Forensic Science

Forensic Science

Current Issues, Future Directions

Douglas H. Ubelaker, Editor

Former President, American Academy of Forensic Sciences
Senior Scientist, Smithsonian Insititution
Department of Anthropology, National Museum of Natural History,
Smithsonian Institution, Washington, DC

A John Wiley & Sons, Ltd., Publication

Blackwell Publishing was acquired by John Wiley & Sons in February 2007. Blackwell's publishing program has been merged with Wiley's global Scientific, Technical and Medical business to form Wiley-Blackwell.

Registered office: John Wiley & Sons Ltd, The Atrium, Southern Gate, Chichester, West Sussex, PO19 8SQ, UK

Editorial offices: 9600 Garsington Road, Oxford, OX4 2DQ, UK
The Atrium, Southern Gate, Chichester, West Sussex, PO19 8SQ, UK
111 River Street, Hoboken, NJ 07030-5774, USA

For details of our global editorial offices, for customer services and for information about how to apply for permission to reuse the copyright material in this book please see our website at www.wiley.com/wiley-blackwell

Library of Congress Cataloguing-in-Publication Data

Forensic science : current issues, future directions / Douglas H. Ubelaker, editor.
p. cm.
Includes index.
ISBN 9781119941231 (cloth)
1. Forensic sciences. I. Ubelaker, Douglas H.
HV8073.F5836 2012
363.25dc23

2012026190

A catalogue record for this book is available from the British Library.

Wiley also publishes its books in a variety of electronic formats. Some content that appears in print may not be available in electronic books.

Set in 10.5/12.5pt Times by Thomson Digital, Noida, India.
Printed and bound in Malaysia by Vivar Printing Sdn Bhd

First Impression 2013

Contents

List of contributors

Douglas H. Ubelaker, PhD (editor) was the 2011–2012 president of the American Academy of Forensic Sciences and is a fellow in the Physical Anthropology Section. He received his PhD degree in 1973 from the University of Kansas and is currently a curator and senior scientist at the Smithsonian Institution's National Museum of Natural History in Washington, DC. He has published extensively in the general field of human skeletal biology, with an emphasis on forensic applications.

Adam Aleksander, PhD is a forensic engineer and a fellow and 23-year member in the Engineering Sciences Section of the American Academy of Forensic Sciences. He received his PhD degree in 1995 from Texas A&M University, and is the President of Aleksander & Associates P.A., Boise, Idaho, and Vice President of Precision Energy Services Inc., Hayden, Idaho. He specializes in forensic engineering issues in product liability, safety engineering, warnings and investigations of industrial and energy system issues. Dr. Aleksander practices nationally and internationally, including some 30 overseas assignments.

Susan Ballou, MS is the program manager for forensic science in the Law Enforcement Standards Office (OLES) at NIST. She has managed this program since 2000, targeting the needs of the forensic science practitioner by identifying and funding research at NIST in areas such as latent print analysis, burn patterns, computer forensics and material standards. Her forensic crime laboratory experience spans over 27 years and includes working on case samples in the areas of toxicology, illicit drugs, serology, hairs, fibers and DNA. She has Diplomate Certification with the American Board of Criminalistics (ABC) and is a fellow of the American Academy of Forensic Sciences (AAFS) and a recipient of the AAFS Criminalistics Section's Mary E. Cowan Outstanding Service Award.

David W. Baker, MFS is a fellow of the AAFS, and served as the secretary of the Digital and Multimedia Sciences Section from 2008–2010, and as chair of the section from 2010–2012. He received a Master of Forensic Science degree from the George Washington University in 1994. David is a principal information security engineer at the MITRE Corporation in McLean, Virginia, and is a member of the organizing committee of the Digital Forensics Research Workshop (DFRWS). He is a retired Special Agent of the US Army Criminal Investigation Command, and was the command's principal forensic science advisor during his last assignment. David has been working in the field of digital and network forensics for 19 years.

Robert Barsley, DDS, JD is the 2012–2013 president of the American Academy of Forensic Sciences and a fellow in the Odontology Section. He received his DDS degree in 1977 from Louisiana State University Health Science Center and his JD degree in 1987 from Loyola University School of Law in New Orleans. He is a professor at the LSUHSC School of Dentistry in New Orleans, Louisiana, where he serves as a consultant to multiple medicolegal entities. He has published and lectured in the field of forensic odontology.

Stephen B. Billick, MD is in full-time private practice of clinical child, adolescent and adult psychiatry and forensic psychiatry. Dr. Billick is clinical professor of psychiatry at New York Medical College, clinical professor of psychiatry at New York University School of Medicine and lecturer in psychiatry at Columbia University College of Physicians and Surgeons, the former associate chair for faculty development at St. Vincent's Hospital/New York Medical College, past president of the American Academy of Psychiatry and the Law (AAPL), past president of the American Society for Adolescent Psychiatry (ASAP) and past president of the New York Council on Child and Adolescent Psychiatry (NYCCAP). He is a member of the board of directors of the American Academy of Forensic Sciences (AAFS) and past chair of the Section on Psychiatry of the New York Academy of Medicine (NYAM).

C. Michael Bowers, DDS, JD received his DDS degree from the University of Southern California. He is a deputy medical examiner in Ventura, California and has written numerous peer review articles on bitemark evidence and dental identification methods. He also is a licensed lawyer (CA) and has contributed to the legal literature regarding the scientific issues related to forensic odontology's role in erroneous criminal convictions.

Samuel I. Brothers, BBA is a digital forensics analyst working for US Customs and Border Protection. He has earned over 20 different certifications in the field of computers and computer forensics. He is currently working at CBP for Laboratory and Scientific Services developing a nationwide digital forensics program. In his free time he is a magician and marathon runner.

Mary Bush, DDS is the 20112012 president of the American Society of Forensic Odontology and a fellow in the Odontology Section. She received her DDS degree in 1999 from the State University of New York at Buffalo and is currently an assistant professor of restorative dentistry and director of the Laboratory for Forensic Odontology Research at that university. She has published extensively in the area of forensic odontology and is currently a member of the editorial board of the *Journal of Forensic Sciences*.

Peter Bush, BS is director of the South Campus Instrument Center at the State University of New York and adjunct professor of Art Conservation at Buffalo State

College. He is a member of the General Section of the American Academy of Forensic Sciences. He has published extensively in a variety of scientific fields, focusing more recently on victim identification and bitemark analysis in forensic odontology.

John Clement, PhD is the inaugural chair of forensic odontology at Melbourne Dental School in the University of Melbourne. He is also a visiting honorary research fellow at the Forensic Institute of the Defence Academy of the UK at Cranfield University. Prof. Clement is past president of both the British and Australian Associations/Societies for Forensic Odontology (BAFO and ASFD), a founder member of the International Association for Craniofacial Identification (IACI) and Dental Ethics and Law Society (IDEALS). Prof. Clement has had practical hands-on experience of working in forensic odontology, especially mass disaster victim identification, since the 1970s.

Cecelia A. Crouse, PhD is the crime laboratory director and forensic biology manager of the Palm Beach County Sheriff's Office Crime Laboratory. She received her PhD degree in 1988 from the University of Miami Department of Microbiology and Immunology and conducted her post-doctoral thesis at the Bascom Palmer Eye Institute. She has been a member of the National Institute of Justice Technical Working Group, the FBI Scientific Working Group on DNA Analysis and Methods, and many other committees for the advancement forensic DNA testing. She has published in the field of virology and forensic DNA analysis and is currently a member of the Inter-Agency Working Group on Accreditation and Certification.

Gregory G. Davis, MD is a fellow in the Pathology/Biology Section of the American Academy of Forensic Sciences. He received his MD degree in 1987 from Vanderbilt University. He is a professor of pathology at the University of Alabama at Birmingham, where he serves full-time as an associate coroner/medical examiner in the Jefferson County Coronary/Medical Examiner Office. His publications concentrate on death associated with drug abuse and on interacting with attorneys and court.

Robert Dorion, DDS is a past President of the American Board of Forensic Odontology, the Canadian Society of Forensic Science, and is a distinguished fellow of the American Academy of Forensic Sciences, presently serving as a member of the board of directors. He received his DDS from McGill University in 1972, where he is currently director of the forensic dentistry program, and is director of forensic dentistry at the Laboratoire de Sciences Judiciaires et de Médecine Légale, Ministry of Public Security for the Province of Quebec, Montreal, Quebec, Canada. He has written for and edited the first stand-alone comprehensive forensic textbook on bitemarks, entitled *Bitemark Evidence* in 2005, with a second edition in 2011.

Janet Barber Duval, MSN is a forensic nurse and a fellow in the General Section. She received her baccalaureate degree in nursing from the University of Cincinnati in 1963 and a master's degree in Nursing Education from Indiana University in 1965. She retired from the United States Air Force Nurse Corps in 2001 with the rank of colonel. During the last decade, Barber Duval has served as a clinical nurse consultant for Hill-Rom Company and has been active in journal and textbook editing. She has developed curricula and has taught continuing education courses in forensic nursing at the University of Texas, the University of New Mexico, and the University of California, Riverside. She is currently an adjunct associate professor at the Indiana University School of Nursing in Indianapolis.

Mary P. Fitzgerald, MS is a fellow in the Criminalist Section of the American Academy of Forensic Sciences and a member of the Midwestern Association of Forensic Scientists. She received her BS and MS from the University of Illinois, Chicago in 1977 and 1983 respectively. Mary is currently employed by the IRS National Forensic Laboratory as a physical scientist in the Questioned Document Section. She has published papers in the field of drug chemistry and ink analysis.

ARW Forrest, LLM received degrees in pharmacology and medicine at Edinburgh University and a law degree from University College of Wales. He received postgraduate education in clinical and analytical chemistry. He combines his legal and medical educations with his scientific background, having held positions as professor of forensic chemistry at the University of Sheffield, professor in the Faculty of Health & Welfare at Sheffield Hallam University and assistant deputy coroner in the jurisdictions of South Yorkshire (West) and the North Riding of Yorkshire & Kingston upon Hull. A fellow in the Royal College of Physicians, Royal College of Pathologists, Faculty of Legal & Forensic Medicine of the Royal College of Physicians and the Royal Society of Chemistry, he is also a fellow in the Jurisprudence Section of the AAFS. Dr. Forrest is a past president of the Forensic Science Society (UK).

Adam Freeman, DDS is a fellow of the American Academy of Forensic Sciences in the Odontology Section. He received his DDS degree in 1992 from Columbia University's College of Dental Medicine, where he now serves as an assistant clinical professor. Dr. Freeman is currently on the board of directors of the American Board of Forensic Odontology and is a past president of the American Society of Forensic Odontology.

Robert Gaffney, MFS is on the 2011–2012 American Academy of Forensic Science board of directors from the General Section. He has a master's in forensic science from George Washington University and is certified as a senior crime scene analyst. He is a special agent with the US Army's CID and the operations officer for the Forensic Technology and Training Division, USACIL in Forest Park,

Georgia. He has published in the *Journal of Forensic Science* and is co-editor of the *CID Crime Scene Handbook.*

Zeno J. Geradts, PhD is a fellow in the Digital and Multimedia Sciences Section and was elected in 2010 for the three-year position as director at the board of directors of the American Academy of Forensic Sciences. He received his PhD degree in 2011 from the University of Utrecht and is currently senior forensic scientist at the Netherlands Forensic Institute in the field of image analysis and biometrics, and R&D coordinator for the Digital Technology and Biometrics Section. He is also chairman of the European ENFSI Forensic IT Working group and published in the field of pattern recognition and image analysis, as well as forensic implications of identity management systems.

James R. Gill, MD is the deputy chief medical examiner for Bronx County of the New York City Office of Chief Medical Examiner, and a clinical associate professor in the Department of Forensic Medicine at New York University School of Medicine.

Scott Grainger, PE is a licensed fire protection engineer and licensed civil engineer in numerous states, and a licensed land surveyor in Arizona. He is a fellow member of the Society of Fire Protection Engineers and senior member of the National Academy of Forensic Engineers.

Judith A. Gustafson, BS, is a member of the American Academy of Forensic Sciences Questioned Documents Section and the Midwestern Association of Forensic Scientists. She received a BS degree in 1982 from Western Illinois University and a BFA degree in 1998 from The School of the Art Institute of Chicago. She began her career with the Illinois State Police, Bureau of Forensic Sciences in 1985, and is currently employed by the US Treasury Department, National Forensic Laboratory, Questioned Documents Unit in Chicago, Illinois.

Sarah Hainsworth, PhD is a professor of materials engineering at the University of Leicester, UK. She received her PhD degree in 1993 from the University of Newcastle upon Tyne, UK and is a chartered engineer, chartered scientist and fellow of the Institute of Materials, Minerals and Mining. She has worked in the area of forensic engineering relating to materials failure investigations, and in forensic science related to stabbing and dismemberment. She has published in the areas of failure analysis of materials, forces required for stabbing with a variety of weapons and analysis of toolmarks.

H. Theodore Harcke, MD received his doctorate of medicine in 1971 from the Pennsylvania State University and has 40 years experience in academic radiology. He is a colonel in the United States Army and currently serves as the forensic radiologist for Armed Forces Medical Examiner System. An American Academy of

Forensic Sciences member in the Pathology/Biology Section, he is co-author of a recently published textbook on forensic imaging.

Max Houck, PhD is an internationally-recognized forensic scientist who specializes in anthropology, trace evidence and education. A former FBI scientist, Dr. Houck has worked for Oxford Instruments, the Tarrant County (Texas) Chief Medical Examiner and West Virginia University, where he directed the Forensic Science Initiative, a multi-million dollar resource to the nation's forensic laboratories. He is currently Director of the District of Columbia Consolidated Forensic Laboratory.

Julie Howe, MBA is a fellow in the American Academy of Forensic Sciences General Section. She is a medicolegal death investigator for Franklin, Jefferson and Saint Charles Counties in Missouri. She also serves as the executive director for the ABMDI, secretary/treasurer for SOMDI, and is a member of the Scientific Working Group for Medicolegal Death Investigation, chairing the Accreditation, Certification, Education and Training committee. Ms. Howe has been involved with national training of medicolegal death investigators for 15 years. She has a master's in business administration from Saint Louis University and is a registered diplomate with the ABMDI.

RT Kennedy, JD is former chair of the AAFS Jurisprudence Section, in which he is a fellow. He was awarded the JB Firth Medal from the Forensic Science Society (UK) in 2003. Judge Kennedy has been an adjunct professor at the University of New Mexico School of Law teaching forensic evidence, and has received an award from the New Mexico State Bar for his continuing education series 'Skeptically Determining the Limits of Scientific Evidence.' He is a judge on the New Mexico Court of Appeals, and has participated in all levels of New Mexico courts and on the bench of the Jicarilla Apache Nation in his 23-year judicial career. He received his JD degree in 1980 from the University of Toledo College of Law.

Douglas S. Lacey, BSEE is a fellow in the Digital and Multimedia Sciences Section of the American Academy of Forensic Sciences. He is the section secretary for 2011–2012 and is the chair-elect for 2012–2014. He earned his BS degree in electrical engineering in 1996, from the audio engineering program at the University of Miami, and currently is a forensic audio/video examiner in the private sector, after having worked as an examiner at the FBI's forensic audio/video laboratory from 1996 through 2003. He has published over a dozen peer-reviewed papers in the fields of forensic audio and video analyses.

John J. Lentini, BA of Scientific Fire Analysis, LLC presents fire scene investigation and laboratory analysis of fire debris; he is one of a handful of people certified to conduct both fire scene investigations and fire debris analysis. He has personally conducted more than 2,000 fire scene inspections and has appeared as an expert witness on more than 200 occasions. He is an active proponent of standards for fire and other forensic investigations.

Jim Lewis, DMD is the chair of the American Board of Forensic Odontology, Dental Age Estimation Committee and the 2012–2013 president of the American Society of Forensic Odontology. He received his dental degree from the University of Alabama School of Dentistry, is a fellow of the Odontology Section, currently serves as an odontology consultant to the Alabama Department of Forensic Sciences and is an adjunct faculty member of the Fellowship in Forensic Odontology at the Center for Education and Research in Forensics and the Southwest Symposium on Forensic Dentistry.

Jane A. Lewis, MFS is a fellow in the Questioned Document Section of the American Academy of Forensic Sciences. She received her Master of Forensic Science degree from the George Washington University in 1985. She is currently a unit leader in the forensic document unit at the Wisconsin State Crime Lab-Milwaukee. She also has a private practice. She has made numerous presentations at forensic science meetings and has published several articles on forensic document examination.

Laura L. Liptai, PhD received a PhD and MS in biomedical engineering as well as a second master's specializing in robotics from the University of Southern California. Dr. Liptai holds a bachelor's of science in mechanical engineering. Specializing in kinetics and kinematics of trauma/injury to the human body, she specializes in the analysis of the mechanics, quantity of forces/accelerations and human factors in traumatic incidents. This specialty of biomedical engineering uses the signature of evidence within human tissues in the context of the incident environment to derive causal factors. Dr. Liptai has received the Andrew Payne award and Founder's award for her exemplary contributions to the forensic engineering sciences. She compiled and edited the AAFS Reference Series that is the largest collection of research and case study proceedings worldwide. She serves four boards: the University of California at Davis, BioMedical Engineering external advisory board; the American Academy of Forensic Sciences; the Society of Forensic Engineers and Scientists (also serves as vice-president) and the International Board of Forensic Engineering Sciences.

Barry K. Logan, PhD is the 2011–2012 treasurer of the American Academy of Forensic Sciences, and a fellow of the Toxicology Section. He is board certified by the American Board of Forensic Toxicology. He received his PhD degree from the University of Glasgow in Scotland, and is currently national director of Forensic Services with NMS Labs in Willow Grove, Pennsylvania, and executive director of the Center for Forensic Science Research and Education at the Fredric Rieders Family Renaissance Foundation. He has faculty appointments at Indiana University, Bloomington, Indiana and Arcadia University, Glenside, Pennsylvania. He has over 80 publications in the field of forensic toxicology.

Ryan Loomba, BS is a diversified biomedical engineer that specializes in complex system design, electrical circuit design, microfluidics and mechanical engineering applications. Ryan currently is an associate engineer at Nanomix in Emeryville, California where he is helping to design a point-of-care biomedical device utilizing carbon nanotube technology. Ryan received a BS in biomedical engineering with a specialization in electrical engineering from the University of Southern California.

Daniel A. Martell, PhD is a fellow in the Psychiatry and Behavioral Sciences Section of the AAFS. He is also a fellow of the American Board of Forensic Psychology and the National Academy of Neuropsychology. He obtained his PhD from the University of Virginia and is currently on the clinical faculty at the Semel Institute for Neuroscience and Human Behavior of the David Geffen School of Medicine at UCLA. Dr. Martell's research interests include brain damage, mental disorder, and violent criminal behavior.

Edward L. Mazuchowski, MD, PhD is a reviewer for the *Journal of Forensic Sciences* and a diplomate of the American Board of Pathology in Anatomic, Clinical and Forensic Pathology. He received his doctorate of medicine from the Uniformed Services University of the Health Sciences in Bethesda, Maryland and his doctorate of philosophy in bioengineering from the University of Pennsylvania in Philadelphia, Pennsylvania. Currently he is a lieutenant colonel in the United States Air Force and a deputy medical examiner with the Armed Forces Medical Examiner System, which has the responsibility of providing comprehensive medicolegal death investigation services to the Department of Defense and other federal agencies.

Kara L. Nance, PhD is professor and chair of the Computer Science Department at the University of Alaska Fairbanks and runs a computer security consulting firm. Her research interests include digital forensics, data systems, network dynamics, visualization and computer security. She is the founder and director of the Advanced Systems Security Education, Research and Training (ASSERT) Center, which is a multi-disciplinary center to address computer security issues and provides an isolated networked computer environment suitable for computer security education, research and training that is used by institutions around the world. She serves on a senior-executive advisory board for the Office of the Director of National Intelligence and is a frequent author and speaker on cyber security as it relates to national security.

Skip Palenik, BS has had a lifelong fascination with the microscope that started when he received his first instrument at the age of eight. Since then he has devoted himself to increasing his knowledge of analytical microscopy and microchemistry and applying it to the solution of real-world problems, especially those of forensic interest. He was fortunate in having worked closely with his mentor, Dr. Walter McCrone, for over 30 years and to have studied forensic microscopy with Dr. Max

Frei-Sulzer of Zurich, a disciple of Dr. Edmond Locard of Lyon. Skip has been teaching analytical microscopy to forensic scientists for more than 30 years and has published numerous scientific articles and book chapters on the applications of chemical and forensic microscopy. He established Microtrace in 1992 to provide a resource for organizations and individuals in need of scientific services involving the analysis of microscopic trace evidence. His special research interests are the identification of single small particles, small amounts of complete unknowns and tracing dust and soil back to their origins. He is the 2009 recipient of the Paul L. Kirk Award, the highest award given by the Criminalistics Section of the American Academy of Forensic Sciences, and the 2003 Distinguished Scientist Award from the Midwestern Association of Forensic Sciences. He is listed in American Men and Women of Science. In 2010 he was awarded the Chamot Medal in chemical microscopy by the State Microscopical Society of Illinois.

Iain A. Pretty, DDS, PhD is a full-time academic at the School of Dentistry, University of Manchester, England. He has published extensively on odontological matters, concentrating on establishing metrics of reliability and validity for forensic techniques. His work continues to develop the theme of application of robust scientific methods to forensic dentistry and he has a particular interest in wrongful convictions and what can be learned from them.

William M. Riordan, BA is a director on the American Academy of Forensic Sciences board of directors and a fellow in the Questioned Document Section. He is also a diplomate of the American Board of Forensic Documents Examiner, Inc. and is currently a director on that board. He received a bachelor of arts degree from Roosevelt University, Chicago in 1975 and was trained in forensic document examination in the Chicago Police Department Crime Laboratory under Maureen Casey Owens. In 1986, after 13 years with the Chicago Police Department Crime Laboratory, he was employed as a forensic document analyst by the Department of the Treasury in the Internal Revenue Service National Forensic Laboratory, Questioned Document Unit, where he is currently employed.

Jeri D. Ropero-Miller, PhD holds a 2011–2014 position on the American Academy of Forensic Sciences board of directors and is a fellow in the Toxicology Section. She received her PhD degree in 1998 from the University of Florida and currently is a principal investigator and senior research forensic scientist at RTI International in Research Triangle Park, North Carolina. She has published in the fields of forensic toxicology and clinical chemistry with an emphasis on drugs of abuse in post-mortem testing, drugs of abuse in alternative matrices such as hair, and laboratory technological evaluations and advancements.

Daniel J. Ryan, JD is a professor at the Information Resources Management College of the National Defense University where he teaches cyberlaw, intelligence, information security and computer forensics. His research is in the areas of risk management, digital forensics and admissibility of scientific, technical or specialized evidence. He received a *Juris Doctor* degree from the University of Maryland and has published one book and numerous journal articles on law, business, and technology.

John E. Sammons, MS is an assistant professor at Marshall University, where he teaches digital forensics, electronic discovery, and information security. He received his Master of Science degree in 2008 from Mountain State University. He is the founder and director of the Appalachian Institute of Digital Evidence. He is an investigator with the Cabell County Prosecuting Attorney's Office and former Huntington police officer. He is a member of the West Virginia FBI Cyber Crime Task Force. He is an associate member of AAFS.

David Senn, DDS is a fellow in the Odontology Section and a past President of the American Board of Forensic Odontology. He received his DDS degree in 1969 from the University of Texas Dental Branch at Houston. He is currently the director of both the Center for Education and Research in Forensics and the Fellowship in Forensic Odontology at the University of Texas Health Science Center at San Antonio (Dental School). He has published in the field of forensic odontology.

Claire Shepard, MS is a professor and the program coordinator for the forensic science program at Southern Crescent Technical College in Griffin, Georgia. She received her master's degree in forensic science from the University of New Haven and earned a bachelor of science degree from Millsaps College. Within the American Academy of Forensic Sciences, she served as program chair for the General Section, was the 2003–2004 president of the Young Forensic Science Forum, and has served on numerous committees. She is also a certified senior crime scene analyst (CSCSA) by the International Association for Identification and a member of DMORT.

Jay A. Siegel, PhD is currently chair of the Department of Chemistry and Chemical Biology at Indiana University Purdue University, Indianapolis, and a professor of forensic science and analytical chemistry. He is a distinguished member of the American Academy of Forensic Sciences. He holds a PhD in analytical chemistry from the George Washington University and received the Distinguished Alumni Scholar Award from there in 2009. He has published many articles in research in forensic chemistry and is co-author of two textbooks in forensic science and co-editor-in-chief of the *Journal of Forensic Science Policy and Management*.

Dawnie Wolfe Steadman, PhD, D-ABFA is the director of the Forensic Anthropology Center at the University of Tennessee, Knoxville. She is a diplomate of the American Board of Forensic Anthropology and a fellow of the American Academy of Forensic Sciences. She received her PhD from the University of Chicago in 1997.

Her areas of teaching and research include forensic anthropology, bioarchaeology and forensic human rights investigations.

Peter Stephenson, PhD is a cyber criminologist, digital investigator and digital forensic scientist at Norwich University (Vermont). He is a writer, researcher and lecturer on information assurance, digital investigation, and forensics. He has lectured throughout the world on digital investigation and security and has written, edited or contributed to 17 books and several hundred articles. He teaches network attack and defense, digital forensics and cyber investigation at Norwich University, where he also is the chief information security officer, and is director of the Norwich University Center for Advanced Computing and Digital Forensics. Dr. Stephenson obtained his PhD at Oxford Brookes University, Oxford, UK, and he holds a Master of Arts degree (cum laude) in diplomacy with a concentration in terrorism from Norwich University.

Yingying Tang, MD, PhD is director of the Molecular Genetic Laboratory at the NYC OCME. Dr. Tang received her MD from China Medical University, Shenyang, China, her PhD from Columbia University, and completed her post-doctoral fellowship in clinical molecular genetics and biochemical genetics in the Department of Human Genetics, Mount Sinai School of Medicine. She is certified by the American Board of Medical Genetics in clinical molecular genetics and clinical biochemical genetics, and by New York State Department of Health as laboratory director in genetic testing. She has published on mitochondrial genetics and its role in human disease, and on genetic testing in sudden death. Dr. Tang has been an investigator or co-investigator on four National Institute of Justice research grants.

Morris Tidball-Binz, MD is the forensic coordinator of the Assistance Division of the *International Committee of the Red Cross* (ICRC), based in Geneva, Switzerland. He received his medical degree from the National University of La Plata, Argentina (1989). He did his postgraduate training in forensic anthropology and in forensic medicine, including an honorary research fellowship at the Department of Forensic Medicine at Guy's Hospital, London, UK, between 1993 and 1996. He specialized in the application of forensic sciences to human rights and humanitarian investigations, has carried out forensic casework in over 30 countries and been actively involved in training and dissemination of applied forensic sciences in all regions (Americas, Africa, Asia, Europe, Middle-East and Pacific). Dr. Tidball-Binz co-founded and was the first director of the Argentine Forensic Anthropology Team (NGO), Buenos Aires, Argentina (1984–1990); assisted the Grand-Mothers of Plaza de Mayo (NGO) in the establishment of a national genetic databank for the identification of missing children in Argentina (1984–1989); was researcher for Chile, Mexico, Venezuela and Central America at the Americas Department of Amnesty International's International Secretariat in London, UK (1990–1995) and then head of that department (1996–1997).

He directed the Regional Programme for the Prevention of Torture of the Inter-American Institute for Human Rights in Costa Rica (1998–1999), founded and was first director of the Latin-America Regional Office of Penal Reform International (NGO), also in Costa Rica (2000–2001), directed the Human Rights Defenders Office of the International Service for Human Rights (NGO), Geneva, Switzerland (2001–2003) and also directed that organization during 2003, before joining the ICRC in 2004.

Jan Unarski, PhD graduated in car mechanical engineering at the Technology University of Cracow in 1977, earning a PhD engineering degree in 2002. Employed by the Institute of Forensic Research in Cracow (Ministry of Justice) for 33 years, accident reconstruction became a specialty in 1985. Head of the Accident Reconstruction Department, Dr. Unarski became the chairman of the Road Accident Analysis Working Group (RAA WG) of ENFSI (European Network of Forensic Science Institutes) in 1998, serving until 2001. Dr. Unarski is a member of the Presidential Board of EVU (European Association for Accident Research and Analysis).

D. N. Vieira, PhD, MD is the current president of the International Academy of Legal Medicine and of the European Council of Legal Medicine, and a fellow in the Pathology and Biology Section. He is past president of the International Association of Forensic Sciences, of the World Police Medical Officers, of the Mediterranean Academy of Legal Medicine and of the Latin-American Association of Medical Law. He is full professor of forensic medicine and forensic sciences and of ethics and medical law at the University of Coimbra, and invited professor in several European and South-America universities. He is also the director of the National Institute of Forensic Medicine of Portugal and a member of the Portuguese National Council of Ethics for Life Sciences. He has published extensively and he has been awarded 11 scientific prizes and 12 honorary fellowships from scientific associations from European, Asian and Central and South American countries. He has participated in many international missions as forensic consultant, especially in the field of human rights.

Frank Wright, DMD is fellow in the Odontology Section of the American Academy of Forensic Sciences, serving as a past Odontology Section program chair, secretary and section chairman. He is the immediate past president of the American Board of Forensic Odontology, a 1984 graduate of the University of Kentucky College of Dentistry and he currently serves as the forensic dental consultant for the Hamilton County Coroner's Office in Cincinnati, Ohio. He has lectured and published extensively in all areas of forensic dentistry, with an emphasis on visible and non-visible light photography.

Acknowledgements

Since this volume represents a team effort, many acknowledgements are due. Rachael Ballard of Wiley-Blackwell extended the original invitation to consider this book effort and offered continuous encouragement and support during production. Our 2011–2012 AAFS board of directors unanimously approved the volume concept and was instrumental in selecting chapter authors, with some members serving as authors themselves. Of the AAFS staff, Anne Warren contributed considerable support, especially regarding contract negotiation. Kimberly Wrasse maintained communication with authors regarding manuscript production deadlines and format issues.

During the manuscript editing phase, Smithsonian Institution colleague Kristin Montaperto worked closely with me in the evaluation and formatting of each contribution. An additional Smithsonian colleague, Keitlyn Alcantara, assisted with various editing issues, along with my son Max Ubelaker and daughter Lisa Ubelaker. Lauryn Guttenplan of the Smithsonian Institution was instrumental in the formulation of contract language.

Wiley-Blackwell editor Fiona Seymour offered prompt advice on format issues during our initial work and then took charge of production once the volume manuscript was formally submitted. Finally, I thank all of our authors who gave so generously of their time and expertise in writing the chapters. Our authors met the somewhat formidable challenge of the rapid production schedule, as well as the goal of squeezing so much information about our beloved forensic sciences into this volume.

1 Introduction

Douglas H. Ubelaker

Department of Anthropology, National Museum of Natural History, Smithsonian Institution, Washington, D.C.

The forensic sciences represent the application of knowledge and methodology in various scientific disciplines toward the resolution of legal issues. The approaches employed in case applications are evidence-driven. Practitioners must evaluate what problems need to be addressed in evidence analysis, then utilize the most appropriate scientific methodology available. Such procedures call for broad understanding of the underlying science and full awareness of the appropriate available techniques, technology and databases, as well as their limitations.

Many areas of science can be applied to problems in legal contexts, sometimes in unexpected ways. In some scientific disciplines, legal applications are relatively uncommon. For example, a zoologist specializing in the taxonomy and natural history of a particular genus of rodent may work his or her entire career without ever dealing with a forensic issue. However, if key evidence relating to the classification and behavior of that rodent genus becomes important in the investigation of a homicide, the zoologist may be needed to provide forensic perspective and to present that evidence in court.

In contrast, some areas of science are utilized routinely in the analysis and interpretation of evidence. With a growing history of forensic perspective, experienced practitioners in these areas have honed methodology specifically geared to address evidentiary issues. They have developed laboratory procedures designed to ensure the security of evidence and to maximize the information that can be extracted from it. Research has been organized and scientific approaches developed to deal with specific problems revealed through forensic practice.

This volume presents key perspective on the forensic sciences from experienced practitioners in the core areas of this endeavor. It attempts to provide a general overview of the main activities in the forensic sciences, some historical perspective on forensic science origins, international factors and current central issues. However, the volume does not represent a practical, how-to guide to forensic practice. Due to limited space, not all forensic applications are addressed.

Forensic Science: Current Issues, Future Directions, First Edition. Edited by Douglas H. Ubelaker.

The idea for this volume surfaced during a meeting I attended with our publisher, Wiley-Blackwell, on February 24, 2011, at the annual assembly of the American Academy of Forensic Sciences (AAFS) in Chicago. Recognizing the strong and growing academic and public interest in forensic topics, the publisher's representative, Rachael Ballard, expressed interest in pursuing quality book projects in this scholarly area. Of course, many of our more than 6,000 AAFS members have published well-received books in their areas of expertise. As book review editor for the *Journal of Forensic Sciences*, also published by Wiley-Blackwell, I was keenly aware of the quality of those products. However, at this meeting, the publisher specifically was seeking perspective from AAFS leadership on what more could be done. I attended the meeting as AAFS President-Elect, along with then President Joseph Bono and Executive Director Anne Warren.

The inquiry strongly attracted my interest for multiple reasons. I agreed with the publisher that both public and scholarly interest in forensic science remained strong. However, much of the public interest has been met and shaped by television and popular fiction writing, in which analyses are conducted almost immediately and forensic scientists are portrayed as action figures with wide-ranging (and largely inappropriate) involvement in the case investigation. Although most of the public with whom I have discussed this issue understand that reality is likely different from that depicted by fiction, some misinformation persists among those entering our classrooms and jury pools.

For the last several years, the broader forensic science community has discussed and debated a variety of issues related to the quality of current forensic practice. Much of this discussion has focused on factors of objectivity, accreditation, certification, reliability, error analysis and the need for, and nature of, sustained focused research. Critical review has taken place at presentations during the annual meetings of the AAFS and other forensic science organizations, in the pages of our journals and in gatherings of scientific working groups.

The topics also have attracted attention on Capitol Hill in Washington, DC, where both the legislative and executive branches of the United States federal government have formed initiatives directed toward shaping forensic science reform. Along with many colleagues from the forensic, general science and legal scholarly communities, I have been an active participant in many of these dialogues. This participation has impressed on me the need for clear, comprehensive information on the current international practice of forensic science. Particular needs involve the clarification of central issues, ongoing efforts (including research) to advance the field, and thoughts on future developments from those engaged in the practice of forensic science. This perspective positioned me to respond positively to the Wiley-Blackwell inquiry.

Following the meeting with the publisher, I spent the remainder of the 2011 AAFS gathering discussing with colleagues my idea of producing an AAFS-sponsored book. After receiving a uniformly positive response to those discussions, I formally presented the concept to our Board of Directors on the final day of the conference week. A plan was formulated to create a volume that would include perspectives from each of our 11 sections and would include international coverage. The section representatives on our board of directors would serve as the primary contacts on the project and would work with me to select the appropriate authors. Although this involved a challenging amount of work for all, the group embraced the idea with unanimous approval.

As a result of the subsequent positive review of the book proposal and formal publisher approval, author selection moved forward. Authors were chosen for their prominence in the profession, skill at writing and meeting deadlines, availability, and their knowledge of the areas needed to be covered in each particular chapter. Reflecting the structure of the AAFS and the section representation on our board of directors, chapters were organized into 11 areas of the forensic sciences. The variability within these areas, regarding the diversity of their scholarly endeavors and their slant on the forensic sciences, is reflected in the nature of author selection for the individual chapters. The research and other activities of the authors selected for the individual chapters speak to the diversity of the scholarly endeavors within the 11 sections of AAFS.

The reviewers of the original book proposal to the publisher focused particularly on the stated need for international coverage. While such perspective had been planned for the individual section contributions, reviewers felt it needed to be strengthened to ensure a global view of the central issues. Accordingly, invitations were extended to the international colleagues who had agreed to serve as plenary speakers at the 2012 AAFS meeting in Atlanta, Georgia. This invitation not only added the recommended international strengthening of the planned volume, but also provided these future speakers with a publication outlet for many of the thoughts they were developing for the conference presentations.

To encourage continuity within the volume, guidelines were provided to those writing the section contributions. The following paragraphs discuss areas and topics planned to be covered, based on these guidelines. These general goals were presented to the chapter authors at the initiation of the project.

- *In regard to practice in the forensic sciences, the authors should explore the underlying science in various types of forensic cases, the necessary training and education of the forensic scientist and the information indispensable to the public/non-specialist knowledge base. The volume also should address the origin and historical development of the forensic sciences, including its early pioneers and main contributors. Each section chapter should relate its own topic-specific history, as well as the key relative developments within the AAFS.*

- *Contributors should review key issues and how they are being addressed. Emphasis should be on scientific developments but other fundamental topics, such as accreditation, certification, objectivity and training, can be included when appropriate. Objective analysis of controversies presented with different points of view can contribute to discourse surrounding forensic science issues.*

- *The authors should discuss the future directions of the field, including major research efforts underway that are likely to have high impact, as well as additional needed research particularly deserving of funding. Furthermore, expectations for future advancement and major breakthroughs should be presented.*
- *As an aid for future professionals, this volume should also review beneficial training and coursework for young scholars, preparing them for entry into this field. In addition, supplemental reading should be recommended by topic for those desiring added detail and perspective.*

Although the above guidelines were communicated to all authors, the goal was not to limit them to these specifics. These suggestions represented my concept of the key factors that needed to be addressed and conveyed in the planned volume. The points mentioned also provided the authors with some detail on the general goals of the volume. In addition, they emphasized that the book was not going to be about specific laboratory procedures and practical methodology. In short, this volume was structured to present the thoughts of the authors on the topics laid out above. A central goal was to provide the reader with an in-house forensic look at the major disciplines represented in the global practice of forensic science.

The American Academy of Forensic Sciences is ideally positioned to provide the perspective conveyed in this volume. In 2011, with 6,260 members from the United States, Canada and 62 other countries, the AAFS can, indeed, be characterized as an international forensic organization. Founded in 1948 [1], the AAFS consists of a professional society dedicated specifically to the forensic sciences. The Academy promotes its educational and scientific objectives through publication of the internationally recognized *Journal of Forensic Sciences*, its newsletter *Academy News*, an annual scientific meeting, and conferences, workshops and policy relating to important forensic issues. Headquartered in Colorado Springs, Colorado, the AAFS represents its membership to the public, serving as a key outlet for information on the forensic sciences.

Academy members are organized into 11 sections that are represented in this volume. These sections consist of Criminalistics, Digital and Multimedia Sciences, Engineering Sciences, General, Jurisprudence, Odontology, Pathology/Biology, Physical Anthropology, Psychiatry and Behavioral Science, Questioned Documents and Toxicology. Applicants for membership must meet the general AAFS requirements as well as those specific to their section of professional interest.

Each year, in February, the AAFS holds its annual meeting. At this time, over 800 scientific papers, workshops, breakfast seminars and other relevant events are presented. The 2011 annual meeting in Chicago included 3,851 attendees, including 512 representatives from countries other than the United States and Canada. For additional details on the American Academy of Forensic Sciences, consult the official website (www.aafs.org).

This volume provides background information, international perspectives and discussions of key issues for many of the scientific disciplines engaged in forensic applications. This effort contributes to the ongoing dialogue regarding challenges to the forensic sciences and the path toward future growth. The points of view expressed represent those engaged in the practice of forensic science, and who are well-positioned to recognize real solutions and advancement.

Reference

1. Field KS. *History of the American Academy of Forensic Sciences: 1948–1998*. West Conshohocken, PA: American Society for Testing and Materials; 1998.

2

General forensics – no one else starts until we finish

Julie Howe[1], Janet Barber Duval[2], Claire Shepard[3] and Robert Gaffney[4]

[1]*Division of Forensic Pathology, Saint Louis University, St. Louis, Missouri, USA*
[2]*Indiana University School of Nursing, Indianapolis, Indiana, USA*
[3]*Southern Crescent Technical College, Griffin, Georgia, USA*
[4]*Forensic Technology and Training Division, United States Army Criminal Investigation Laboratory, Forest Park, Georgia, USA*

2.1 Introduction

The General Forensics Section of the American Academy of Forensic Sciences is not made up of generalists; rather, the section is comprised of forensic experts in 18 accepted forensic sub-disciplines. The term 'general' is derived from the first titles given to 'floating', 'at large' or 'general' members of the early Academy's accepted disciplines. The expertise, training and education of these individuals were highly valued within the Academy, but they did not have a section of their own. 'General' members were associated with various disciplines such as pathology and toxicology. Although the need for a 'General' Section was identified as early as 1953 and the concept accepted by the Academy in 1956, the General Section was not created until 1968 [1].

The General Forensic Section members believe they are the present and future of the Academy [2]. They have moved from 'at large' members in 1953 to the 'gate keepers' of the American Academy of Forensic Sciences today. It is through the General Section that forensic expertise is first identified, vetted and accepted after a critical review of the forensic specialties to protect the integrity of the Academy. The ultimate goal of these new experts is to become a separate section.

The Academy acknowledged in 1990 that the General Section was the 'Mother of the Academy' because section members took their responsibilities as a serious trust in establishing new forensic disciplines [2]. This process has been repeated several times, with the first being forensic anthropology in 1973, followed by odontology in

Forensic Science: Current Issues, Future Directions, First Edition. Edited by Douglas H. Ubelaker.

1974, engineering sciences in 1981 and the latest, digital and multimedia sciences, in 2008.

Globally, General Forensic Section members are involved in forensic research as forensic researchers strive to improve analysis and methodologies. There are ongoing projects in accelerating DNA results, which will benefit all labs dealing with a backlog in DNA cases. There are also projects in latent print development and firearms analysis, to improve methodologies and to aid the examiner in analysis. These and other research projects will impact on forensics around the globe, as the focus is on smaller and faster analysis without compromising the evidentiary value of forensic material.

Unfortunately, all of the sub-disciplines within the section cannot be highlighted in this volume. Each would require a chapter. Medicolegal death investigation, crime scene investigation and forensic nursing were selected, as these disciplines are the largest in membership. The General Forensics Section is comprised of accountants, administrators, educators, firearms examiners, nurses and researchers, with a myriad of consultants and investigators. Members are providing forensic expertise in countries around the globe such as India, Australia, Canada, France and the United Kingdom. Section members can be found on the battlefield in Iraq and Afghanistan. Each has a common thread, which is the identification, collection and preservation of forensically relevant evidence for further analyses by other forensic experts to seek truth and justice. No other forensic discipline starts until we finish!

2.2 Medicolegal death investigation

2.2.1 Definition of the field

Local laws define which deaths are investigated but, typically, death investigations occur when there is a sudden unexpected, unexplained, suspicious or violent death. Death investigations are performed by coroners, medicolegal death investigators, law enforcement representatives working on behalf of medicolegal jurisdictions, and forensic pathologists.

The qualifications for coroner vary tremendously among the many countries who have adopted coroner systems. Qualifications of coroner range from legal and medical education, to no requirement other than being 18 years of age and residing in the jurisdiction for which one is elected. Medical Examiners, on the other hand, are typically physicians who are board certified in forensic pathology. They are appointed to state or county positions. Lay medicolegal death investigators are utilized by medicolegal systems to represent coroners and medical examiners at death scenes, receive initial death notifications, perform scene investigations, collect and process evidence, obtain medical and investigative information, and interact with families and outside agencies.

The majority of cases handled by a medicolegal office, regardless of location, are natural deaths that occur outside of a medical facility. These deaths do not require a death scene investigation, but instead require investigation into the decedent's medical history, documenting and verifying prescription medications found on scene, interviewing medical personnel and obtaining and reviewing medical records. Therefore, the medicolegal death investigator should have basic knowledge in medicine, in order to correlate the information obtained at the scene where the decedent is discovered with the story being provided. The investigator should also have knowledge of local statutes that apply to medicolegal authority.

Medicolegal death investigators also respond to death scenes of non-natural causes to view the body in the context of its surroundings. They determine the scope and extent of death investigations, according to budgets and mandated procedures based on statutory requirements. They are generally the first point of contact for law enforcement, determining the details of death and deciding the extent of investigation needed, with knowledge and objectivity. In many instances, the scene investigation and medical, psychiatric and social background of the decedent will provide the necessary information for the medical examiner or coroner to certify the cause and manner of death correctly. For example, the sleep environment of an infant death can depict accurately why the child was found unresponsive, whereas the autopsy will be negative.

The current global death investigation system requires a multidisciplinary approach and involves ancillary professionals, including administrators, autopsy assistants, crime laboratory technicians, law enforcement personnel, radiologists and toxicologists, to name a few.

2.2.2 History of medicolegal death investigation

The Chinese were the first to perform the duties of the coroner [3]. In 1975, a tomb in China dating from the period of 475–221 BC was found that contained bamboo slips with instructions to judges on how to detect bruises, wounds and general post-mortem changes [4]. There is evidence that the role of the coroner existed in China in 1618. These individuals were the equivalent of judicial officers or magistrates, and they had rigid routines for investigating deaths. The first systematic book on forensic medicine in any civilization is the *Xi Yuan Ji Lu* ('The Washing Away of Wrong' or 'Collected Cases of Injustice Rectified'); this was written by Son Ci (1186–1249), who is honored as the greatest figure in the history of Chinese forensic medicine [5].

According to a number of authors, there is evidence of a coroner in England as early as 871 AD [6]. However, Hunisett, a professional historian specializing in the study of the medieval coroner, disagrees, stating that while the duties of the coroner had been done earlier by other officials, with the office of coroner actually being established in September, 1194 [6]. The English Articles of Eyre were written at that time, requiring that three knights and a clerk attend each death, but the duties of the coroner were not clearly defined. Coroners were tasked with ensuring that a portion of the decedent's assets were secured for the Crown, as well as determining the cause and circumstances of death while enforcing *lex murdrorum*, a law prohibiting homicide.

The name 'coroner' originated from these officials' duties as 'keepers of the pleas of the Crown'. English coroners were often physicians, attorneys or legal magistrates. In the late 1800s, the English recognized the need for physicians with medical knowledge to respond to death scenes to obtain medical information related to the death.

Today, the English coroner is part of both the executive branch of government, providing ministerial services, and the judicial branch, acting as judge. The Ministry of Justice oversees coronial law but has no responsibility for jurisdictional operations. In order to qualify as a coroner, one must have a degree in law or medicine (e.g. criminology or bio-medical science). Coroners continue to retain the services of physicians to perform autopsies.

English colonists brought the coroner system to America when they settled in the 1600s. American coroners were often farmers, craftsmen, or undertakers, as opposed to their English counterparts, who were judicial officials. Coroner duties during the colonial period included inquests, arranging for burials and requesting that physicians examine bodies and perform autopsies. Coroners were initially appointed by state governors, before becoming an elected position placed under the influence of the political party in the mid-19th century. Because many coroners were elected based on their political influence, in 1877, Massachusetts became the first state to develop a medical examiner system, as a response to the corrupt coroner systems in existence. Other large cities and states eventually followed, establishing medical examiner systems. The current United States medicolegal death investigation system relies upon medical examiner systems, coroner systems and mixed medical examiner/coroner systems.

Coroners are responsible for holding inquests into deaths of non-natural causes. The inquest is an inquisitorial proceeding, not an accusatorial process, to help determine the manner of death, as the cause of death is usually obvious, based on medical evidence and the facts surrounding the case. Inquests are led by the coroner and consist of jurors residing within the county where the victim died. Testimony is recorded by a court reporter and open to the public. After the jury has deliberated, the coroner will rule on a final cause and manner of death. Coroners in the United Kingdom, Australia and other countries also make recommendations for safety practices that may prevent deaths, in addition to holding inquests and other investigative duties.

In the United States, coroners are not required to have a medical and/or legal background in the majority of the 28 states having coroner systems. Only four states require the coroner to be a physician (Kansas, Louisiana, North Dakota and Ohio) [7]. Medical examiners are appointed to their positions by governing bodies of the jurisdiction, either at county or state levels. Only three states (West Virginia, Wisconsin and Vermont) allow non-physicians to be appointed as medical examiners [7].

While forensic pathologists who are board certified by the American Board of Pathology are the most highly educated and trained individuals to conduct medicolegal death investigations and perform forensic examinations and autopsies, there is a critical shortage of these individuals. Medical examiners direct the investigative process in larger metropolitan areas and/or perform autopsies ordered by the coroner or authorized by state statute. Twenty-two states are medical examiner systems, eleven are coroner systems and eighteen have mixed systems, with medical examiners in the metropolitan areas and coroners in rural or suburban counties.

The position of the lay medicolegal death investigator in the United States is relatively new, with the majority of positions being created after 1970. According to most state statutes, the medicolegal death investigator has jurisdiction of the decedent, whereas law enforcement or Crime Scene Investigators have jurisdiction to process the scene. Therefore, medicolegal death investigators work in concert with CSIs at death scenes as well as follow-up investigations. The duties of the position require a person to have skills in investigation and basic medicinal knowledge. Historically, medicolegal jurisdictions have hired investigators who are retired law enforcement officers beginning a second career, but this trend is diminishing as the profession recognizes the necessity to possess skills sufficient to obtain pertinent histories of the deceased, in order to assist in proper determination of cause and manner of death.

Canada has coroner and medical examiner systems, whose role is to make and offer recommendations to improve public safety and prevention of death for unnatural, unattended, unexpected or unexplained deaths [8]. Coroner services fall under the jurisdiction of the provincial or territorial government. A chief coroner is supported by a team, including other coroners, medical examiner physicians and investigators who are appointed by executive council. By law, coroner jurisdictions in Alberta, Manitoba, Ontario, Nova Scotia, Newfoundland and Labrador and Prince Edward Island are physicians, and therefore they are considered medical examiners, not coroners [8]. Coroners are not necessarily physicians in other Canadian provinces, but many have background in law or medicine.

In Japan, there are two different systems of death investigation: criminal inspection and judicial autopsy from a criminal justice standpoint; and, from a public health standpoint, administrative inspection and either administrative or consent autopsy. As a result, it is sometimes unclear which system is responsible for investigating a death [9]. Japanese coroners are police detectives who have significant field experience, with the majority holding the rank of captain or higher and having studied forensic medicine. The Japanese system does not concentrate on preventing loss of life in similar circumstances, which most other coroner systems consider.

Hong Kong coroners are judicial officers who oversee many of the same responsibilities as US coroners, including determining when an autopsy must be performed, ordering inquests and issuing certificates of fact of death. If the coroner determines that a death should be investigated, the police carry out the investigation and submit a report outlining the details of the death to the coroner.

The Chinese modern medicolegal death investigation system was established in the 1930s, pioneered by Professor Lin Ji. Since that time, the system has evolved into one that is based on the administrative divisions of China, due to the vast territory and dense population. At present, medicolegal investigations are conducted by forensic medical experts mainly within five relatively independent agencies [10]:

1. police organizations;
2. prosecutor's offices;
3. departments/divisions of forensic medicine or science within medical colleges and universities;
4. the Institute of Forensic Science in the Ministry of Justice; and
5. the government or private forensic societies.

Each agency is responsible for specific investigations. The police organizations investigate homicides, suicides, unintentional injuries, drug-related deaths and deaths with suspicion. The prosecutor's office investigates deaths in police custody, in prison and in correctional institutions. The remaining agencies investigate cases sent to them by the police, prosecutor's office, courts or any other that the decedent's family requests [10]. Many of these cases are civil in nature and result from therapeutic misadventures.

The Italian system does not employ medicolegal death investigators. If a non-natural death occurs, or the manner of death is natural but is related to medical or nursing malpractice, the autopsy and medical records examination is performed by Legal Medicine Specialists, who are the equivalent of forensic pathologists. However, there is no law that limits the competencies of the specialist who performs the external body examination or the anatomical dissection. It is not surprising, therefore, that specialists in other medical disciplines are required by prosecutors to perform forensic activities, even though the absence of forensic methodology and competencies appear later at the cross-examination [11].

Italian medical examiners are required to respond to crime scenes but there is no law to state the competencies of those who perform external body exams or autopsies. Prosecutors do not require that a forensic pathologist respond to crime scenes, even though their participation is considered essential and mandatory. Crime scene investigations are sometimes conducted by the forensic pathologist and police officers together; at other times, police officers conduct scene investigations without the assistance of the forensic pathologist, resulting in inadequate procedure ending with controversial trials. In fact, crime scene investigators commonly collect evidence and reconstruct the dynamics of the event without input from the medicolegal representative globally, with the same problematic ending [11].

2.2.3 Education, training and certification

There is no baccalaureate degree specifically for medicolegal death investigation in the United States, although several forensic programs do include a course or two on this topic. To compensate for the lack of formal education, there are a number of basic training courses conducted across the country that teach investigative techniques and procedures in how to perform a thorough, competent medicolegal death investigation, as well as specialized courses that teach specific topics to enhance knowledge in ancillary forensic sciences, such as blood stain pattern analysis, entomology, etc. However, most medicolegal death investigators receive on-the-job experience instead of formal training. This form of training presents several deficiencies, including an absence of criteria to establish levels of competency.

As mentioned, qualifications for coroners still vary significantly by jurisdiction. Some states require training in order to take the position, but this is usually minimal, with no requirements for continuing education. Training requirements range from 8–80 hours of instruction and are typically provided within the specified state by coroner or medical examiner associations/offices.

Because jurisdictional qualifications and skills for those employed by medicolegal systems vary widely, the quality of global death investigations differs significantly. It is advantageous for the medicolegal death investigator to possess knowledge and skills to screen cases in a 'scientifically defensible manner'. Properly trained and educated medicolegal death investigators are capable of deciding how to proceed with a case after the initial death report, screening phone calls and following up with completed medical, psychological and social histories.

College and university programs offer certificates and bachelor degrees in forensic science. Greater opportunities for training have also emerged in recent years. Online training has been made available through funding provided by National Institutes of Justice (NIJ) grants. However, training conducted by local offices or jurisdictions may not be accredited by any professional organization. Professional accreditation is an accepted, structured process, requiring specific documentation to ensure that the training meets rigorous standards. A goal would be to ensure standardized training so that everyone receives the same level of competency in their education.

In other countries, almost all coroners are trained in law and/or medicine. Of those who are medically trained, however, few are current practitioners of medicine and they rarely perform as forensic pathologists.

2.2.4 Key issues

The medicolegal death investigation system overall has greatly improved in recent decades, but there are still significant weaknesses and obstacles worldwide. Systems remain inconsistent and fragmented, with low budgets and wide variances in best practices, policies and reporting laws. Qualifications for American coroners continue to be deficient in the majority of cases, as there is no education or training required. Individuals continue to be elected based on their political influence instead of skills and knowledge.

The position of the medicolegal death investigator continues to be plagued by low pay, lack of educational funding, enforceable death scene protocols such as those put forth by the National Institutes of Justice 1999 publication *Death Investigation:*

A Guide for the Scene Investigator and mandatory certification. The fact that many investigators have different job skills and titles results in a professional identify crisis [12].

There is great variance in the scope, extent and quality of current investigations worldwide, as politics continue to greatly influence the Chinese medicolegal system and United States coroner systems, for example. Global governments are attempting to standardize existing systems by establishing guidelines, best practice policies and regulations for medicolegal death investigation.

A lack of resources and budgets to support the growing number of personnel needed to perform well-defined investigations creates inefficiency in many countries. On many occasions, investigations are conducted with inadequate operations, contributing to problems in quality control mechanisms across the board.

Professional certification provides official recognition by an independent certification organization that an individual has acquired specialized knowledge and skills in the standards and practices necessary to properly perform job duties. The American Board of Medicolegal Death Investigators (ABMDI) was established in 1998 by veteran, practicing medicolegal death investigators in the United States, to certify death investigators nationally. In response to the need for international certification, the ABMDI has expanded its certification program to include other countries that can meet the qualifications and maintain certification with the required continuing education component.

The ABMDI established uniform criteria, ethical conduct and standardized practices for individuals investigating deaths, as well as requiring 45 hours of approved continuing education within a five-year period as part of the recertification requirements. There are two levels of certification: registry and board certification. Currently, there are 1,454 registered diplomates and 194 board certified fellows representing 49 US states, Canada, Puerto Rico and Australia. Pakistan and Kosovan medicolegal death investigators have applied and are in process.

2.2.5 Research and the future

Approximately one percent of the United States population dies annually, amounting to over two million deaths. Of those deaths, about 50% are referred to medical examiner/coroner offices, and complete death investigations are conducted approximately 20% of the time [13]. This estimate can be extrapolated to other countries, resulting in similar numbers. Death investigations aid civil litigation, criminal prosecution, and they are critical for public health practice, safety and research. Medicolegal death investigators frontline efforts provide closure for survivors and provide documentation and information to families to complete necessary insurance and estate issues.

Despite the type of system in which medicolegal death investigators are employed, their role is necessary for effective, efficient and complete death investigations. A proper evaluation of the death scene provides the best opportunity to obtain accurate, thorough and timely information surrounding the events of that death. Drs. Prahlow and Lantz stated: ‘Indeed, the death investigator is perhaps the most crucial element in the entire death investigation process’ [14]. Until mandatory standards of practice are enforceable, however, the quality of death investigations will continue to vary significantly.

Budgets also play an important role in determining the extent of the investigation. Increased funding for medicolegal jurisdictions is greatly needed in order to provide high quality services to the communities they serve.

With the advent of media programs and attention on the role of the medicolegal death investigator, the pool of qualified applicants has increased. Many have baccalaureate degrees in forensic science or even master’s degrees in areas like anthropology, increasing the overall education level of those seeking positions in the field. It is not uncommon to have tens of applications for one position, allowing employers the opportunity to choose the most qualified individual to meet the office needs.

Generally, medicolegal death investigators are not involved in national or international organizations. This is partly due to the fact that they often have to fund their own annual dues and/or travel to meetings. However, participation in these activities helps to elevate the status of the position. Presentations and peer-reviewed published papers also assist in professionalizing the field. Thus, medicolegal death investigators should make more of an effort to be involved in national or international organizations and to give presentations, even if they have to fund meeting attendance themselves. Death investigation, like any other profession, requires an individual to take pride in their work product and interest in furthering their education, even if they have to bear the expense themselves.

The International Association of Coroners and Medical Examiners (IACME) has been in existence for more than 70 years to offer educational opportunities and assist coroners and medical examiners in their duties. The National Association of Medical Examiners (NAME) has an affiliate status which allows medicolegal death investigators who are sponsored by a medical examiner to participate. The American Academy of Forensic Sciences (AAFS) includes death investigators and coroners in their General Section. The Society of Medicolegal Death Investigators (SOMDI) was established in 2011 to provide a much-needed voice specifically for death investigators to the forensic science community, interdisciplinary agencies and local, state and federal governments.

It appears that the perception of medicolegal death investigators has evolved from an undefined, uneducated field with no recognized standardization, to one where standards do exist. Professional status and respect is increasing as others observe improvements within the field. Stringent job requirements now exist, including mandatory certification and continuing education. The competitive job market will continue to increase the educational level of those working in the field.

2.3 Crime scene investigation

Crime Scene Investigation (CSI) is arguably one of the most diverse fields in forensic science, due to the variability in laboratory facilities, the qualifications and caseloads of investigators and the types of crimes handled. Accordingly, it is also one of the most important fields in forensic science, because the forensic investigation begins with the crime scene.

2.3.1 Definition of the field

A crime scene is the place where a crime has occurred and/or a place where evidence from a crime is or has been located. However, crime scenes can be further defined by describing them based on their environment (indoor, outdoor), their location (primary, secondary), the offense (homicide, burglary), their size (macroscopic, microscopic) and the criminal activity (active, passive). Additional information such as this is often needed to obtain the necessary equipment and people adequately to process and investigate the scene.

Once on scene, the crime scene investigator will employ a variety of investigative and documentation techniques based on scene characteristics. The scene investigation begins with the investigator conducting a walk-through with the first responder to determine what happened. Subsequently, a variety of documentation techniques will be used, based on agency protocol and the scene characteristics. Those include video recording to give a three-dimensional view of the scene; photography to give a detailed visual account of the scene; and sketching, which gives accurate measurements of the evidence and scene location. After the documentation is complete, a search for additional evidence is conducted, as items in the scene should not be disturbed before they are documented.

Next, the processing of specific items is conducted. This could include fingerprint processing, performing presumptive tests for biological fluids, chemical enhancement of pattern evidence or casting footwear impressions and tool marks. Other advanced techniques may also be employed, including the reconstruction of a bloodstain pattern or determining the trajectory of a projectile. The techniques required will be based on the scene characteristics and will vary widely, depending not only on the scene itself but also on availability of equipment, the education and training of investigators and agency protocol. Finally, the evidence is collected, using proper evidence collection and preservation techniques to ensure its integrity and its ability to be processed further in the crime laboratory. While many believe the CSI process ends here, this only concludes the scene investigation.

The investigators must subsequently upload their photographs to their case file, complete a final sketch of the scene and prepare an investigative report. In addition, the evidence collected must be documented on a chain of custody form, providing a detailed description of the evidence collected and all individuals who have been in contact with that item of evidence. Items collected from the crime scene may require further processing in the CSI lab, which could include specialized photography, chemical processing of fingerprints or enhancement of bloodstain evidence.

A case file, either in hard copy or electronic format, will then be constructed, consisting of the CSI reports, notes, photographs, sketches and other supplemental materials. And as is the case with any forensic investigation, the CSI could later be called to court to testify not only about the case at hand, but also about the scientific nature of the techniques used to gain results.

While the use of all the documentation and processing techniques listed previously are not used on every scene, they are typically required on larger scenes, especially homicides. However, in the majority of instances, CSIs are actually responding to lesser crimes such as burglaries and automobile thefts. In these cases, a video or sketch of the scene may not be necessary and will be determined by agency protocol or by the investigator. In addition, the types of crime scenes handled by a unit will vary, based on geographical location and crime rate. For example, a large metropolitan police department may handle thousands of cases per year, with each investigator averaging 40 plus cases a month, including 70–100 homicides per year and thousands of property crimes. However, a more rural CSI unit may average one homicide a year, handle far fewer calls and spend the majority of their time on lesser crimes.

2.3.2 Education, training and certification

Since there are numerous services provided by the different crime scene units, there is also a variety when it comes to each agency's requirements for education and training. The minimum qualifications required range from a high school diploma to a master's degree, while some require investigators to be certified peace officers (Peace Officer Standards Training or POST certified). While some civilian job postings may list a high school diploma as the minimum qualification, many will not hire an individual without an associate's or a bachelor's degree, as the agencies are receiving more applicants (and more qualified applicants) due to the popularity of the field and the increase in educational programs. The degree field varies as well, with most agencies requiring a degree in a natural science or forensic science, while others accept a degree in criminal justice. Hiring decisions are typically agency-driven when it comes to CSI.

Many agencies also require certification as a crime scene investigator by the International Association for Identification (IAI) within a certain time period. Other agencies may substitute a state-approved certification for the IAI certification. In addition, a set number of hours of continuing education are typically required each year to ensure that investigators are aware of the latest trends in the field.

Many individuals have become interested in the field of CSI but, before entering this field, one should know that while television shows do a good job of portraying some aspects of CSI, they do a poor job in the portrayal of other aspects. Thus, the term 'the CSI Effect' was coined to highlight these misnomers. For example, investigators on television are rarely seen accomplishing the often daunting and always time-consuming task of completing paperwork, although that is a large portion of any CSI's job. They are portrayed responding to exciting crime scenes, mainly homicides, which are riddled with forensic evidence, when in reality the majority of crime scenes are property crimes. Finally, the field is not glamorous. CSIs typically wear police-style uniforms due to the adverse working conditions, which can be environmental, structural, biological or chemical – and, since these are typically government jobs, extensive background checks are required for all employees. Thus, those with questionable criminal, financial and social backgrounds are not likely to be employed.

2.3.3 History of crime scene investigation

Due to the recent popularity of the field of crime scene investigation, it may appear to be a relatively new career field. However, its beginnings can be traced to as early as 1248, when it was recognized by the Chinese that the post-mortem examination of the body could give information regarding the cause of death. This led to a publication in 1250 that focused on issues such as blunt force trauma, drowning and examination of fire victims.

While this may seem to be primarily rooted in pathology, CSIs are responsible for using various disciplines to conduct their investigations. For this reason, many of the pioneers in other fields are also considered relevant to the development of the crime scene investigation process. Sir Robert Peel, who formed the first police force in London in 1829, and Allan Pinkerton, who was the first detective at the Chicago Police Department, were instrumental in the development of the investigative practice.

During those early investigative years, it became increasingly apparent that scientific research was critical to investigations. Thus, much of the research was focused on identification and, through this, Alphonse Bertillon created the first known scientific system of identification, known as anthropometry. While his system of using body measurements for identification was being used, research into the use of fingerprints continued to be conducted. This resulted in the 1880 suggestion by Henry Faulds that fingerprints from crime scenes could be used to identify criminals. His publication, along with ones by Sir Francis Galton, identifying a basic classification system for fingerprints based on minutiae, were the turning points in fingerprint research, which became widely used in 1903 when two individuals in Leavenworth Prison were found to have the same Bertillon measurements but different fingerprints.

In addition to fingerprint evidence, biological evidence has also played a large role in the history of CSI, with Karl Landsteiner's discovery of blood groups in 1901 and Leone Lattes' subsequent determination of the A, B and O blood groups from a dried bloodstain in 1915. Perhaps the most significant advance came in the 1980s, when Alec Jeffreys developed DNA fingerprinting and specified that individual identification could be made with certainty based on DNA [15,16].

There are also pioneers specific to crime scene analysis and reconstruction, including Hans Gross, Luke May and Henry T.F. Rhodes. Gross was one of the first people to speak of crime scene reconstruction when, in 1898, he stated that a crime scene reconstruction should be based on careful examination of evidence and the facts surrounding the case, thereby stating that a theory should have hard evidence to support it. May, in his 1933 writings, encouraged investigators to compile seemingly unrelated facts from a case in order to construct a theory free from personal conjectures. Many of his writings focused on searching for evidence to construct an accurate theory, rather than forcing evidence to fit an already devised hypothesis, thus reminding investigators to remain unbiased in reconstruction efforts. Finally, Rhodes, also in 1933, was the first person to suggest using the scientific method for reconstruction, and to tout the crime scene investigative process as a scientific process.

Other more modern day pioneers include Charles O'Hara, who, in his 1965 publication on criminal investigation, emphasized the importance of the scientific and objective analysis of physical evidence; and Dr. Henry Lee, who mentioned the importance of the scientific crime scene investigation process in several publications, including those in 1992 and 1999. Ross Gardner and Tom Bevel have also been pioneers in modern day crime scene investigations, with publications spanning decades, introducing new terminology and techniques such as event analysis and crime scene analysis [17].

Any history of crime scene investigation would be remiss if it failed to mention the Locard Exchange Principle and Paul Kirk. The Locard Exchange Principle, published by Edmund Locard in 1920, is the basis for crime scene investigation, as this principle states that evidence is transferred with every contact. In addition, in 1974, Paul Kirk postulated that physical evidence is always present at crime scenes, and the only way it can be compromised is through human error. Both of these historical events are significant, as they illustrate the importance of recognition of evidence at a crime scene, and they showed remarkable forethought by professionals [16,18].

Thus, the historical progression in crime scene investigation is based primarily on the pressures of science, causing a transition to modern police and forensic science agencies which are rooted in valid and reliable techniques, performed by educated and well-trained professionals.

2.3.4 Key issues within crime scene investigation

One of the main issues in CSI is the lack of standardization in the field, as outlined in the 2009 National Academy of Sciences (NAS) report on *Strengthening Forensic Science in the United States*. As previously mentioned, there are no national educational or certification standards for crime scene investigators. While that does not mean that there are no standards or protocols in the field, it does mean that, unlike other fields, the educational level of an individual is not obvious by their title. One investigator may be a police officer with little to no science background, while another may have a master's degree in forensic science. Furthermore, certifications by the IAI or other such bodies are not required by all agencies, so a level of continuity across the field is not endorsed. The importance of standardization lies with public perception; most agencies are likely not improperly operating but, without standards in place, no one can be sure.

The most significant scientific issue in crime scene investigation is the use of the Scientific Crime Scene Investigation model, which challenges investigators to go beyond the traditional step-by-step approach to CSI and employ the use of the scientific method. While the methodology of CSI is certainly important, investigations are lacking if the investigator simply goes through the steps, does not use inductive and deductive reasoning and does not analyze the entire scene.

The scientific approach extends beyond processing of the evidence to processing of the scene. In other words, the investigator should determine the question, gather data/evidence, form a hypothesis, perform experiments to test the hypothesis and then analyze the data to form a conclusion or a new hypothesis. Using this method not only produces better results, but also demonstrates the scientific approach to investigations and the use of the scientific method to arrive at a feasible conclusion or theory [17,19].

In other countries, standardized techniques are the norm but variation is also seen in regard to personnel qualifications. For example, in Australia, most of the Scenes of Crime Officers (SOCOs) are police officers, but in some states they do hire civilians. In Queensland, police officers wishing to be SOCOs may apply to do so after three years of service and, if accepted, they are required to undergo an intensive six-month training program followed by an eleven-month probationary period, at which time they are awarded a Diploma of Public Safety in Forensic Investigations. These officers are then qualified to handle basic crime scene processing, while the more intensive processing is handled by Scientific Officers, who not only have more training but also are required to have a science degree and continue their education to obtain a master's degree (Leah Tigchelaar, 2011, email communication).

In Canada, those wishing to be Forensic Identification Apprentices with the Royal Canadian Mounted Police (RCMP) must first be officers and then apply to be accepted to the Forensic Identification Apprenticeship Training Program (FIATP). Certain forensic science-based competencies must be met for acceptance to the program, to satisfy the Forensic Identification Suitability Assessment. A three-year commitment to service is also typically required once the apprenticeship is complete [20].

2.3.5 Research and the future

While the standard methodology, along with the processing, documentation and collection procedures, remains relatively unchanged in the CSI field, technological progress has allowed for the implementation of advanced techniques. Nevertheless, this research is typically conducted in other fields of study and then adapted to CSI. For example, the research that created digital photography was primarily done in the photography industry, but CSIs were then able to use that technology to convert from film to digital cameras. Along the same lines, research primarily conducted in the architecture and drafting fields to create computer-aided drawing programs changed the way crime scene sketches were done, just as total station systems have changed the way that measurements can be recorded.

Accordingly, DNA research resonated with CSIs. It was once necessary to collect a biological sample about the size of a quarter but, with the advent of PCR, it was only necessary to collect a sample about the size of a dime. With further advances in STRs, a sample that is not even visible can be used to obtain a DNA profile. Obviously, these advances changed the way that crime scene investigators recognize and collect biological evidence, just as advances in alternate light sources and chemical enhancement methods have changed those procedures. Current research is focused on the ability to capture crime scene data quickly by scanning the scene with a computer system developed for battlefield forensic investigations [21]. This technology would be useful in scenes that are hazardous or need to be worked quickly for other reasons. Therefore, while there is some research conducted specific to crime scene investigation, much of the research originates in other fields and is applied to CSI.

In the future, CSI practitioners are likely to see more standardization in the field, specifically a required industry certification and a minimum degree level requirement. It is likely that CSI will be similar to most medical fields, requiring an appropriate degree and board certification prior to, or immediately following, employment. Not only will this provide more continuity within the field, it will also be necessary, with continued advancements in technology, to have well-educated and trained individuals to perform such techniques.

The popularity of the field will also likely level off at some point but, in the meantime, not only have educational programs in crime scene investigation continued to increase, the funding for the field has increased as well. These advances will likely lead to better-funded CSI units and more educated and trained employees, which should result in the continued betterment of the CSI practice. As the first generation of graduates from newly developed educational programs are only beginning to emerge, one can only imagine the effects on the field as a whole. Perhaps, with more educated employees and more of a devotion to scientific techniques, we will continue to see technological advances and be performing the techniques seen on television in the real world. In the future, fingerprint and DNA database searches may even be able to be conducted from the crime scene.

2.4 Forensic nursing

2.4.1 Definition of the field

Forensic nursing science is a discipline that combines the knowledge, skills and techniques of nursing with those of forensic science. The primary role of the forensic nurse focuses on the investigation of traumatic injuries and the recovery of medical evidence in living and deceased individuals, or investigates practices and staff performance within healthcare organizations that are, or potentially could be, associated with negative patient outcomes.

Subjects of a nurse's forensic investigation may be victims, suspects, healthcare organizations and known perpetrators of violent or unlawful acts. The majority of forensic nurses are engaged with cases involving interpersonal violence, such as sexual assault, child and elder abuse or neglect and domestic violence. Undoubtedly, the focus on these types of victims is closely related to the historical foundations of forensic nursing in the United States, which are deeply rooted in the forensic examination of rape victims. Consequently, the earliest champions and financial support for forensic nursing organizations and staff positions in hospitals and clinics stemmed from groups invested in assisting and protecting victims of violence.

In the official publication *Forensic Nursing Scope and Standards of Practice*, the International Association of Forensic Nursing (IAFN) defines the domain of forensic nursing as 'nursing globally when health and legal systems intersect'. The document further explains that forensic nurses address the needs of both victims and perpetrators of violence, and 'care for the physical, psychological, and social trauma that occurs in patients who have been assaulted or abused'. Such nurses possess a 'specialized knowledge of the legal system, and collect evidence, provide testimony in court, and provide consultation to legal authorities' [22].

2.4.2 History of forensic nursing

The earliest roots of forensic nursing can be traced to forensic medicine and pathology. However, forensic roles for nurses became more realistic with the advent of the concept of 'living forensics'. In 1988, Harry C. McNamara, Chief Medical Examiner for Ulster County, New York, conceived a special forensic role which involved proper processing of evidence associated with living victims of trauma. He was the first to recognize and publicize the potential role of nursing in evidence collection within healthcare settings, especially the Emergency Department [23].

In 1991, The American Academy of Forensic Sciences (AAFS) recognized forensic nursing as a unique discipline within the General Section. However, professional nursing did not formally acknowledge its status until 1995, when the American Nurses Association's Congress of Nursing Practice approved the Scope and Standards of Practice and a Code of Ethics submitted from the newly formed organization, the International Association of Forensic Nurses [23].

The IAFN was founded in Minneapolis, Minnesota, in 1992. Dr. Linda Ledray, a well-known sexual assault examiner, had convened the group in order to establish a formal organization for this nursing specialty. Virginia Lynch, an attendee at the meeting, encouraged the group to expand its vision to include a broad range of other viable forensic opportunities for nurses, such as death investigation, forensic psychiatric nursing, correctional care and the management of patients who have sustained trauma as a result of interpersonal violence. These original nurses adopted an international vision for their newly conceived organization and, on that day, the International Association of Forensic Nurses (IAFN) became a reality [18].

Most of the original forensic nurses in the US were either sexual assault examiners (SANES), or participated in investigations of interpersonal violence involving intimate partners, children or the elderly. However, although other nurses became enticed with death investigation roles, obtaining the required education and practical training posed a significant challenge. Some worked with medical examiners, or served as nurse coroners or forensic autopsy assistants, while others became field investigators and worked with forensic toxicologists or with medical examiners. Additional specialty interests that were gaining attention included forensic psychiatric nursing, correctional nursing, clinical forensic investigation and legal nurse consulting.

The *Journal of Forensic Nursing* (*JFN*), IAFN's official peer-reviewed journal, published its inaugural issue in 2005. The 2010 IAFN Annual Report indicates that *JFN* reaches approximately 3,500 readers in 20 countries [24]. The first comprehensive textbook for forensic nursing was published in 2006 and is now in its second edition. At least eight additional specialty books have been authored by forensic nurses. Twelve forensic nurses hold the rank of fellow in the American Academy of Forensic Sciences.

2.4.3 Areas of practice

The realm of practice for forensic nurses has been expanded to include several specialties in which the acumen of nursing can be applied. Among these are:

- death investigation;
- forensic psychiatric and correctional nursing;
- accident and incident reconstruction; hospital risk management;
- healthcare fraud investigations;
- human trafficking and exploitation;
- medical device failures resulting in injury or death.

Specialized skills of nurses which are applied in such roles include:

- systematic physical assessment and trauma evaluation;
- focused interviewing and documentation that can withstand legal scrutiny;
- forensic photography;

- interpersonal communication and behavioral analyses;
- retrieval and use of digital data stored in medical equipment and monitoring devices;
- collection of biological specimens and other evidence in healthcare settings;
- providing fact or expert witness testimony by deposition or courtroom testimony.

Forensic nurses are valuable resources within hospitals and in any community where there is an interface of healthcare and the law. Among the caseload of today's forensic nurses are:

- patients presenting with injuries associated with interpersonal violence;
- victims of human neglect and abuse in hospitals, nursing homes, correctional facilities or the community;
- incidents of healthcare fraud or sub-standard practices that endanger the lives of patients;
- sudden, unexpected deaths in hospitals or domiciliary institutions;
- failure or malfunction of medical devices that injure patients or result in death;
- identification of victims associated with human trafficking and other social-cultural exploitations; and
- protecting the rights of those in custody by application of psychiatric assessment skills to determine their competence to withstand judicial proceedings.

Although these are some examples, this is not an exhaustive list of all types of forensic cases in which the unique expertise of forensic nurses can be utilized. However, they illustrate the versatility of this unique human resource, both within healthcare and the community.

2.4.4 Education, training and certification

Forensic nurses must be registered professional nurses, preferably with a baccalaureate degree in nursing. Specialized roles, such as sexual assault nurse examiner or death investigator, typically mandate additional formal training and certification. Forensic nurses engaged in research or education usually have earned a master's degree or a doctorate to qualify for their positions. In most states, advanced practice roles and clinical specialization require a master's degree.

Forensic content is now being integrated into some progressive undergraduate nursing curricula, but most aspirants still develop their acumen by relevant clinical experiences with a forensic science mentor, attendance at workshops and conferences and continuing education programs sponsored by forensic-related professional organizations. During the last two decades, nurses have relied heavily upon online education for both basic and advanced study as well as continuing education.

In the United States, 2011 data reflect approximately ten schools of nursing which offer curricula leading to a master's degree in forensic nursing, and at least two others which offer a post-graduate Certificate in Forensic Nursing. One additional school permits students to pursue a Minor in Forensic Nursing [23]. There are some universities that provide limited opportunities for students to enroll in open electives offered within its several academic curricula. Online courses for college credit and continuing education units are also widely available. Professional organizations and commercial enterprises offer yet more alternatives for nurses who want to explore the specialty or advance their acumen in the forensic sciences.

Educational opportunities have not kept pace with the demand from students, primarily due to financial constraints and curricular limitations within schools of nursing, as well as a dearth of qualified faculty resources. Although post-master's and doctoral studies in forensic nursing emerged in 2002 in the United States, there are only about 150 forensic nurses with earned doctorates. Few of these individuals have full-time forensic nursing faculty positions, however. Existing curricula in most programs still rely heavily upon part-time or volunteer lecturers from related disciplines, such as criminal justice, law enforcement, psychiatry or psychology, forensic pathology or schools of law.

Certification standards for Sexual Assault Nurse Examiners (SANEs) were developed by the Forensic Nursing Certification Board of the International Association of Forensic Nurses. These original standards were approved and published in 1996. Approximately 1,200 nurses now hold certification credentials for performing adult (SANE-A) and/or pediatric (SANE-P) sexual assault examinations.

In the 1990s, some of the original forensic nurses who aspired to being death investigators received special training at the Dade County Florida (Miami) Injuries and Death Investigation Course, designed especially for nurses by Dr. Joseph Davis, Chief Medical Examiner. Others attended the Medical/Legal Death Investigators Training course at the Saint Louis University School of Medicine. Several of these nurses have become certified by the ABMDI and work in various settings as death investigators or nurse coroners.

In 2010, the IAFN published Forensic Nurse Death Investigator Guidelines. However, certification examinations have not yet been developed and, unfortunately, both training and employment opportunities remain limited for nurses in this specialty.

2.4.5 Key issues within forensic nursing

A primary concern for forensic nursing is its strong historical ties to victims and their rights, as opposed to operating in an unbiased framework, seeking truth solely on the basis of evidence and application of scientific methods. Many early positions for sexual assault nurse examiners were funded by the local Office of the Prosecutor, and much of the work of forensic nursing has been underwritten by grant support from organizations whose mission relates to the protection of human rights or women's advocacy initiatives, such as the Office of Violence against Women. These associations, although helpful in supporting the growth of forensic nursing, have given rise to criticisms that forensic nurses are 'victim-oriented' and are not unbiased scientists who give equal attention to the forensic examinations of suspected offenders or perpetrators of interpersonal violence who may be falsely accused.

In past decades, some sexual assault nurses have been subject to disciplinary action, and even employment termination, for performing suspect examinations or testifying for defendants. However, in recent years, there is increasing anecdotal evidence of jurisdictions and agencies permitting forensic nurses to perform evidentiary examinations for suspects as well as for victims, and to allow them to testify on behalf of their client without impunity. It is imperative that forensic nurses be viewed as 'truth-seekers', and that their scientific activities and testimony be unbiased, whether representing the Prosecution or the Defense.

2.4.6 International scope of forensic nursing

There are international representatives among the nursing membership of AAFS and IAFN. Forensic psychiatric facilities and correctional settings, especially in the United Kingdom have employed forensic nurses for many decades. However, other forensic nursing roles are slower to develop outside the USA. There are many influences in certain countries that curtail the progress of forensic nursing, including economic constraints, gender bias and how human rights of women and children are perceived in some cultures.

Several nursing fellows within the AAFS have been active overseas, training physicians and nurses in forensic examination procedures. Although many of these initiatives have related to training individuals in sexual assault examinations, other endeavors have related to human trafficking, child abuse, women's rights and war-related mass casualty death investigations.

The first international meeting for IAFN was held in Calgary, Alberta, in 2000. In 2001, reports were presented about work in Kosovo at the AAFS Annual Scientific Meeting, and multiple forensic nursing papers reporting on overseas activities were proffered at additional professional meetings. Current projects staffed by AAFS forensic nurse members are based in Zimbabwe, Kenya, Peru, Central America, Portugal and Italy. In 2007, the University of Bari (Italy) graduated its first class of 42 students with master's degrees in forensic nursing, and plans are now underway there for the development of doctoral study curricula [18].

2.4.7 Practice trends

The Joint Commission for the Accreditation of Healthcare Organizations has acknowledged the value of forensic nurses within hospitals. Although many hospitals have offered sexual assault nurse examiner services for two or three decades, more facilities are broadening forensic services to include investigation of cases involving intimate partner violence, human abuse and neglect, medical device failure and cases of unexpected deaths within the domain of risk management.

The US Army, Navy and Air Force, and the Department of Veterans Affairs, now have special initiatives and career opportunities for forensic nurses. Public health organizations are seeking the contributions of forensic nurses in the US and in international programming. Perhaps the most important indicator of the acceptance of forensic nurses is that hospitals have now acknowledged the value of their understandings and skills in forensic science, and are extending the roles of the sexual assault nurse examiner into other areas of patient care [25].

2.4.8 Research

Research in forensic nursing has been steadily increasing as more nurses pursue post-graduate study. There are also grants available to support studies related to interpersonal violence, especially sexual assault, domestic violence and human abuse of vulnerable subjects such as children, the elderly and those in custody. Nurses are often involved with other forensic scientists in research endeavors, and published research illustrates the value of these partnerships.

References

1. Field KS. *History of the American Academy of Forensic Sciences, 1948-1998*. West Conshohocken, PA: ASTM; 1998.
2. Freed RA. A Revisit to the Beginning: History of the General Section. *Proceedings of the American Academy of Forensic Sciences*; 1998 Feb 9–14; San Francisco, CA: American Academy of Forensic Sciences, 1998; 94.
3. Thurston G, Knapman P, Powers M. *Thurston's Coronership: The Law and Practice on Coroners*. 3rd ed. London: Barry Rose Publishers Ltd; 1998.
4. Knapman P. The Crowner's quest. *Journal of The Royal Society Of Medicine* 1993; 86:716–720.
5. Pen Z, Pounder D. Forensic Medicine in China. *American Journal of Forensic Medicine and Pathology* 1998; 19(4):368–371.
6. Hunnisett RF. *The Medieval Coroner*. London: Cambridge University Press; 1961.
7. Hanzlick R. *Death Investigation Systems and Procedures*. Boca Raton, FL: CRC Press; 2007.

8. The Coroner System. www.usw.ca.program.content.3179.php (accessed October 1, 2011).
9. Fujimiya T. Legal Medicine and the Death Inquiry System in Japan: A comparative Study. *Legal medicine (Tokyo, Japan)* 2009 April; 11 (Suppl 1): S6–S8.
10. Chang L, Zhang JD, Baosheng Y, Yan P, Fowler D, Li L. Current Medicolegal Death Investigation System in China. *Journal of Forensic Sciences* 2011; 56(4):930–933.
11. Di Vella G. *The Italian Tradition of Legal Medicine*. National Association of Medical Examiner's Interim Meeting presentation. 2011 Feb 22; Seattle, WA.
12. Jentzen J. Ernst M.F. Developing Medicolegal Death Investigator Systems in Forensic Pathology. *Clinics In Laboratory Medicine* 1998; 18(2):279–322.
13. Board on Health Promotion and Disease, Prevention. *Medicolegal Death Investigation System: Workshop Summary*. Washington, DC: National Academies Press; 2003.
14. Prahlow J, Lantz P. Medical Examiner/Death Investigator Training Requirements in State Medical Examiner Systems. *Journal of Forensic Sciences* 1995; 40(1):55–58.
15. Becker RF. *Criminal Investigation*. Boston: Jones and Bartlett; 2009.
16. Dutelle A. *An Introduction to Crime Scene Investigation*. Boston: Jones and Bartlett; 2011.
17. Gardner RM, Tom B. *Practical Crime Scene Analysis and Reconstruction*. Boca Raton: CRC Press; 2009.
18. James S, Nordby JJ. *Forensic Science: An Introduction to Scientific and Investigative Techniques*. 3rd ed. Boca Raton: CRC Press; 2009.
19. Lee HC, Palmbach T, Miller M. *Henry Lee's Crime Scene Handbook*. San Diego: Academic Press; 2001.
20. Forensic Identification Apprentice – Lateral. Advertisement. Infoweb Job Opportunities; 2011 May 11. http://infoweb.rcmp-grc.gc.ca/job/job_bulletins/1-11-50-276-eng.htm (accessed October 1, 2011).
21. Tamburini E. *Battlefield Forensics. PowerPoint file*; 2010.
22. International Association of Forensic Nurses (IAFN). http//iafn.org/displaycommon.cfm?an=1&subarticlenbr=137 (accessed September 24, 2011).
23. Lynch VA, Barber Duval J. *Forensic Nursing Science*. 2nd ed. St. Louis: Elsevier; 2011; 4–5.
24. Annual Report. Arnold, MD: International Association of Forensic Nurses; 2010.
25. Goryl C (Executive Director). International Association of Forensic Nurses. www.iafn.org (accessed September 12, 2011).

Further reading

Chang L. A Research on Expert Opinion and Preventing Post Decision Dispute. (Chinese) *Evidence Scientific Journal* 2009; 5(7):629–633.

Chang L. Discussion on the Ten Major Issues in the Reform of Judicial Examinations. (Chinese) *Evidence Forum* 2007; 13:171–176.

Clark SC, Ernst MF, Haglund WD, Jenzten JM. *Medicolegal Death Investigator: A Systematic Training Program for the Professional Death Investigator*. Big Rapids, MI: Occupational Research and Assessment Publishing; 1996.

Clark SC. Jentzen JM. Medicolegal Death Investigator Training and Certification. *American Journal of Forensic Medicine and Pathology* 1996; 17:112–116.

National Research, Council. *Strengthening Forensic Science in the United States: A Path Forward.* Washington, DC: The National Academies Press; 2009.

Wagner S. *Death Scene Investigation: A Field Guide*. Boca Raton: CRC Press; 2009.

3

Criminalistics: the bedrock of forensic science

Susan Ballou[1], Max Houck[2], Jay A. Siegel[3], Cecelia A. Crouse[4], John J. Lentini[5] and Skip Palenik[6]

[1]*Law Enforcement Standards Office, National Institute of Standards and Technology, Gaithersburg, Maryland, USA*
[2]*District of Columbia Consolidated Forensic Laboratory, Washington, DC, USA*
[3]*Department of Chemistry and Chemical Biology, Indiana University Purdue University, Indianapolis, Indiana, USA*
[4]*Palm Beach County Sheriff's Office Crime Laboratory, West Palm Beach, Florida, USA*
[5]*Scientific Fire Analysis, LLC , Big Pine Key, Florida, USA*
[6]*Microtrace, Elgin, Illinois, USA*

3.1 Introduction

In 1953, and in the second edition released in 1974, Paul L. Kirk's *Crime Investigation* publications preached about the proper collection and interpretation of physical evidence. Dr. Kirk stated 'Physical evidence cannot be wrong; it cannot perjure itself; it cannot be wholly absent. Only in its interpretation can there be error' [1]. Words that were true then still ring true today.

Analysis of evidence in the late fifties was conducted with the understanding of science and technology at that era. The interpretation of the evidence in the 1950s was offered through testimony with conviction, supported by the sound underpinnings of science and bench laboratory research. Over the years, evolvement of technology and breakthroughs in research have unlocked doors, allowing the analyst to see beyond what was accepted in the past. DNA analysis is one example of a breakthrough that has dramatically changed the interpretation of physical evidence. DNA technology has shown past interpretations of physical evidence to be incorrect.

Does this imply that earlier experts were wrong in their conviction of the interpretation of the case evidence that was offered in testimony? We offer the response that these experts were correct in their conviction; that the science was

Forensic Science: Current Issues, Future Directions, First Edition. Edited by Douglas H. Ubelaker.

astutely applied according to the belief and understanding for that timeline in science. Science continues to evolve, and new technology continues to be discovered. The forensic analyst must accept these discoveries and keep an open mind to the changes that occur throughout the world.

Amidst these changes there is one area that maintains its status quo. Dr. Kirk stated: 'Law Enforcement officers probably make more errors in the collecting and subsequent handling of physical evidence than in any other phase of their work' [1]. He further stated: '[e]ven the laboratory investigator, who should appreciate the value of such evidence more than would any ordinary police officer, frequently destroys it quite casually' [1].

Although these statements were made half a century ago, they are just as accurate today as when Dr. Kirk made them. New recruits for the police department have a limited time in training. The collection and preservation of physical evidence is of low priority in comparison to learning the workings of the criminal justice system, self-protection, safety and crowd control measures. As the profession migrates through its promotional tiers, there is little time to emphasize the attributes and delicacies of physical evidence. The end result is that physical evidence suffers and the detective or investigator wonders why the experts failed in their analysis of the evidence. As a result, there continues to be evidence handling, collection and packaging training targeted to the crime scene technician, investigator, property custodian and novice analyst.

However, there is a misconception that only physical evidence training at the crime scene technician level is needed. Adhering to this position is tantamount to immediate failure. Understanding the movement of physical evidence from its collection to its analysis demonstrates the need to respect this as a team effort. Consider a team on the playing field – a superlative winning season is demonstrative of a well-oiled program. The same holds true for the management of physical evidence – each aspect of its movement through the system must be done by individuals trained the same way. Until this is recognized and enforced, slight deviations from proper collection technique, packaging and storage actions will irrevocably affect the evidence. Although the expert may suffer professional criticism for failing to provide the evidence that may solve a case, ultimately the victim is the one to suffer irrevocable consequences.

This chapter offers a look at these concerns in the world of criminalistics. Although the term 'criminalistics' includes an expanse of fields such as DNA, glass, paint, tape, illicit drugs, fire debris, biology, fibers and hair analysis, and applied instrumentation such as GC/MS, hyperspectral imaging, Raman and robotics, due to space restrictions only a few of these are highlighted in this chapter. The highlighted fields in this chapter – management, illicit drugs, DNA, fire debris and trace – discuss historical aspects to include the belief of the science and the limitations associated with that belief when new technologies came onto the horizon. As mentioned above and explained further in this chapter, the newly identified limitations of past science do not imply improper actions by the analyst – just the recognition that each analyst's conviction is justifiably based on the knowledge of the time.

This chapter looks at training objectives, new research requirements and other pertinent issues such as accreditation and certification. Significant research efforts being undertaken today that will impact these fields of study in the future are also discussed. The chapter will also address expectations for advancement and what training, education and coursework will be required by the next generation of students who are preparing to enter the dynamic field of forensic science.

3.2 Managing forensic services for quality performance

Forensic science managers, like all managers, need to strive for the best possible outcomes from the resources readily available to them. The success of a manager depends on the sound performance of his or her duties at a reasonable cost, which is a typical approach on how a manager is evaluated. This form of evaluation may lead managers to focus on short-term quantitative issues, such as cost, scheduling and production, rather than longer-term qualitative issues, such as quality and performance. For a manager to be effective, both the quantitative and qualitative issues must be balanced to produce the desired outcome.

Efficiency, the extent to which time and effort are used to produce the desired outcome, is often confused with effectiveness (the *capability* of producing a desired effect), but they are intertwined (one could fill a bathtub with water using a teaspoon – effective but not efficient!). The need for training and support in forensic laboratory management has been recognized for many years, but not enough has been done to transition the tools of business management to the forensic laboratory environment. However, efforts are continually expanding to overcome and counteract this deficiency.

3.2.1 The forensic industry

From a historical perspective, one reason there has been a lack of managerial development in the field of forensic science is due to the origin of forensic laboratories. Starting in the 19th century and into the 20th century, the field of scientific criminal investigation was in its infancy in the US. Most crime laboratories during this period were connected directly to state or local law enforcement agencies. Following the end of the Second World War, the field of forensic science became more widely recognized for its value in crime scene detection by both law enforcement and the public. Throughout the second half of the 20th century, numerous organizations were created or expanded in the different areas of expertise comprising the field of forensic science.

Currently, most forensic service providers are essentially governmental, non-profit, production-oriented organizations staffed largely by specialized knowledge workers. Forensic scientists take evidence and data of uncertain quality and origin, converting them into knowledge in the form of reports and testimony, which need to be of the highest level of quality in order to be of any intrinsic value to the criminal justice system and other outside third parties. A key efficiency in producing such reports is the fact that forensic scientists specialize in different fields of expertise and can therefore simplify and clarify the presentation of evidence and data for the benefit of the criminal justice system, alleviating investigators and attorneys from having to find separate individuals to conduct the specific examinations required for a case.

It is specialization of knowledge, not simply generalization of a subject matter, that is the most critical and important element available to a forensic knowledge worker. The more specialized the knowledge, the more useful it is. It should be noted that increased specialization does not imply that such knowledge will become more 'applied'. For example, many knowledge workers with highly specialized knowledge conduct very basic research, as with high-energy particle physics.

The forensic industry encompasses many types of specialties and sciences (as detailed in this book) that are not limited to 'the laboratory'. So long as the costs of providing such services to external agencies do not exceed the internal cost of production, then the forensic organization can prosper. If a government laboratory's costs are greater than the cost generated from a private laboratory providing the same services, then the need for evaluation, cost comparison, and managerial review is required.

A discussion that will likely persist for years to come concerns the mechanism for providing forensic services, that is, are they a public good, like police and fire services, or can they be best provided through the free market on a fee-for-service basis ('privatized')? With the problems faced by the United Kingdom's Forensic Science Service (FSS), which went from a wholly governmental agency to a near-fully privatized system and is now scheduled to be shut down in 2012 due to financial losses, this begs the question: can forensic science be turned into a commodity and sold like breakfast cereal or hamburgers, or is it better off as part of the political whirlwind? As governmental budgets tighten, these questions will be looked at more seriously and will require not only forensic professionals but economists, accountants and other business professionals to help study the forensic industry.

3.2.2 Forensic service processes: scene to court

Forensic services occur at many levels throughout the investigation of a crime, starting at the scene, through one or more laboratories or service providers, to reporting and ultimately testifying in a court of law. What complicates providing these services is the patchwork nature of jurisdictions at the city, local, state or province and national levels. Overlaid is the complex nature of all of the various forensic services, some of which are provided by more than one agency within a jurisdiction.

Crime scenes may be processed by one agency's employees – some of whom are sworn as officers, while others are civilians – who hand off evidence to one or more laboratories, who might also need to wait for evidence from other service providers such as toxicology laboratories or morgues. Every professional in this network of case involvement (a system of systems, ultimately) may be subpoenaed to testify. Thus, which agency performs what services for whom can become a tricky proposition – one that requires negotiations, diplomacy and no small amount of paperwork.

An issue related to this is outsourcing – one entity paying another entity to do tasks that are too costly (in time or money) for the first entity to do themselves. For example, is it more effective and efficient for an agency to have a staff of toxicologists, instrumentation and its upkeep and other attendant costs to work some number of toxicology samples, or should the agency just send them to a laboratory that specializes in toxicology and pay them per sample instead? This can be a difficult question to answer, and one that requires a well-honed understanding of costs, time, and process – how to budget and charge for testimony, for instance. Again, a business mindset is required, but one attuned to the requirements of civil service and a non-profit orientation.

3.2.3 Management vs. leadership

Historically, management and leadership were seen as two separate things; today, their relationship is seen as being far more intertwined. Still, the distinction persists that *things* are managed while *people* are led, with the goal of making the specific strengths and knowledge of each employee productive. Organizing a group of people to achieve that goal is considered to be the definition of leadership. Forensic managers are only now incorporating the lessons of management to their profession; leadership skills are also desperately needed in the forensic enterprise.

Communication and leadership are deeply intertwined with organizational effectiveness. Unless the organization's vision and goals can be clearly articulated to all relevant audiences, the group will not be effective. Ethics is clearly central to an organization's values, as this delineates how that group wants to behave towards each other and towards others as they achieve their goals in pursuit of the vision. The relationship between goals and vision are simultaneous, dynamic, and evolving. If these qualitative aspects are absent, an organization will not seem worthy of respect and integrity.

3.2.4 Scientists managing scientists

The education of scientists rarely includes preparation to become a laboratory *manager*; the common assumption is that management skills will come through on-the-job experience. Added to this situation is that scientists who manage laboratories see themselves primarily as scientists and secondly as managers. They tend to downplay or ignore the managerial aspects of their jobs. Forensic laboratory managers are no different, but the stakes are higher. given the importance of quality science to the criminal justice system.

Managing knowledge workers is far more complex than managing typical industrial or manufacturing workers. To manage knowledge workers properly, collaboration and professionalism must be the main emphasis. Conversely, incentives and performance measures should be de-emphasized; knowledge workers should be committed to their work and the organization's vision for its own sake. Relationships between managers and knowledge workers should be based on professionalism and mutual respect, not on a structured hierarchy with incentive schemes or production quotas. All of these factors relate to forensic service providers in their work.

3.2.5 Metrics for success

An agency's strategic value is a function of benefits as well as costs, and managers frequently will focus on the wrong measures. Comparatively, non-profit (such as government) and for-profit (private businesses) organizations are similar in some ways (money is an input for both), yet different (money, in the form of profits, is an output only for the private sector). Non-profits must, by definition, measure success in other ways, such as 'low-cost', 'cost-effective' or 'success'. Forensic service providers and their parent organizations use terms such as 'cost-effective' vaguely, without reference to other disciplines which use these as well-defined technical terms in evaluative phrases or formulae.

Despite the great concern and administrative angst over forensic service providers' 'performance' and 'capacity', these metrics go undefined as industry standards. Despite (or perhaps because of) this lack of explicit understanding, forensic service providers are held accountable by one or more agencies for their performance, based on non-standard, ill-defined or non-existent criteria. Successes and improvements go unrecognized, and opportunities for advancing the mission and goals of the organization are squandered.

The comparison of values between entities is called benchmarking. Comparisons may be made between units within a business, between businesses, or may be idealized or industrial standards. The most efficient and effective performers set the industry's best practices, usually in terms of time, cost or quality. Called 'best practice benchmarking', this process can provide invaluable benefit for process improvement and strategic planning for an organization.

Benchmarking can be a single event, but is typically a periodic (quarterly or annually) part of a continuous quality process for an organization. Benchmarking can help with assessments of efficiency (short-term gains) that lead to greater effectiveness (overall performance). A best practice benchmarking project for forensic service providers has been funded for several years by the National Institute of Justice through West Virginia University's Forensic Science Initiative and the College of Business and Economics. All of the agencies participating in the project, called FORESIGHT, submit their data under standardized definitions.

Without this requirement, any data would be incommensurable and perhaps misleading. For example, if Laboratory A reported a cycle time (or turnaround time, the amount of time required to complete a case) measured from the date of the first submission of evidence to the date that the report was issued (signaling the start of the casework), but Laboratory B reported their cycle time as the date of the last submission of evidence to the date of reporting (reasoning that the casework cannot be complete unless *all* of the evidence has been submitted), then the two numbers would be very different and could not be compared.

After an initial examination of mission, vision and values of the participating laboratories in the FORESIGHT study, several common themes emerged that led to a listing of goals. The key performance indicators (KPIs) related to these goals permitted the evaluation of performances by individual laboratories to fall into a series of categorical measures. Analysis is conducted by faculty at the WVU College of Business and Economics, in collaboration with the participating laboratories.

Forensic organizations across the world participate in the FORESIGHT project, which is free to any forensic service provider who submits data. Participant agencies already have seen the benefit of benchmarking and, with additional data, time series and trends can be calculated. It is hoped that FORESIGHT becomes an industry standard and will assist forensic service providers in improving their quality and effectiveness.

3.2.6 Quality is your product

Forensic services providers are considered high reliability organizations (HROs) in that, much like the military, the police, and other professions like airline pilots, a drop in quality or even a lack of service can have devastating results. Quality is determined by the end-users of the service or product. Do not confuse 'expensive' with high 'quality': low-cost services or products can have very high quality (like paper clips). The quality of a forensic service provider's work directly relates to the quality of justice provided by a society, and it is therefore of paramount importance. Assuring quality processes and results through the systematic monitoring and evaluation of the various aspects of all aspects of a process creates the highest probability that minimum standards of quality are met.

Two guiding principles lie at the heart of quality assurance:

- 'Fit for purpose': the result or product should be suitable for the intended purpose.
- 'Right first time': mistakes should be reduced or eliminated.

Quality assurance involves the oversight and documentation of the quality of all incoming materials that are used to produce a product. In manufacturing, this means raw materials, components, assemblies and intermediate processes; in forensic services, it means the quality of the evidence submitted, the validity of the methods employed, the proficiency of the forensic professionals examining the evidence and the accuracy of reports and testimony. Arguably, these last two are the *physical* products that a forensic agency 'produces'. The actual outcome, however, must be *quality*.

3.3 Illicit drugs

3.3.1 Introduction

In many forensic science laboratories today, more than half of all submissions by law enforcement agencies involve drugs. This has been going on since President Nixon began the 'War on Drugs' in the 1970s. As drug enforcement activities become better coordinated and funded, and as arrests are made, the drug evidence is submitted to a forensic science laboratory. Today, the critical question is 'what actually is a drug and what is not a drug?' Are the terms 'illicit drugs', 'illegal drugs', 'abused drugs', 'seized drugs' and 'controlled substances' synonymous with one another? Does how we label the drug matter in this context? This section will discuss the nature of drugs, some of the legal history of drug control in the USA, the analysis of drugs and some of the recent issues that surround drug control.

A drug can be defined as 'any substance, natural or artificial, other than food, that, by its chemical nature, alters structure or function in the living organism' [2]. This begs the definition of food, but that usually comes into play when one is considering ethyl alcohol, which is often used as a food but which, like many drugs, has psychoactive effects. If one considers how the federal government controls drugs in the US, alcohol is not included as a drug.

Clearly, there are thousands of substances that fit the definition of a drug, but only a relatively small number of these are of interest to law enforcement agencies worldwide. Certain drugs have qualities that make them the subject of legal controls, whereby people who possess, use, manufacture and/or distribute these drugs are subject to legal sanctions, including arrest, heavy fines and imprisonment, because of the notion that use of such drugs causes harm not only to individuals but to society as a whole. People would not ingest these substances unless there were some psychological benefits to doing so. All of the drugs that are subject to legal sanction confer some psychoactivity that is, in some way, pleasurable to the user. This desirable psychoactivity is the key to determining which drugs are the targets of law enforcement today.

There are a number of terms that are used to describe which drugs are subject to law enforcement in this country. In this section, the term 'illicit' will be used. It, like other terms, connotes something that is unlawful to possess or use. Other terms include 'illegal', seized, abused and controlled (substances).

Today, in the United States, the government uses two criteria to determine if and how a drug should be controlled: whether or not the drug has an accepted medical use in this country, and its potential for abuse. The latter can be viewed as a proxy of the degree of psychoactivity, how addictive the drug is and how easy it is to obtain.

3.3.2 Legal control of drugs

The control on the proliferation of illicit drugs in the United States has a rich and constantly changing history. What began as a means to raise revenue through taxation then progressed through a reform and medical model and, today, is centered more on interdiction and law enforcement than on rehabilitation. The control of drugs in the US can actually be traced back to the 1791 Whiskey Rebellion, when the federal government tried to levy an excise tax on alcoholic beverages. The philosophy behind the enactment of this tax was characterized as a promotion of 'the public good'.

The reformism movement of the mid-1800s was the first significant major effort to control drugs in this country. At the time, many substances that are now heavily regulated were more or less freely available to the public. Many patent medicines contained narcotics such as morphine and opium and other 'dubious' substances. Opium smoking in 'dens' was very popular, and 'morphinism' was recognized as a medical issue as its addictive nature became known. Cocaine became readily available in the United States, starting in the 1880s as an ingredient in many patent medicines and even by mail order. These drugs were often pedaled from town to town by hucksters in covered wagons. They made outrageous, unsubstantiated claims about the benefits of 'Hostelter's Bitters', which contained 44% alcohol, or Briney's Catarrh Cure, which contained 4% cocaine (3, p. 56).

The result of all of this was the enactment of the 1906 *Pure Food and Drug Act* [4], which regulated interstate commerce in misbranded and adulterated foods, drinks, and drugs. It defined a drug as: 'any substance or mixture of substances intended to be used for the cure, mitigation, or prevention of disease' [4]. This Act referred to Cannabis Indicus (as marijuana was known back then), cocaine, alcohol, morphine, opium, heroin and other substances. Interestingly, there were no restrictions on buying or selling any of these concoctions, as long as they were properly labeled. In one way, the Pure Food and Drug Act can be viewed as a kind of consumer protection law.

By 1914, the federal government took a different approach to its attempt to control drugs by deciding to tax them. The focus behind this taxation was to control opium and cocaine and thus, in 1914, the Harrison Act became the law of the land. This Act was described as: 'An Act to provide for the registration of, with collectors of internal revenue, and to impose a special tax upon persons who produce, import, manufacture, compound, deal in, disperse, or give away opium or coca leaves, their salts, derivatives or preparations, and for other purposes' [5]. However, since there were no prohibitions on individuals possessing or using these drugs, the Act was, for all intense and purposes, a tax law raising revenue for the government. Following enactment of the Harrison Act, many opium users were still able to get their opium from physicians legally. Law enforcement emphasis then turned on arresting drug smugglers.

The next few decades saw additional efforts to control drugs. In 1919, narcotics enforcement was folded into alcohol enforcement. This began an era when drug abuse became centered on users and distributors, and a view emerged of drug abuse as a legal problem rather than a revenue issue. In 1932, a Bureau of Narcotics was established in the Treasury Department. In 1937, a marijuana tax act that mirrored the Harrison Act was made into law. In 1965, the Bureau of Narcotics became the Bureau of Narcotics and Dangerous Drugs (BNDD) and hallucinogens, including LSD and prescription drugs, were added to the government's portfolio of what constituted an illegal drug.

In 1970, a new sweeping federal law was passed that changed the entire landscape of law enforcement of illicit drugs in the US. This is known as the *Comprehensive Drug Abuse Prevention and Control Act of 1970* [6], and is often referred to as the *Controlled Substances Act*. Although it has been modified many times since then, the Controlled Substances Act remains essentially in place today and is the major drug enforcement tool of the federal government. The major features of the Controlled Substances Act are as follows:

- All previous statutes that controlled illicit drugs were repealed and replaced by this Act.
- Whereas previous laws were mainly concerned with the interstate commerce of drugs, the Act swept in all illegal drug activity and trumped many state laws.
- In addition to controlling illegal drug use, distribution and manufacture, the Act also had provisions for drug abuse prevention and treatment programs.
- The BNDD was replaced by a new Drug Enforcement Administration (DEA), which was placed within the Department of Justice, rather than the Department of the Treasury, signaling its emphasis on drug enforcement rather than taxes.
- Illegal drugs were placed in five Schedules of Controlled Substances [6]. A drug was placed in a schedule by virtue of whether or not it had an accepted medical use in the United States, as determined by the US Food and Drug Administration (FDA), and its potential for abuse. The potential for abuse is determined by such factors as cost, availability, intensity of psychoactive effects, the degree of addiction and/or dependence caused by the drug and other similar factors. Only the Congress could add, reschedule or remove a drug from its schedule. For example, heroin was placed in Schedule I because of its lack of approved medical use (all controlled substances with no accepted medical use are placed in Schedule I) and its high potential for abuse. Schedules II through V all contain drugs with a medical use whose potential for abuse generally decreases with increasing schedule number. Schedules I and II contain the strongest penalties for possession and for manufacture or distribution.

The scheduling system in the Controlled Substances Act was created as a model law that could be easily adopted by individual states. Many states adopted this system completely, while other states modeled their laws after the federal statute but changed the number of schedules and/or placed drugs in different schedules.

The most recent major reform to the Controlled Substances Act was the enactment of the *Omnibus Drug Act of 1988* [7]. This Act put more emphasis on controlling possession of drugs and created a forfeiture provision, whereby any property that is involved in a drug enterprise such as sale or manufacture could be seized and turned over to law enforcement. For example, if a large boat is used to transport drugs, it can be seized. The Act also controlled for the first time, major precursors that are used in the manufacture of drugs. Paraphernalia such as 'hash pipes' also came under this Act. The Act also established a *National Office of Drug Control Policy* and a 'Drug Czar' to set overall drug enforcement policy.

3.3.3 Analysis of illicit drugs

Illicit drugs can come in a variety of forms, such as solids, liquids and even, on occasion, as vapors. Some are naturally occurring substances found in plants, such as THC in marijuana or cocaine in the coca plant or morphine in the opium poppy. Others are pure or adulterated solids that have either been manufactured or extracted from, or prepared from, plant material. Heroin is chemically converted from morphine, and methamphetamine is manufactured from precursors. Some illicit drugs are themselves liquids at room temperature, such as isoamyl nitrite, or are commonly encountered dissolved in a liquid. An example of this would be codeine in any number of cough medicines. Glue sniffing used to be fashionable; even though the glue is a liquid, it is the vapors that are inhaled.

Whatever the form of the drug, it is axiomatic that the analytical protocol chosen in a particular case will be driven by what form the actual illicit substance is in. It is a rare illicit drug seizure that consists of a single, pure drug. Drugs are often cut with diluents, which are inert fillers such as sugars or carbohydrates, or even powders like cocoa. They may also be diluted with excipients, which are themselves drug substances that mimic or enhance the effects of the illicit drug. An example of this would be cocaine cut with lidocaine, another topical anesthetic. Because of these cutting agents, in the vast majority of cases, some sort of separation or extraction must be accomplished as part of the analytical scheme.

Another major factor that drives the analytical scheme in a particular case is the amount of the sample that is available for analysis. There are two situations that arise here: the case where there is insufficient sample for a complete analysis that preserves sample for reanalysis by the other party in the case; and the case where there is a very large exhibit that contains hundreds or thousands of samples. These will be considered in more detail later.

The above considerations of form and quantity must be taken into consideration when designing a scheme for the analysis of illicit drugs. Even with these variables, however, there are several guiding principles that should be followed, which transcend particular issues:

1. The analyst should be mindful of the laws that control the various illicit drugs. Most states in the USA place drugs in various schedules based upon their potential for abuse and whether or not they have an accepted medical use (based on the Federal Uniform Controlled Substances Laws), but there are local variations that may need to be considered.
2. The scheme must lead to an unequivocal identification of the illicit drug(s) present in the sample. There is no room in the forensic analysis of drugs for a conclusion that is less than an identification. One cannot be 90% sure that the drug is present. One cannot state that the drug is probably present but could not be confirmed. If a conclusion of identification cannot be made, then the conclusion must be that no drug was identified.
3. The scheme of analysis should proceed from screening tests to confirmatory tests. That is, they should increase in specificity as the scheme proceeds.
4. Notwithstanding #2 above, if the amount of sample available is a consideration, then non-destructive tests should be performed first.

3.3.4 The SWGDRUG Approach

In recent years, federal agencies in the Department of Justice have set up a series of Technical Working Groups (TWGs) and Scientific Working Groups (SWGs). These consist of groups of peer experts in particular areas of forensic science that meet and develop guidelines for various types of scientific evidence, concentrating on analytical methods. The National Institute of Justice sponsors the TWGs and joins the FBI in sponsoring the SWGs, with one exception – the Drug Enforcement Administration sponsors and staffs the Scientific Working Group for the Analysis of Seized Drugs, or SWGDRUG. This is a truly international SWG, made up of nearly two dozen drug experts with approximately one-third from outside the United States. SWGDRUG began its work in 1997 and continues to this day. To date, they have developed recommendations on:

- a code of professional practice;
- education and training of drug chemists;
- methods of analysis;
- quality assurance.

This section will be concerned only with methods of analysis of illicit drugs. It should be noted that the overall SWGDRUG philosophy is international in scope.

It recognizes that legal requirements, culture, prevalence of drugs, education and training of analysts, and availability of instrumentation, vary widely among countries and jurisdictions within countries. Rather than proscribe a particular protocol for quality assurance, education and training of drug analysts, sampling and analysis of drugs, SWGDRUG proposes several ways of accomplishing the goal of reliable drug identification.

SWGDRUG provides a number of guidelines concerning sampling of large numbers of exhibits within one submission. These include sampling strategies, descriptions of various types of sampling strategies and a flowchart sampling scheme. These will not be discussed in this chapter, but the reader is invited to consult the SWGDRUG recommendations on sampling.

In the SWGDRUG section on 'Methods of Analysis/Drug Identification', the Introduction indicates that the purpose of this section is to recommend *minimum standards* for the identification of drugs. Further, it states that the correct identification of a drug depends on employing a scheme based on *validated methods*. This concept is developed further in the SWGDRUG documentation. The analytical approach is to require the use of multiple orthogonal (uncorrelated) techniques. The basic framework is a set of three categories of analytical techniques that are based upon their *maximum potential discriminating power*. Table 3.1 is reproduced from the SWGDRUG recommendations document below [8]:

Based on the techniques in the three categories, SWGDRUG recommends that a scheme of analysis can be built from judicious use of these techniques. For example, if a Category A technique is used, it should be supplemented with at least one other technique from any category. If a Category A technique is not used, at least three other techniques shall be used, and at least two of them shall be uncorrelated techniques from Category B.

Table 3.1 SWGDRUG methods of analysis/drug identification. Provided are three categories of analysis techniques for consideration when analyzing drugs. The classification is according to the technique's maximum potential for discrimination. The selection of a combination of these techniques that are uncorrelated, according to the SWGDRUG recommendations, provides a valid scheme of analysis

Category A	Category B	Category C
Infrared spectroscopy	Capillary electrophoresis	Color tests
Mass spectrometry	Gas chromatography	Fluorescence spectroscopy
Nuclear magnetic resonance spectroscopy	Ion mobility spectrometry	Immunoassay
Raman spectroscopy	Liquid chromatography	Melting point
X-ray diffractometry	Microcrystalline tests	Ultraviolet spectroscopy
	Pharmaceutical identifiers	
	Thin layer chromatography	
	Cannabis only:	
	Macroscopic examination	
	Microscopic examination	

In recognition that most cannabis exhibits consist of plant material, a separate set of recommendations are proposed. Both macroscopic and microscopic techniques are used in the identification of cannabis and they are considered to be uncorrelated Category B techniques. However, if they are used, then the botanical features of the plant material must be documented. A third technique from any category must also be used. In the case where the exhibit is an extract or residue of cannabis, such as hashish, where there are few or no documentable botanical features, then the exhibit must be treated as any other submission, and tetrahydrocannabinol (TCH – present in all cannabis plants) must be identified.

In addition to cannabis, there are other illicit drugs that occur in plant material form. These include mescaline, opium poppy, psilocybin and others. In these cases, where the plant material is the exhibit, a competent, trained botanist can identify the material by elucidating the relevant botanical features alone. If the active ingredient in the plant is to be identified, the scheme shall follow the same recommendations as for other drug exhibits.

In recognition of the fact these drug analyses are being performed for adjudicative processes and the possibility that the other party in the crime or dispute may wish to review the analysis of an illicit drug exhibit, SWGDRUG recommendations include the provision that all schemes of analysis must produce reviewable data. This applies to all Category A methods, and also to Category B methods if a Category A technique is not used. Examples of such reviewable data include charts, spectra, chromatograms, photographs and photocopies, as appropriate.

There are a number of other important considerations in the analysis of drugs using the SWGDRUG protocol. There are times when a test on a suspected drug exhibit will yield a negative result; this most often occurs with field screening tests. A negative result can be useful for eliminating the presence of a particular drug or group of drugs (at least to the degree of sensitivity of the test), but can never be used as evidence of the presence of a drug.

Several of the techniques used in forensic drug analysis are so-called 'hyphenated' techniques. The prime example of this is, of course, gas chromatography-mass spectrometry. When a hyphenated method is used, it can count as two uncorrelated tests, but only if the results of both are used. In the GC-MS case, the major useful data from the GC test is the retention time of the drug, whereas the MS yields data about the molecular weight of the drug as well as the pattern of fragmentation of molecules and ions that occurs in the test. Both types of data must be used and reported for GC-MS to be counted as two tests. Finally, the recommendations include the use of quality assurance measures, including two separate samplings, proper sample ID procedures, good laboratory practices, etc.

The SWGDRUG approach to the analysis of illicit drugs is, of course, not the only set of recommendations. Other organizations, such as the American Society for Testing and Materials (ASTM) publish guidelines for the analysis of various types of forensic evidence, including drugs. In many cases, they use guidelines developed, at least in part, from SWGs, including SWGDRUG. Many forensic science laboratories have their own protocols for the analysis of drugs. The advantage of using SWGDRUG is that it is a rational, consistent set of tests that require validation, reviewable data, sufficient testing and scientifically grounded sampling methodologies. It is also a truly international effort, which extends its reach and influence.

3.3.5 Reporting and court testimony

Regardless of which methods or protocols are chosen for the analysis of illicit drugs, the goal is always to provide an unequivocal identification of all drugs in the seized exhibit. There must be no reasonable doubt as to the identity of the drug(s) present. If quantitative analysis of the drug(s) is necessary, then this must be accomplished to measurable, defensible limits of accuracy.

The next step after the analysis is to commit the scientific findings to a report. ASTM has published standards for laboratory reports and has recently revised them for forensic science laboratory reports, but a specific standard for laboratory reports for drug analysis does not exist, and there is no mechanism for mandating that forensic science laboratories adopt these standards.

It has been all too common for forensic science drug laboratory reports to be little more than certificates of analysis. Such a report includes the demographic information surrounding the case, such as the name of the accused, the submitting law enforcement agency, the location of the offense, the type of evidence being submitted and the requested examinations. The report includes a description of what evidence was received and the results of the examinations (what drugs were present, the weights of the exhibits and, if required, the amount of drug(s) present). There is no information concerning what analytical tests were run, the results obtained for each test, how these results lead to the conclusion of the identity of the drugs present or any possible sources of uncertainty or error.

These reports also do not normally contain any charts, graphs, spectra, chromatograms, etc., that were generated during the analysis. Such a report hardly qualifies as a scientific report. All of this information is standard practice in scientific reports that are generated by scientific agencies, universities and other such entities.

Among the reasons given for the lack of completeness in forensic science drug reports are:

- The audience – judges and attorneys do not understand science and would not know what to do with all of this scientific information. The prosecutor, defense attorney and the court need to know if an offence has been committed – are there controlled drugs present and, if so, what drugs are present? The rest can be obtained by calling the analyst into court to testify and be cross-examined.
- Some scientists complain that providing all of this information to the adverse party (usually the defense attorney) just gives the adverse party ammunition for questions when cross-examining the scientist in court. This is an entirely specious

argument. A competent forensic scientist has nothing to hide and should be willing to make all data available to either side without hesitation.

A drug chemist, like all forensic scientists, will be offered and qualified as an expert prior to testifying in court. There is no 'grandfathering'. It is often useful for the expert to supply a bulleted list of qualifications to the attorney prior to testifying, so that the qualification process goes smoothly. Once the witness's qualifications have been recited and the witness is offered as an expert to the court, the opposing side may wish to challenge these credentials or qualifications in an effort to keep the witness from being admitted as an expert. The judge makes the final determination on whether a proffered expert is qualified.

If the drug chemist is testifying on behalf of the people (prosecution), the direct examination will usually consist of a series of questions that will establish the chain of custody of the evidence in the laboratory, describe the evidence received, the tests performed and the results of the analysis. Any mitigating factors should also be brought out at this time. During cross-examination, the chemist may be asked to describe the tests done and how they contributed to the conclusion. The data may be examined and questioned. The witness is only finished testifying when released by the judge.

3.3.6 A sampling of current issues in drug analysis

The analysis of illicit drugs is a fairly settled issue in forensic science. The currently used methods of analysis have been employed for many years and have been extensively validated. Standard methods and protocols have been developed and published; the aforementioned SWGDRUG guidelines are a good example. The ASTM has also developed some standards for the analysis of drugs. Still, science is always in flux and there are also societal and political pressures that must continually be addressed. A few of the more important recent developments are presented below.

Sampling

As law enforcement agencies focus on major drug operations, large-quantity seizures increase. This has raised questions about how much of a large quantity of drugs should be tested. This issue arises in two contexts, the first occurring when there are a large number of small samples in a large seizure and the laboratory wishes to report out the characteristics of the entire seizure without having to analyze every sample. The issue of representative sampling and of what constitutes a sufficient number of samples then arises. Various strategies have been devised to address both of these issues. SWGDRUG has devoted a great deal of effort into presenting scientifically sound methods for ensuring that sampling is both representative and sufficient to allow inferences about a large number of samples from analysis of a subset.

The second context for sampling of large seizures occurs in those jurisdictions where the penalties associated with drug distribution or possession vary with the quantity involved. Often the best strategy is to analyze enough of the samples so that the weight threshold has been exceeded and then stop. This removes the necessity of having to infer the characteristics of the entire seizure from the analysis of a subset. Research is still being carried out on the optimal methods for sampling large seizures of illicit drugs.

Over-analysis of drug seizures

SWGDRUG is a rarity among the forensic science SWGs in that it is international in scope. Approximately one-third of its members have traditionally been from countries outside the USA. During the development of the guidelines for the analysis of seized drugs, it was noted that other countries do not always take the same approach to the analysis of certain drugs as is done in the US. This may be due to a different philosophy concerning what constitutes sufficient proof of identity; or it may be due to a lack of analytical resources, forcing a laboratory into settling for less in the way of the most probative tests; or it may be due to a lack of human resources.

For example, in some countries, marijuana is identified solely by macroscopic and microscopic analysis. In some other countries, the major mode of identification is by microscopy, especially for plant materials and characteristic crystal structures. Is this situation problematic? Clearly, the protocols that have been developed in countries other than the USA are acceptable to the various actors (judges, attorneys) in the criminal justice system and to the forensic scientists, or they would not be used. It would, of course, be preferable for the global community to coalesce around a common set of protocols, such as SWGDRUG, but cultural and financial realities must also be considered. It may very well be that drug laboratories in the US that insist on mass spectrometry to confirm every identification may be over-analyzing.

Tracking marijuana by genetic analysis

DNA typing in forensic science is not limited to human DNA from crime scenes. Clearly, it is possible to perform genetic analysis on any living organism, and such analysis is being applied increasingly to cases where there are major seizures of marijuana plants. Prosecutors see this as an effective technique in tracking marijuana growers who grow vast fields of marijuana, using seed stock from one or a very few plants. Theoretically, having evidence of this type would make a case of conspiracy and/or manufacture easier to develop and prosecute. However, this practice also raises questions of cost-benefit. Most forensic science laboratories are not equipped to perform DNA typing on plants, and case backlogs preclude this from being a consideration. Thus, such analysis must be done by outside, private laboratories. This is bound to be expensive, and its benefits may be dubious when the cost and evidentiary value are factored in.

The war on drugs

For the past 40 years, the US government has spent more than $2.5 trillion on the War on Drugs, which has targeted smuggling, manufacture, distribution and use of illegal drugs. Despite all of the advertising campaigns, increases in penalties for drug activity, forfeiture laws and increases in incarcerations for drug offenses, as of 2009 almost 20 million Americans used illicit drugs. Much of the drugs reach the United States through Mexico, including over 90% of cocaine and a large quantity of marijuana.

The term 'War on Drugs' was coined when President Nixon created the DEA in 1973, and it has spawned many initiatives. These include First Lady Nancy Reagan's 'Just say no' campaign aimed at children and teenagers, the DARE program that started in Los Angeles and spread nationwide, and recent efforts to curb drug trafficking through Mexico. Most recently, in June of 2011, a Global Commission on Drug Policy released a report on the War on Drugs that was highly critical, calling the war a failure.

The War on Drugs has had effects on the entire criminal justice system, resulting in increased arrests, trials, and jail and prison sentences. Even forensic science laboratories have been impacted in several ways, some positive and some negative. Increased activity by the criminal justice system has resulted in major increases in the number of illicit drug cases submitted to forensic science laboratories. This has contributed to case backlogs, but has also resulted in large increases in the number of drug chemists that have been brought into forensic science laboratories.

Many laboratories have also been the recipients of drug forfeiture money, as well as federal and state/local grants to purchase analytical instrumentation to increase drug analysis capabilities. Since much of this instrumentation has uses throughout the laboratory, the War on Drugs has benefited whole laboratories, not just drug sections. The increased attention to the analysis of illicit drugs has also accelerated the development of standard methods of analysis, and has thus driven the efforts of SWGDRUG and other organizations.

3.3.7 Conclusion

The analysis of illicit drugs has been, and remains, one of the most active areas in forensic science. In many forensic science laboratories, illicit drugs comprise more than half of all cases submitted. Because forensic drug analysis is essentially an exercise in the identification of organic compounds that exhibit psychoactive properties, it is an excellent example of applied analytical chemistry. The methods of analysis of drugs are mostly well recognized and thoroughly validated. They run the gamut from non-specific field tests for presumptive identification to sophisticated confirmatory tests that leave no doubt as to the identity of the drugs.

The most unsettled segment of drug analysis remains the vexing sampling issues that accompany very large seizures containing either large quantities of drugs in a single or a few submissions, and/or a large number (often thousands) of small packages of drugs that add up to a large quantity. Challenges are presented by the need to analyze representative samples of large seizures, the need to avoid time-consuming analyses of all of the exhibits in a large seizure and the need to respond to many state and local laws that tie penalties for possession and distribution to the quantity of drug present. Of all of the SWGs sponsored by the federal government, SWGDRUG has been among the most successful in terms of the comprehensiveness of its recommendations and its international representation. There is every reason to believe that forensic drug analysis will continue to be a major component of, and a force in, forensic science.

3.4 Forensic DNA Analysis: a Primer

3.4.1 Introduction

The importance of using cellular DNA to discriminate one human being from another was realized in 1985, when a British immigrant's son was accused of not being related to his mother and was ordered to be deported [9]. The defendant's attorney read a serendipitous article in a local newspaper recounting a new scientific discovery in Dr. Alec Jeffrey's laboratory at Leicester University that could prove maternity. DNA testing was conducted, the relationship verified, and the science of DNA testing was accepted by the British Home Office. The breakthrough that an individual's genetic composition could be visualized as unique discrete patterns ushered in the age of forensic DNA analysis.

The purpose of forensic DNA analysis has remained the same over the past 25 years – the comparison of one DNA profile to another DNA profile for the purpose of establishing an association. During this time period, further to achieving this stated purpose, DNA methods have been in a continuous, and often rapid, state of development, with improvements in testing reagents, automation techniques, characterizing DNA genetic markers, customized laboratory consumables, manufacturing sensitive DNA detection instrumentation, mandating national and international laboratory standards and using analytical software to aid in DNA interpretation.

3.4.2 Historical perspective

The critical difference between the uses of DNA analysis in a forensic laboratory versus a research laboratory is that the end-user of forensic DNA data is the justice system. The judicial system generally accepts the underlying scientific principles used for forensic DNA testing in the United States. It is the laboratory's protocols and practices and analysts' training that are often the reason for challenges to DNA results.

An analyst's sworn opinion is an amalgam of their education, appropriate training, continuing experience and the laboratory's validation studies. Forensic DNA laboratories have been fortunate to have had government, academic and private industry partners since the inception of DNA profiling that have allowed scientific reliability and judicial acceptance. The 1994 DNA Identification Act formalized the authorization of the FBI to operate a national DNA database, using the Combined DNA Indexing System (CODIS), and to set national standards for forensic DNA analysis which was the impetus for the establishment of uniformity for forensic laboratories [10].

As a result, since 1998, with updates as recently as 2011, the *FBI Quality Assurance Standards (QAS) for Forensic DNA Testing Laboratories and Convicted Offender DNA Databasing Laboratories* document has been used to assess the quality of DNA analysis in laboratories [11]. The FBI QAS Standard 5 provides for mandatory education, training and continuing education requirements for forensic biology personnel, so that qualified analysts have solid science academics with classes covering biochemistry, genetics, molecular biology, statistics and/or population genetics, and that a laboratory must have a documented, comprehensive training program.

The FBI QAS Standard 8 provides that a laboratory must use validated methods and procedures before implementing any analysis on casework. There are three major components of a credible validation study: the design of the study; selection of the appropriate number and type of samples from which data will be generated; and applicable interpretation guidelines based on validation data results. The laboratory's validation studies allow an analyst to make critical, supported, scientifically based decisions when conducting DNA testing. External scrutiny of a laboratory's validation studies, training of analysts, and all policies and procedures is conducted by the accreditation process and is critical to supporting a laboratory's scientific credibility.

Nobel Prize awards offer an important chronology for DNA discoveries that have become the foundation for the general acceptance of forensic DNA analysis in the scientific community [12]. These include: in 1962, James Watson and Francis Crick, who co-authored the first published structure of DNA; Arthur Kornberg in 1959 for providing the first model for cellular DNA replication; H. Gobind Khorana in 1968 for his description of the genetic code; Frederick Sanger in 1980 for developing a method for the sequencing of DNA; and Kary Mullis in 1986 for his description and research into a process known as the polymerase chain reaction (PCR).

The latter is a very rapid *in situ* production of millions of copies of a specific target DNA sequence. The concept is based on what actually occurs during DNA replication within a cell, and was first described in 1971 by Kjell Kleppe from Khorana's laboratory who wrote [13]:

One would hope to obtain two structures, each containing the full length of the template strand appropriately complexed with the primer. DNA polymerase will be added to complete the process of repair replication. Two molecules of the duplex should result. The whole cycle could be repeated, there being added every time a fresh dose of the enzyme.

PCR, as a viable research method for copying well-defined specific DNA sequences and ultimate visualization and interpretation of the copied DNA products introduced by Mullis's laboratory, ultimately changed the face of how genetic research was conducted. It was the single most important event ever to move the science of forensic DNA analysis forward (Figure 3.1) [14].

The first PCR-based forensic kit exploited internal DNA sequence differences known as the AmpliType HLA DQα kit and the AmpliType PM-HLA DQα kit, commonly called the 'blue dot tests' because of the colored dots of the data strips (Figure 3.2) [15]. The test required a small sample size and was a relatively quick

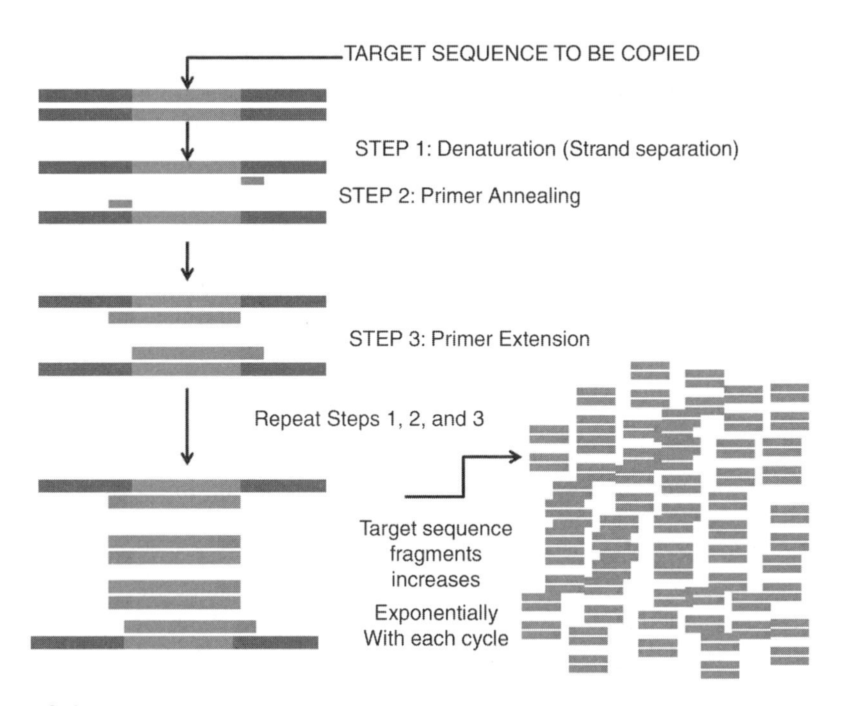

Figure 3.1 The polymerase chain reaction.

Step 1: The process begins by heat denaturation of the DNA strands to form single-stranded DNA.
Step 2: Region-specific primers hybridize to specific selected target sequences.
Step 3: The primers extend the target region by copying the sequence.

These steps are repeated through a series of temperature-dependent cycles, providing millions of copies of the target region.

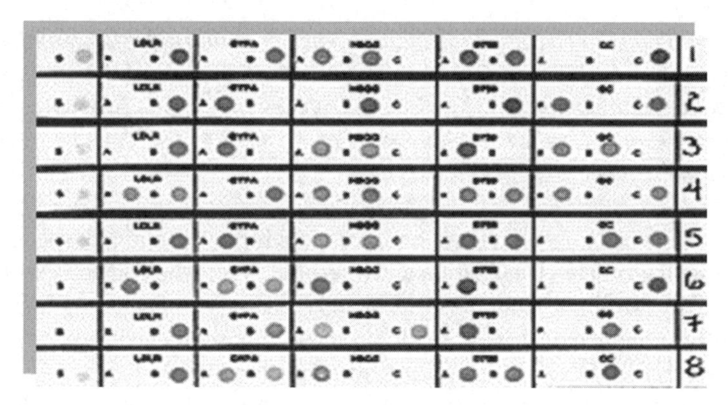

Figure 3.2 The AmpliType PM kit, also known as the polymarker or the 'blue dot kit', tested six sequence polymorphisms detected through hybridization to allele-specific oligonucleotides, including LDLR, GYPA, HBGG, D7S8, GC and HLA DQA1 (not shown).

assay compared to the laborious DNA testing method Restriction Fragment Length Polymorphism (RFLP) used in the late 1980s to early 1990s, which required a large sample size. The AmpliType PCR-based tests were not without limitations. These single nucleotide genetic markers provided low power of discrimination compared to RFLP results, and the kits were expensive. Additionally, the results were not compatible with desired database search methodologies. In fact, one of the purposes of the DNA Identification Act of 1994 was to develop a process whereby a DNA profile could be shared between national public laboratories.

A national coordinated effort to define the number and type of forensic genetic markers to be analyzed using PCR was realized in April 1996, when the STR Working Group was formed. The Working Group included the FBI and representatives from over 20 forensic laboratories [16], and it evaluated DNA PCR testing methods targeting autosomal nuclear repetitive sequences called microsatellite variable number tandem repeat (Short Tandem Repeat) or 'STR' genetic markers.

The Working Group established criteria for the selection of the forensic STR 'CODIS core loci'. These included well-characterized genetic markers, specifically not related to disease states, containing discrete alleles over a defined base pair range, having relatively low mutation rates, which can be multiplexed for ease of analysis, generating minimum PCR-analysis artifacts and having allele frequencies with a useful power of discrimination. The research conducted by the group concluded with the selection of 13 CODIS STR core loci, including D18S51, D21S11, TH01, D3S1358, FGA, TPOX, D8S1179, vWA, CSF1PO, D16S539, D7S820, D13S317 and D5S818 (Figure 3.3). In order for DNA profiles to qualify for entry into the CODIS database, specific requirements were defined so that manufacturers could conduct developmental validation studies before releasing commercial STR kits.

Forensic laboratories conduct additional internal STR validation studies to critically evaluate the reliability of the kits. Implementation of the 13 CODIS

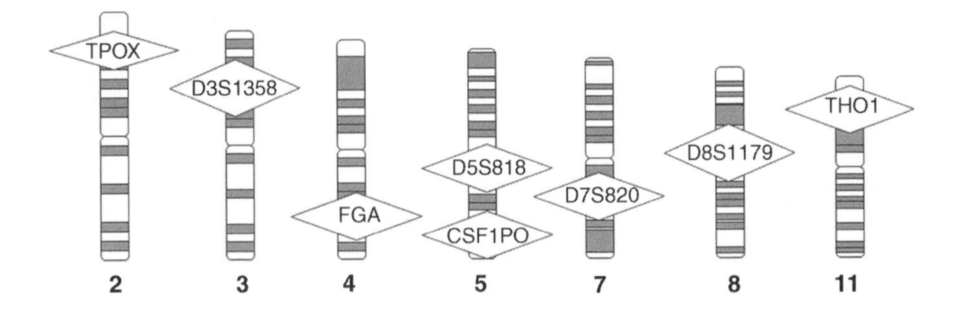

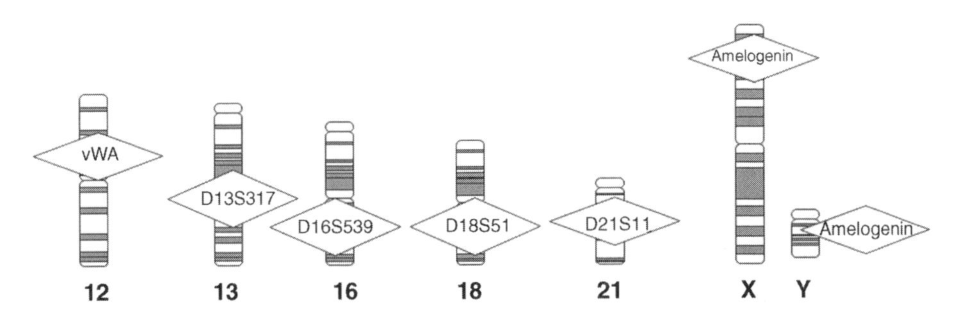

Figure 3.3 Chromosomal location of the US CODIS 13 STR core loci and Amelogenin.

core-loci STR markers in forensic laboratories allows local, state, and national databases to create a repository in which law enforcement agencies populate and share DNA profile information. In 2011, the FBI published the enhancement of the CODIS core loci in the United States to include D2S1338, D19D433, D12S391, D1S1656, D2S441, D10S10S1048, Penta E and the Y chromosome locus DYS391 with TPOX optional [17].

3.4.3 The reality of forensic DNA Typing

'*How long does it take to generate a DNA profile?*' Calculation of the amount of time it takes to generate a DNA profile for a defined number of samples in the laboratory is simple math, based on the time it takes to extract, quantify, amplify and detect the amplified STR products. The question, '*How long does it take to complete a DNA case?*' is much more complicated, and the typical answer is, '*It depends.*' The time it takes on the front and back end of casework testing is often unpredictable, and this is the major reason why there are backlogs in a typical forensic laboratory (Figure 3.4).

Unlike most of the other forensic disciplines the location of a biological sample from a crime scene submission, the type of biological sample present or how many

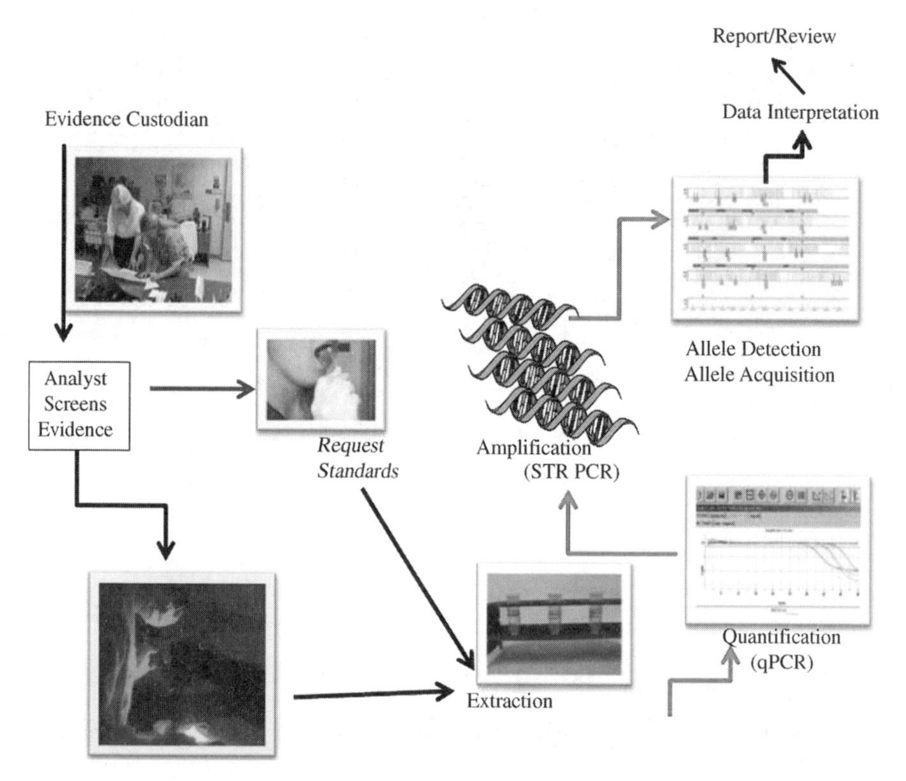

Figure 3.4 The case working schema generally includes obtaining evidence from an evidence custodian, followed by screening the evidence for biological material, conducting DNA analysis (grey arrows) and, ultimately, interpreting the data, writing a report and conducting the review process.

samples will be analyzed is variable. Screening of evidentiary items for human cellular DNA is experience-based and must be learned through extensive training, observation and hands-on casework. The evidence-handling process may take hours or weeks, depending on the complexity of the case.

Case sample number submission policies have been instituted in some laboratories, where law enforcement evaluate the potential informative nature of the crime scene samples and submit only those samples that have the most potential to aid in their investigation. Crimes with a voluminous amount of evidence may require the expertise of the analyst to help prioritize the evidence. This protocol provides faster DNA turnaround times and the analysis of more cases. Most laboratories have a flexible submission policy to accommodate case-related complexities that may occur.

In addition to the unpredictable time necessary for front-end screening of casework evidence, the time it takes post-analysis for an analyst to interpret DNA profiles, write a report and submit the case file for review is also unpredictable. DNA interpretations and case file review may take hours or weeks, as all the

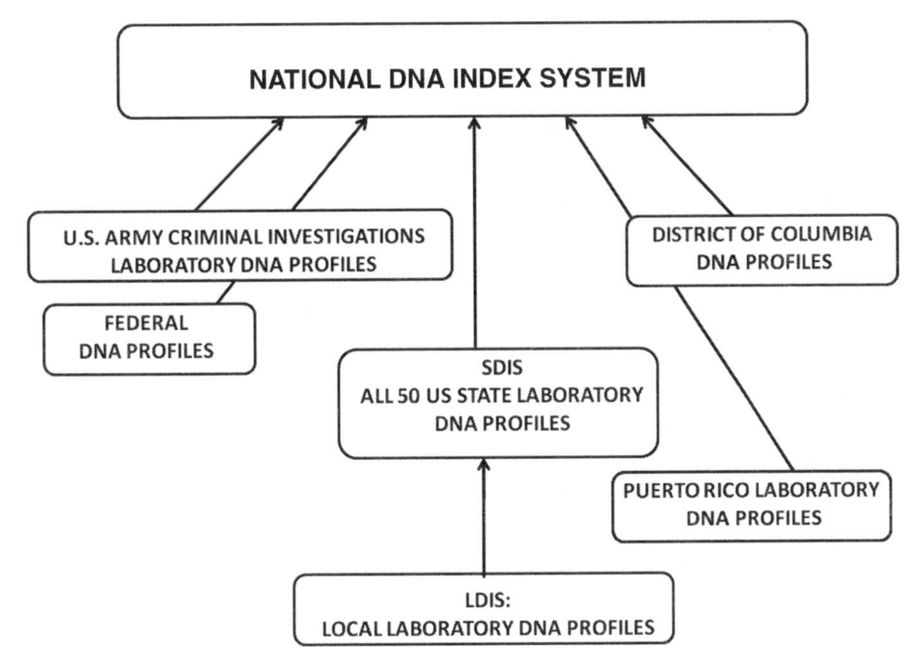

Figure 3.5 The CODIS architecture for US databasing encompasses all qualifying local (city, county), state and federal agencies and, through Public Law 103 322, is maintained by the FBI. The most recent CODIS software will include capabilities for searching forensic DNA profiles against convicted offenders and, in some states, arrestees, as well as Y-STR and mitochondrial DNA profiles for unidentified human remains and the missing persons database.

evidence is not always submitted at the same time. It is important to note that there are two types of forensic DNA laboratories: databasing laboratories are responsible for processing known DNA samples from convicted offenders, arrestees, unidentified human remains or missing persons; forensic laboratories are responsible for processing evidence from crime scenes.

In compliance with the 1994 DNA Identification Act, the DNA database repositories must be maintained by the FBI and individual states (Figure 3.5). Laboratories processing convicted offender samples for the purpose of submitting the known profiles into CODIS do not have the screening or extensive DNA interpretation guidelines and report review protocols used in caseworking laboratories, although databasing analysts must be qualified as defined in the QAS [11]. Samples submitted to a convicted offender DNA databasing laboratory are not evidence, nor is a chain-of-custody required, whereas samples collected through the course of criminal investigations are considered evidence, and a chain-of-custody is maintained to assure the authenticity of the sample. In addition, either a blood draw or, more commonly, a mouth swab is collected from the convicted offender sample, allowing DNA analysis to be automated, whereas the source(s) and type(s) of biological samples collected from a crime scene are unpredictable.

3.4.4 The analytical process

The extraction and purification of DNA from a biological stain is the first analytical step in the DNA profiling process. The condition, size and type of the biological stain are important factors when determining an extraction strategy. The goal of the extraction step is to generate highly purified DNA with an optimum DNA concentration for downstream processing. In every extraction method, the initial step obliterates cellular membranes, denatures proteins and liberates DNA molecules [18]. Removal of cellular debris and potential inhibitors that may co-extract with the DNA is essential to providing a quality DNA profile. Laboratories validate specific DNA extraction protocols for blood, buccal cells, bone, skeletal remains, hair, formalin-fixed or whole tissue, sperm and 'touch' samples. Regardless of the extraction protocol used, the laboratory must process concurrent negative control samples in order to monitor potential contamination issues.

Chelex-100 resin was one of the first manual extraction protocols used on casework evidence that was fast, inexpensive and effective [19]. The disadvantages of Chelex-100 DNA extraction are that the extracted DNA remains in a soup of cellular debris with potential inhibitors, and the resin, if accidentally carried over into downstream processes, may prevent the generation of a DNA profile.

The use of phenol-chloroform, commonly known as an organic, PC or PCIA, is also a manual extraction technique still used in many laboratories, providing more purified DNA than the Chelex-100 method [20]. DNA is released into a buffered solution, which is then vigorously mixed with phenol-chloroform. The centrifugation of the sample results in the separation of two liquid phases – an upper aqueous phase containing the DNA, and a lower organic phase containing the cellular debris. The mixing and separation steps are repeated several times, followed by purification of the DNA through a series of ethanol washes.

The disadvantage of organic extractions is an inability to automate, due to necessary human intervention steps to open and shut the test tubes to add reagents. Special precautions must be used, as phenol and chloroform are both hazardous chemicals and, if the mixing is not efficient, there is a potential to co-extract the phenol, ethanol and PCR inhibitors from the sample, which may affect the downstream DNA analysis.

In fact, there are many well-characterized inhibitors of the DNA process that may be co-extracted with the DNA during the purification step, including the blood component hematin, humic acid found in soil, tannic acid from plant material, melanin pigment present in skin and hair, dyes used in coloring clothing, calcium from bone and collagen from bone or tissue [21]. Extraction and purification reagents such as phenol, guanidium and EDTA, if co-extracted with the DNA, may also affect the quality of the DNA. Therefore, it is important that extraction protocols are evaluated and validated to obtain maximum DNA recovery from a biological stain, and to optimize the generation of purified DNA.

There are various commercialized DNA extraction technologies that are growing in popularity in forensic laboratories because they eliminate the presence of

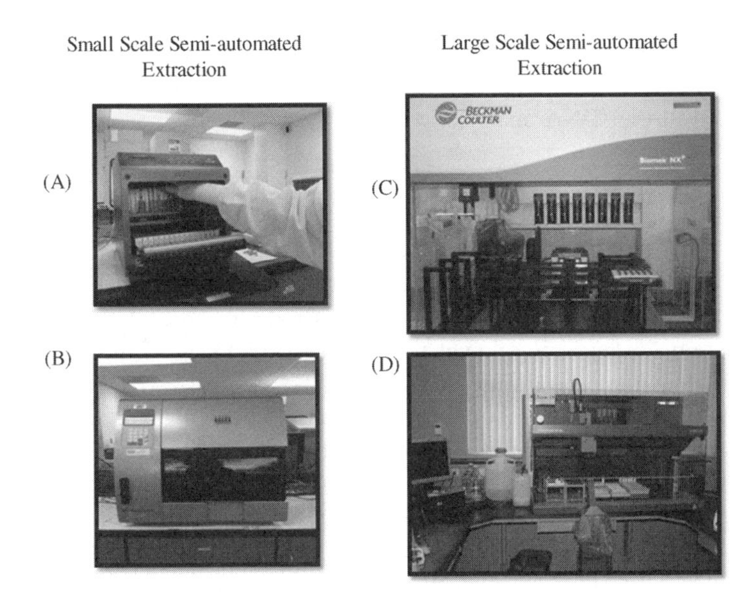

Figure 3.6 DNA extraction automation may be small-scale (4–16 samples simultaneously) such as the Qiagen EZ1 (A) or Promega Maxwell 16 (B), or large-scale liquid handling robots which can process up to 384 samples simultaneously. The Beckman BioMek NX (C) or the Tecan EVO series (D) have large-scale capabilities.

inhibitors and provide highly purified DNA. In general, the overall negatively charged DNA is attracted and reversibly bound to silica-based solid supports, followed by repetitive washing of the bound DNA, which rids the sample of cellular debris and inhibitors. The purified DNA is eluted from the silica support and is available for downstream processing [22].

An advantage of these new DNA solid phase extraction technologies is that the steps are automatable using robotics (Figure 3.6) [23]. The strategy provides high-quality DNA, minimizes human intervention, eliminates the use of hazardous chemicals and, depending on the manufacturer's liquid handler design, dramatically improves the number of samples that may be extracted.

Robotic workstations can process between six and 384 samples at a time. When programmed and used appropriately, an extraction robot conducts the important repetitive purification steps with consistency, precision and accuracy, while preventing potential contamination.

Once the DNA has been isolated and purified, the extracts are either used in the next step of the DNA analysis process or are stored until needed. Although most laboratories do not consider DNA extracts evidence, laboratories usually store the extracts in freezers to maintain the integrity of the samples in case further DNA analysis is required. More recently, laboratories are validating the storage of DNA extracts at room temperature in 96-well plates that have been pretreated with a proprietary matrix to protect the DNA from degradation [24,25]. The liquid DNA extracts are placed in a specific well in the plate, dried down in the presence of the matrix and stored at room temperature. Simple rehydration of the sample allows for recovery and further DNA analysis.

3.4.5 Quantification of extracted DNA

Quantification of DNA extracts provides a crossroad for an analyst where critical decisions are made regarding downstream processing. Quantification of extracted DNA is a mandatory step during analysis of unknown casework samples, as per QAS 9.4, which states: 'The laboratory shall quantify the amount of human DNA in forensic samples prior to nuclear DNA amplification' [11]. Quantification determines how much DNA, if any, can be detected in the extract.

Although there are many quantitative methods, forensic laboratories usually internally validate commercially available quantitative PCR kits (qPCR) or published real-time quantitative PCR systems that specifically measure primate DNA [26,27]. Generally, these kits are multiplexed genetic marker systems capable of generating data for both total autosomal and total male DNA concentrations. The mechanism of qPCR is based on the events during the amplification process, expressed as a graph post-amplification (Figure 3.4).

When amplification begins, the PCR reagents are in excess, and amplification of the DNA target sequence proceeds at a constant, exponential rate. At a point during every reaction, the amplification rate decreases below exponential and enters a linear phase. Eventually, the amplification rate plateaus and little or no more amplified products are generated. The exponential phase is the only time during the amplification when the rate is reproducible and predictable. Comparing the cycle number to the amount of amplified product produced during this time can accurately reflect the amount of template DNA present in the sample.

In a forensic setting, autosomal target sequences are fluorescently tagged during amplification and emit light at a specific wavelength. The Y-chromosome target sequence is amplified using a tag that emits light at a different wavelength than the autosomal PCR products. All light emission data is collected, and quantification data for total DNA and Y-STR DNA concentrations are automatically calculated.

Based on the qPCR results, an analyst can calculate the concentration in a sample and estimate an optimum template DNA concentration for the next step in the DNA process. Additionally, and equally importantly, the analyst can make scientific and valid decisions not to process samples further. Other questions that may be answered by the qPCR data include:

- Does the qPCR data indicate the sample is inhibited or degraded?
- Should the DNA extract be 'cleaned up' before amplification?
- Should the sample be consumed?

- Would it be beneficial to preserve the sample for analysis, in case the sample is outsourced to a laboratory for additional DNA testing?
- Do qPCR results indicate potential contamination in negative controls?

One of the important benefits of qPCR data is that an analyst has the ability to choose which samples are carried on to profiling. 'Touch' samples from items that a perpetrator may have touched are notorious for providing very low DNA concentrations. The qPCR results from such items often indicate sub-optimum concentrations of DNA, and it may be beneficial to concentrate or combine extracts to possibly avoid an incomplete profile or loss of a profile. A laboratory's qPCR validation data is important in order to have confidence in DNA concentration data that truly is a precise and accurate measurement.

3.4.6 Amplification of STR Markers

In the late 1990s, manufacturers were producing STR megaplex kits that could generate data from up to eight STR genetic markers in a single reaction, using fluorescent-based technology [28]. Nowadays, forensic laboratories in the United States typically use kits that simultaneously amplify up to 16 STR genetic sequences which include the 13 CODIS core loci and, depending on the manufacturer, additional non-CODIS genetic markers. Amelogenin, the gender identification or sex-typing marker located on the X and Y-chromosome, is also available in the commercial STR kits. The autosomal STR kits provide the forensic community with consistent, reproducible, comparable and quality DNA data. The international community has a well-defined set of STR markers in which some kit components also analyze some of the US CODIS core 13 loci [28].

A critical component of the amplification process, and why quantitation is so essential, is the DNA template concentration within a sample, which has a clear effect on the quality of the final profile data. Excessive DNA template causes false priming, inability of the PCR process to copy the DNA properly, early cycle depletion of PCR reagents, accumulation of non-specific products and increased PCR artifacts. If the DNA template concentration is too low, this may result in weak signals or high background noise, inability to generate data, drop-in or drop-out of alleles, excessive stutter, allele imbalance and the potential to amplify contaminants. It is critical that validation studies use specific STR kits, laboratory consumables and a thermal cycler instrument to define the range of the DNA concentration necessary during amplification, in order to maximize generation of high-quality DNA profile data.

As described in the quantitation of autosomal and Y-STR target sequences using PCR, autosomal STR products for DNA profiling are tagged fluorescently during the amplification process. The use of fluorescent tags with different excitation and emission wavelengths allows for the detection of the megaplexed amplified STRs within relative base pair ranges (Figure 3.7).

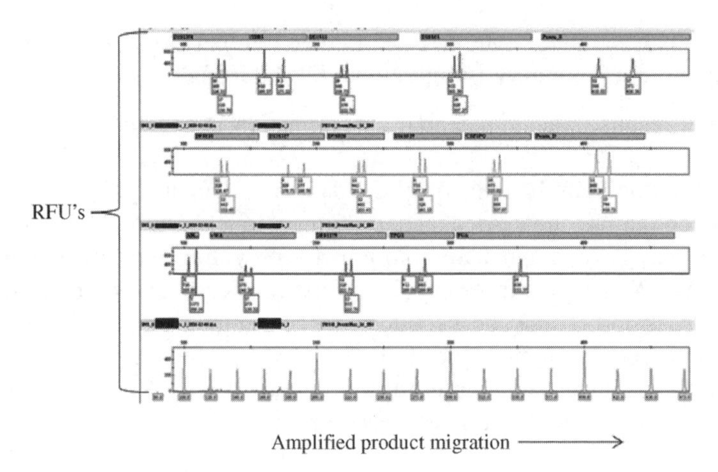

Figure 3.7 Single source STR electropherogram using the Promega PowerPlex16^R System, which includes Penta E, D18S51, D21S11, THO1, D3S1358, FGA, TPOX, D8S1179, vWA, Amelogenin, Penta D, CSF1PO, D16S539, D7S820, D13S317 and D5S818. The peaks are detected by the relative fluorescence units (RFU) emitted by the amplified products.

It is important to note that multiplex STR kits with a large number of loci, in which the fluorescently tagged loci span approximately 100–350 base pairs, provide a very efficient and rapid analysis of DNA samples. However, using the large multiplex kits on DNA extracts from severely degraded DNA samples, or samples that have co-extracted inhibitors, may provide DNA profiles of low quality or a loss of DNA profile data. In these cases, where there is sufficient quantity of DNA but the template is of poor quality, analysis of the sample using an amplification reaction that generates smaller STR fragment lengths from some of the same loci targeted in the larger STR kits has become an attractive alternative. These 'mini-STRs', which generally span approximately 200 base pairs, have been especially useful for unidentified human remains, and the DNA profiles may be searched in the CODIS database [29].

Testing for genetic markers specific to male and female inheritance, called lineage markers, sometimes offer important DNA case information. Lineage DNA markers are often useful when the conventional autosomal STR DNA markers do not provide informative or complete profile data. One of the important differences between autosomal genetic markers and lineage genetic markers is the mode of inheritance. Half of an individual's autosomal markers are maternally inherited and half are paternally inherited. In the case of the Y-chromosome lineage markers, there is only a paternal linkage (father to son), and with mitochondrial DNA inheritance there is only a maternal linkage (mother to sons and daughters).

Caseworking laboratories that conduct Y-STR DNA profiling can use the same laboratory facilities, consumables and thermal cycler instrumentation as used for the megaplex autosomal STRs. Commercial Y-STR kits amplify up to 16 Y-STR loci. Female DNA is not amplified during Y-STR analysis, thus allowing only male DNA data to be generated. This is especially important when there is an overwhelming amount of female DNA and the male contributor cannot be detected. This occurs when female intimate evidentiary samples such as fingernail scrapings, swabs from a victim's skin, or semen low-level/negative vaginal swabs may contain male suspect DNA. Potential informative Y-STR analysis is also routinely conducted on evidence where a male individual may have touched an item and the traditional autosomal STR analysis did not elicit informative DNA data. Y-STR analysis does not provide such discriminating statistical results as autosomal STR analysis, but in many cases the results have the power to exclude individuals. Efforts to provide a comprehensive Y-STR database have been realized with the United States Consolidated Y-STR Database, which allows the upload of Y-STR profiles for the purpose of conducting statistical analysis and mixture deconvolution [30].

There are several forensic laboratories that analyze the maternally derived lineage mitochondrial DNA sequences, which are abundant, well-conserved and useful when analyzing severely degraded DNA samples or samples with low nuclear template DNA concentrations, such as hair shafts. Most state and local forensic laboratories do not conduct mtDNA analysis. There are hundreds of circular copies of mitochondria DNA sequences within the cell and, due to the high mtDNA copy number, special isolated facilities and personal protective equipment are necessary in order to prevent contamination between specimens. The analysis of mtDNA is much more labor-intensive and expensive then autosomal DNA and Y-STR analysis.

3.4.7 Detection of amplified products: allele detection

Separation and detection of the fluorescently tagged STR amplified products is the final step of the laboratory bench DNA analysis. Once the amplified products have been detected, DNA data is electronically acquisitioned and then transferred to an analyst report writing station, where the DNA profile(s) will be interpreted, comparisons and associations made, and a report written and reviewed.

Amplified fluorescently tagged STR products are separated by size using capillary electrophoresis and are then visualized in the form of peaks on a graph called an electropherogram (Figure 3.7). Conventional CODIS STRs, miniSTRs and Y-STRs amplified alleles are analyzed, using the same instrumentation, with the same scientific principles. Mitochondrial amplified products are also separated by a specific size, but the separation is based on the fluorescently tagged internal sequence in which each base of the DNA sequence is detected during separation.

The use of capillary electrophoresis to separate biological fragments by size was first reported in 1930, when Arnes Tiselius showed the separation of proteins in solution using a glass capillary [31]. Over 50 years later, polymer-filled capillaries were routinely used to separate extremely small DNA molecules and, shortly thereafter, DNA fragments were separated using a sieving polymer [32]. The use of capillary electrophoresis (CE) to separate DNA STR alleles by sequence length eventually became the forensic separation method of choice, both in the United States and internationally.

The separation of fluorescently tagged amplified products by size involves the migration of the overall negatively charged DNA fragments deposited at the negative electrode through a polymer to the opposite positively charged electrode. The scientific principles from which this separation occurs are complex, involving:

- the preparation of the amplified DNA prior to migration through the capillary;
- optimization of the buffer and matrix environment in the capillary in which the fragments migrate;
- the length of the capillary;
- the amount and quality of amplified DNA fragments undergoing the separation process;
- the time it takes for the fragments to traverse through the length of the capillary;
- optimizing resolution capabilities;
- an instruments' ability to capture the different colored fluorescence light signals emitted by the amplified products.

Collection of DNA fluorescent data occurs as the fragments pass through a window located near the positive end of the capillary, where laser light continually passes. When the fluorescently tagged DNA passes through the laser light, the tag is excited and light is emitted from the amplified products. The effective and efficient collection of these fluorescent signals, relative to migration times, provides the necessary data to interpret the DNA profile from an evidentiary sample.

Analytical software is imperative to the interpretation of the data. In order to be assured that the DNA fragment's relative length is accurately calculated, the DNA amplified products migrate with an internal known DNA standard provided in the STR kit. Allelic ladders, which are a mixture of representative allele sizes for each STR marker, are used as a reference to 'call' the size of the unknown DNA alleles in an amplified sample. The calling software compares the migration time of the alleles in the ladder and the internal standard with the migration times of the alleles associated with the specific locus in a sample.

When this method was first adapted for forensic analysis, approximately 35 minutes was required to process a single sample. It would take over 12 hours to separate 32 samples, although this was 'walk-away' time, where the analyst could perform other laboratory tasks or go home for the night. Early issues with the instrumentation were resolved through extensive and unparalleled collaborations between the manufacturers and the forensic laboratories, so that the forensic community could assure law enforcement and the judicial system that only optimized quality results would be provided by the laboratory.

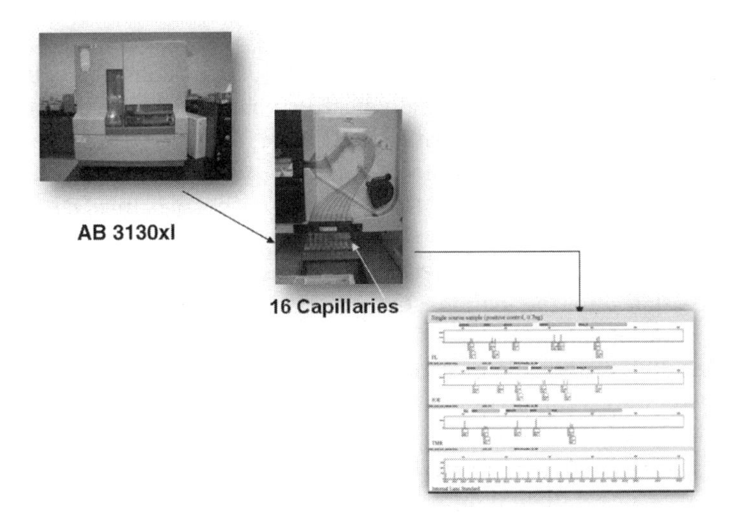

Figure 3.8 The AB 3130xl is an automated multicapillary fluorescent-based instrument used to detect amplified STR products. This instrument is a 16 capillary system in which 16 samples are injected, electropheresed through the capillary, amplified products detected and electropherograms generated.

Since the first available commercial single-capillary CE instrument for forensic DNA testing, 4- and 16-capillary array instruments have been developed, greatly improving the throughput of forensic casework samples (Figure 3.8). There have also been marked improvements in the size of the capillary diameter, the length, type and coating of the capillary, the nature of the electrolytic solutions, optimum voltage conditions, collection of STR fragment migration time and analytical software, all improving amplified product resolution and increased quality DNA profiles.

3.4.8 Interpretation of DNA Profiles

As previously discussed, the science employed in forensic laboratories is accomplished by conducting well-designed, documented, purposeful validation studies to demonstrate that the science of DNA testing is robust and reliable, consistently performs within defined conditions and may be trusted for use on casework evidence. Interpretation of amplified products is heavily dependent on the results of a laboratory's autosomal, Y-STR and mtDNA validation studies, which provide empirical data to support the identification of what is real DNA and part of a profile, what is not part of a profile and what is inconclusive data.

The national and international forensic community defines 'real DNA' in terms of the amount of Relative Fluorescent Units (RFU) that a specific size amplified product emits (Figure 3.7) [33]. Sensitivity, reproducibility and mixture validation studies define the point – called the analytical threshold – at which an analyst can trust that the RFUs from a specific peak are of an appropriate height, size, and morphology to be called an allele, and that the RFU signals are not baseline noise. However, even if a peak is at least the height of the analytical threshold, it cannot be assumed that all the data from an individual is present at the genetic marker, i.e. there may have been drop-out of an allele (Figure 3.9).

Validation studies also define a point, called the stochastic threshold, where the analyst may assume that peaks above this line represent not only true DNA alleles, but also that there has not been any loss of data for a specific profile at the specific locus (Figure 3.9). The empirically derived analytical and stochastic thresholds provide scientific assistance for the interpretation of peaks within an electropherogram.

Interpretation of single-source, complete DNA profiles are usually straightforward, but results for most unknown evidentiary samples in a caseworking laboratory are not always thus. Profiles from alleles generated from very low DNA template concentrations may be challenging, as inefficient amplification may generate amplified products with RFU values that do not meet stochastic threshold requirements. There may be allele drop-out and only a partial profile is generated, thus complicating the interpretation.

The *SWGDAM Interpretation Guidelines for Autosomal STR Typing by Forensic DNA Testing Laboratories* were published in January, 2010 [34]. These were the first comprehensive instructions provided for laboratories regarding mixture interpretation. It is anticipated that these guidelines will provide laboratories with a more uniform process of reviewing their existing mixture validation studies and developing interpretation protocols.

Laboratories in general have mixture interpretation guidelines that are specific for standards and unknown samples, specifically intimate or non-intimate samples. Intimate samples are collected from an individual's body or intimate items, and they are expected to contain DNA from the owner. If a mixture is obtained from an intimate sample, any unknown contributor to the sample may be deduced by subtracting out the known profile where possible.

Non-intimate mixture DNA samples may have two or more unknown contributors. Based on RFU intensities, a major contributor may be determined, and it may also be possible to deduce the minor profile through deconvolution. If the RFU data does not discern a major contributor, a minor profile cannot be called and interpretation becomes more challenging. A mixture of three or more contributors is considered complex, so interpretation guidelines must address the possible issues associated with these types of mixtures (Figure 3.9).

Complicating mixture interpretation is the possible presence of well-documented artifacts that are manifest on the electropherogram [18]. Artifacts may be related to issues with the manufacturing process of the reagents in the STR kits, such as fluorescent dyes that dissociate from the primer and migrate through the capillary, which have an amorphous shape and are easily discernible from an allele peak (Figure 3.10 A). A non-DNA peak may also occur when the different colored dyes are not optimally separated by the software, and one dye's fluorescence is observed in the wavelength of another dye's fluorescence, producing a peak, or 'pull-up' (Figure 3.10B).

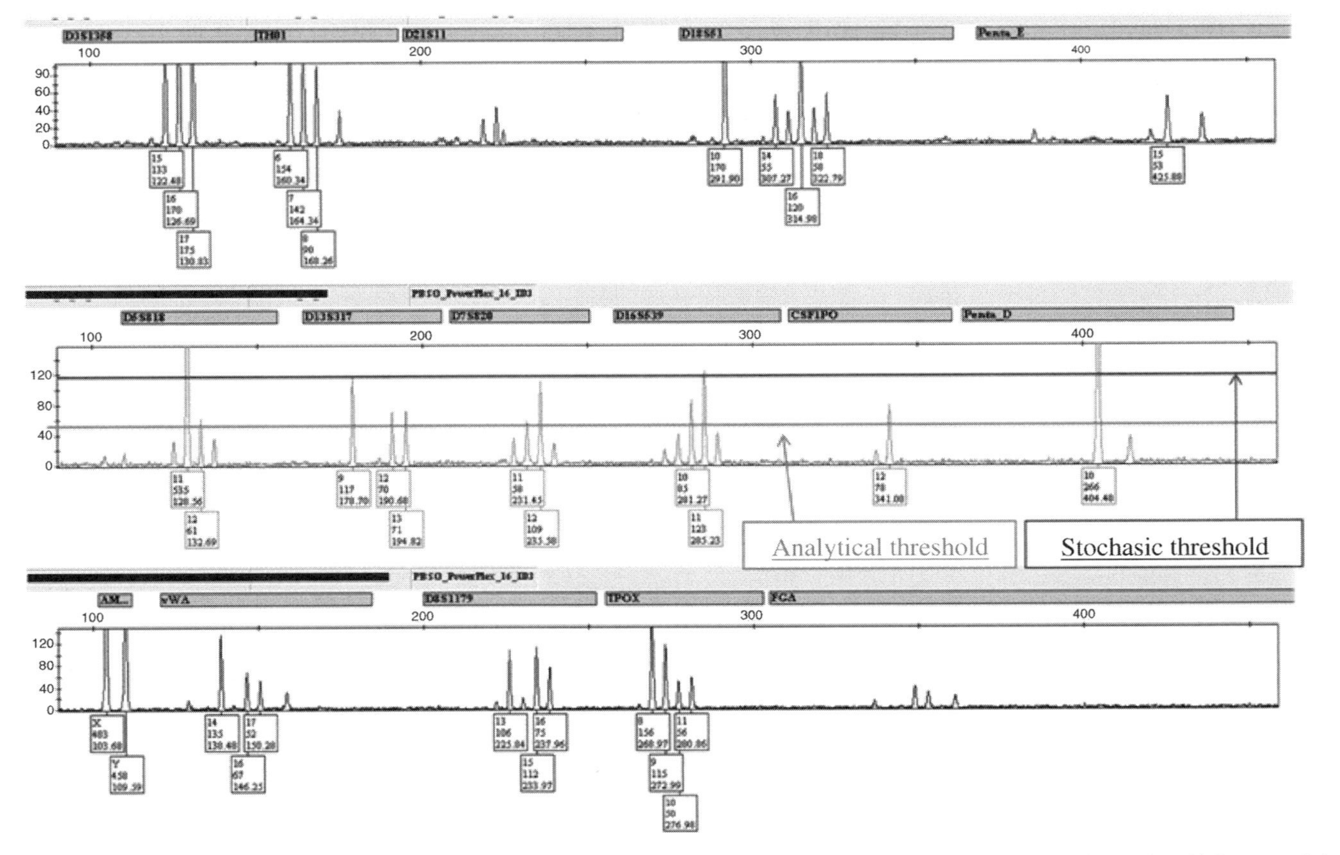

Figure 3.9 Electropherogram of a sample amplified with Promega PowerPlex 16 typing system. This is an obvious mixture in which some of the peaks are not called, because they appear below the analytical threshold (red line), and alleles that are called because the peaks appear above the analytical threshold, the point at which the peaks represent real DNA. The stochastic threshold (blue line) is the point in which an individual's profile is assumed to be present and drop-out has not occurred for that individual, although, in complex mixtures, interpretation is challenging.

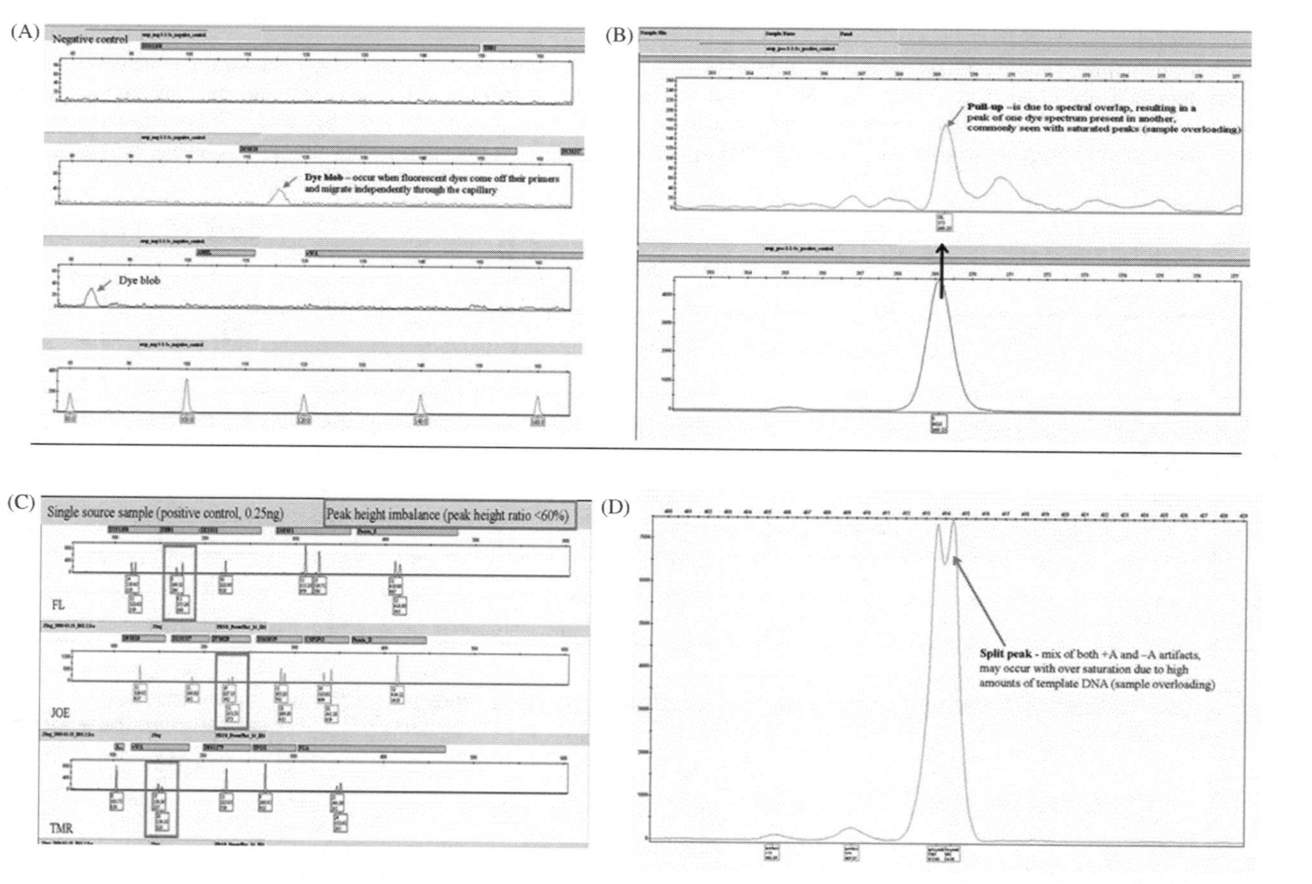

Figure 3.10 (A) Dye blobs from the manufacturer's kit may migrate through the capillary and are easily recognized as they have poor morphology and, following reinjection, the migration pattern cannot be repeated. (B) Spectral overlap, also known as 'pull-up', may occur when one dye can be detected in another dye's spectrum. (C) Peak height imbalance is obvious, as one of the alleles in an individual's profile is amplified at greater efficiency than the sister allele, causing the peaks to have significant RFU value differences. (D) Split peaks may be observed as an apparent single allele with an indentation at the peak, indicating an incomplete synthesis of the true allele.

There are also artifacts that are related to the amplification process. The most common is stutter, defined as a minute peak that is one repeat smaller or, in rare cases, larger than the real DNA peak, generated as a result of inefficient amplification. These stutter peaks are predictable and well-documented in the literature, and usually they can be filtered out by the calling software.

Peak imbalance is a well-documented artifact that occurs when low levels of template DNA are present (Figure 3.10C). There are occasions when the amplification process does not copy the nominal allele completely, thereby causing some amplified products to be a single base pair shorter than predicted. The result are split peaks on an electropherogram, although these are less frequent, as STR kits have been optimized to favor amplification of the larger peak (Figure 3.10D).

The CE instrument can also produce peaks within an electropherogram. These are obvious, and can be ruled out readily as part of a DNA profile as the peaks lack the morphology of a real DNA peak. Generally, these occur in each of the fluorescent dyes and cannot be reproduced upon re-injection. Once an analyst has determined which peaks are real DNA alleles and which are artifacts, the procedure of comparing profiles can begin. Where possible, the DNA profile from the unknown evidentiary samples should be determined before comparisons with the known individual's profiles.

Regardless of the DNA method used in a laboratory, the goal of the laboratory is to provide uniformity in the process, whereby analysts will interpret a DNA profile. The interpretation must be based on data from validation studies, scientific literature and an analyst's education, training and experience. Aside from providing DNA data for the judicial system, laboratories also submit DNA profiles into CODIS [35]. Reliable, qualifying DNA profiles are entered into the local, state or national DNA database and can be searched against database profiles for the purpose of providing investigative leads.

3.4.9 Future technologies

The goals of future forensic DNA analysis are to provide more sensitive and efficient technologies, whereby quality DNA data are generated for the purpose of aiding criminal and non-criminal investigations such as paternity, unidentified human remains and mass disaster identifications. Macro-scale robotics, designed to remove human intervention steps for extraction, sample preparation for qPCR and STR amplification, are becoming more widely used in laboratories.

Dr. John Butler, NIST fellow and Group Leader of Applied Genetics, presents an excellent and comprehensive summary of the new technologies in the 2011 publication of *Advanced Topics in Forensic DNA Typing Methodology*, Chapter 17 [36]. Microfluidic devices, integrated with extraction, amplification and STR allele detection capabilities that will cut the laboratory time from days to a few hours, will be commercially available in 2012.

The use of next generation DNA sequencing, mass spectrophotometry, pyrosequencing and single nucleotide polymorphisms (SNPs) are a potential alternative to amplified fragment analysis, although they most likely will not be replacing STRs in the near future, due to the sheer magnitude of the investment nationally and internationally that has been made to produce the current forensic DNA STR program. The challenges associated with DNA mixture profile interpretation will be met with sophisticated software programs that incorporate the laboratory's validation protocols, which will be based on the SWGDAM mixture guidelines.

Regardless of what the future holds for forensic DNA typing, it is imperative that validation studies are conducted appropriately and interpretation of the studies are accurate. The forensic community will embrace changes that offer improved DNA quality and/or efficiency and still maintain the integrity of the DNA analytical process, while never compromising the fundamental trust of the scientific, public and judicial communities.

3.5 Fire scene investigation and laboratory analysis of fire debris

3.5.1 Introduction

The field surrounding the occurrence of fires is expansive. It covers the investigation of the scene through the chemical analysis of submitted samples for ignitable liquid residues and the examination of potential ignition sources. To address all tasks within the field, many professionals may be required, each with a specialized skill set. These professionals have titles such as fire scene investigator, electrical engineer, metallurgist, fire protection engineer and forensic analyst. There are major differences in the qualifications of the involved personnel. The field investigator may only possess a high school diploma, while the laboratory analyst will possess at least a bachelor's degree and a specialized understanding of instrument mechanics, such as phase separation.

At the outset, the major difference between a fire investigation and the investigation of other types of crimes is the necessity to determine whether a crime has actually occurred. With complex cases such as those involving death, or multimillion dollar losses, there are issues beyond whether the fire was intentionally set or not.

This segment will treat the two facets involved separately. First, the evolution of the science of fire scene investigation will be addressed, followed by a discussion on the techniques for fire debris analysis.

3.5.2 Fire scene investigation: history

The great Chicago fire of 1871 has been attributed to a perfectly placed kick to a gas lantern by the O'Learys' cow. However, in 1893, Michael Ahern, a *Chicago*

Republican reporter, admitted to making the story up as it made great news. The fact of the actual conflagration was that everything was in place for the disaster: dry climate, high wood content in building structures, strong southwest winds and a slow response from the fire department. These elements combined to cause four square miles of destruction. Over the years, there has been speculation about several possible causes of the fire, but nobody really knows. The cause might have been determined if the standardized approach used today had been applied.

Although some progress was made in fire investigation after 1871, most of the important improvements have occurred in the last 30 years. During this time, many of the previously accepted 'rules of thumb' for fire investigation have turned out to have been based on anecdotal evidence or a misunderstanding of the behavior of fire.

In 1977, the Law Enforcement Assistance Administration (LEAA) reported the results of a survey of fire investigators in a publication entitled *Arson And Arson Investigation: Survey And Assessment* [37]. This report contained a compilation of the 'indicators' of incendiary activity used by fire investigators at the time. The authors of the survey warned that none of the indicators had been scientifically validated, and they recommended a series of carefully conducted experiments to identify the value of these indicators.

Three years later, the National Bureau of Standards (NBS – now called the National Institute of Standards and Technology, or NIST) produced a document called *Fire Investigation Handbook* [38]. The NBS handbook was developed with information about origin and cause available at the time and it provided a broad approach to fire investigation, including documenting the scene; information on fire chemistry and building construction; interview techniques; and ignition sources. Prior to publication, the handbook was reviewed by the National Fire Protection Agency (NFPA), the International Association of Arson Investigators (IAAI), the Bureau of Alcohol, Tobacco and Firearms (ATF), the University of Maryland Fire & Rescue Institute, and State and local fire departments or administrations.

The 'Cause and Origin' section, written by Steven Hill and Victor Palumbo, both from the National Fire Academy, limited the content to their interpretation, but many of the arson indicators in that section have since been proved false. The reputation of the NBS for integrity and scientific prowess gave the handbook staying power, when it should have been subjected to re-evaluation in the years after publication. During the 1980s and into the 1990s, many investigators relied on textbooks that referenced some of the misinformation generated from the *Fire Investigation Handbook*.

In 1985, the Standards Council of the NFPA became concerned with the quality of the work product of fire investigators and commissioned the Technical Committee on Fire Investigations. The committee was asked to draft a guideline for fire investigators – a task which took seven years. The first edition of NFPA 921, *Guide for Fire and Explosion Investigations*, was published in 1992 [39]. Automatic acceptance of dramatic changes to established beliefs rarely goes smoothly, and the profession did not immediately embrace NFPA 921. Part of the concern was that if the information within the newly published document was correct, it meant that hundreds or thousands of accidental fires had been wrongly characterized.

A major challenge to fire investigators' traditional thought came in 1996, in *Michigan Millers Mutual Insurance Company v. Janelle R. Benfield* [40]. In that case, a fire investigator failed to document his observations properly and, as a result, he was excluded from testifying. In the appeal against the exclusion of this investigator's testimony, the IAAI filed an *amicus curiae* brief, in which they contended that fire investigators should not be held to a reliability inquiry because fire investigation was 'less scientific' than the scientific testing discussed in *Daubert v. Merrell Dow Pharmaceuticals, Inc.*, 43 F.3d 1311 (9th Cir. 1995).

The Daubert decision emphasized the role of the judge to determine if the expert is, in fact, an expert, whether the information proffered by that expert is relevant, and whether the expert has used sound methodology. Subsequent court rulings, including the decision in the Michigan Millers' case and the unanimous Supreme Court decision in *Kumho Tire v. Carmichael* [41] persuaded the majority of fire investigators to accept the scientific method recommended by NFPA 921.

It is difficult to state exactly when NFPA 921 became generally accepted by the relevant scientific community, but 2000 was an important turning point. That year, the United States Department of Justice released a research report entitled *Fire And Arson Scene Evidence: A Guide For Public Safety Personnel*, which identified NFPA 921 as a 'benchmark for the training and expertise of everyone who purports to be an expert in the origin and cause determination of fires' [42]. That same year, the IAAI, for the first time, endorsed the adoption of the new edition of NFPA 921.

3.5.3 Established through science

NFPA 921 provides guidance that allows an investigator to reach conclusions based on science, instead of basing them on hunches or feelings. Using this guide, along with documented data, sound science and clear reasoning, investigators are establishing stronger justifications for stated conclusions. For example, in the past, investigators were unaware of 'flashover' burn patterns on surfaces and were labeling these burn patterns as patterns made by ignitable fluids. A 'flashover' is a transition phase in the development of a confined fire, in which surfaces exposed to thermal radiation reach ignition temperature more or less simultaneously and fire spreads rapidly throughout the space. This transition from 'a fire in a room' to a 'room on fire' results in the production of irregular fire patterns. It is likely that the understanding of the significance of irregular patterns has reduced the number of cases identified as arson.

Some high-profile criminal arson cases [43] have attracted the attention of the public and the media and have resulted in some interesting studies regarding the prevalence of arson in the United States. Dave Mann of the *Texas Observer* became interested in the study of errors in fire investigation as a result of the cases of Ernest Ray Willis (who was exonerated after 17 years on death row) and Cameron Todd

Willingham (who was executed after 12). Mr. Mann published a study that included a total fire count in Texas versus the number of fires determined to be arson. Those results demonstrated a more than 60% drop in the number of fires determined to be arson between 1997 and 2007 [44].

After reviewing the data from Texas, Jack Nicas, a reporter for the *Boston Globe*, performed the same exercise in Massachusetts, with even more startling results. Between 1984 and 2008, the percentage of fires determined to be arson in Massachusetts dropped from over 20% to less than 2%, despite a net increase in the total number of fires. Nationwide, from 1999 to 2011, the NFPA reports a drop from around 15% to around 8% in the percentage of fires determined to be arson.

There remains a cadre of fire investigators who refuse to accept the fact that post-flashover burning produces artifacts that they can misinterpret. However, they are fighting an uphill battle, as newly revised training modules for fire investigators no longer singly focus on how to 'recognize arson'. Instead, instruction is expanding to include understanding fire patterns, particularly those caused by the effects of ventilation on fully involved compartment fires.

3.5.4 Post-flashover burning

The year 2005 marked another major turning point in our understanding of fire behavior and the accuracy of fire origin determinations. A group of ATF Certified Fire Investigators (CFIs), led by Special Agent Steve Carman, designed a test that mirrored similar experiments that had been conducted (but not documented) at the Federal Law Enforcement Training Center (FLETC) in Glynco, Georgia.

The ATF CFIs set up two 12 × 14 foot bedrooms, set each of them on fire, and allowed the rooms to burn for about two minutes beyond flashover. Fifty-three participants in a Las Vegas IAAI-sponsored fire investigation seminar were asked to walk through the burned compartments and to write down the quadrant in which they believed the fire had originated. In the first bedroom, three participants identified the correct quadrant. When the exercise was repeated on the second bedroom, three different participants identified the correct quadrant. Carman reported that the success rate was 8–10% in the undocumented tests at Glynco, so the error rate in origin determination was not much different [45]. In the two 12 × 14 foot bedroom fires, the participants were not allowed to interview witnesses, collect debris or perform other activities other than a visual observation. It should also be noted that the qualifications of some of the participants were unknown, and that some were participating just to familiarize themselves with fire investigative procedures.

In an attempt to understand the data, Carman and his collaborators at ATF recreated the test fires at the ATF Fire Research Laboratory in Ammendale, Maryland. They modeled the results using a computational fluid dynamics-based program, the NIST *Fire Dynamics Simulator* (FDS), and viewed the results with a companion program called *Smokeview* [46–48]. What came out of these studies was a better, but certainly not complete, understanding of the effects of ventilation in post-flashover fires.

To encourage dissemination of this information, the IAAI has hosted the website www.CFITrainer.net as a means to provide access to free training modules [49,50].

Typically, fire investigators look for the origin of a fire where the *fire plume* has interacted with two-dimensional surfaces such as walls and ceilings. These patterns may cease to be created once the transition to flashover has occurred. In ventilation-controlled fires, patterns will be generated around doors and windows, as well as around any places where air leakage existed before the fire, or where such ventilation was created as a result of the fire.

When flashover occurs and every exposed combustible surface has ignited, the fire can consume almost all of the oxygen in the room, and the burning will diminish once the oxygen is depleted. The fuel at the origin of the fire will not necessarily continue to burn, and much larger patterns not associated with the origin can be generated. These new patterns can obscure the original, plume-generated patterns. Even if the original patterns persist, the new patterns can add confusion to the interpretation of the origination of the fire. Figure 3.11 shows the fire pattern that was mistakenly interpreted as an origin pattern in the Las Vegas tests previously described. The pattern at the actual origin of the fire is shown in Figure 3.12 .

When the fire was modeled, the oxygen concentration was shown to be very low throughout the room, except in the immediate vicinity of the doorway, and only near the floor. This oxygen concentration suggested the existence of a 'floor jet'. This inflowing oxygen collided with the wall opposite the doorway and moved upward, mixing with the unburned fuel, and this resulted in a heat flux of more than

Figure 3.11 Large, conspicuous ventilation-produced pattern on the wall opposite the door incorrectly identified as the origin of the fire by many investigators in the 2005 Las Vegas test. There was no competent ignition source or fuel source at the base of the pattern. The wide base on the pattern may have led some investigators to believe that an ignitable liquid was spread along the wall.

Figure 3.12 Roughly V-shaped pattern on the wall next to the bed, where a test fire was ignited. It is not clear whether this pattern represents a 'clean burn' or whether the dry wall in this area was never sooted because it was too hot for condensation to take place.

150 kW/m^2. Clearly, the total energy impact (heat flux multiplied by time of exposure, expressed as kJ/m^2) on this surface far exceeded the total energy impact on the wall next to the origin.

Besides being incorrect, identifying the wrong origin will lead the investigator to a location where no trace of the ignition source will be found. In this scenario, investigators who fail to understand ventilation could conclude that somebody must have placed some fuel at that origin and ignited it with an open flame. If there is an irregular burn pattern on the carpet in that area, even in the absence of a positive laboratory report, the investigator may conclude that the fire was intentionally set using a flammable liquid. Many investigators have made errors using this kind of 'negative corpus' determination. Finding the correct origin is key to a correct fire cause determination, and it is the most difficult part of the investigation of a fully involved compartment fire.

In 2007, ATF agents refined and repeated the Las Vegas test, this time in Oklahoma City. They set up three burn cells, with identical fuel and identical ventilation but different points of origin. The cells were allowed to burn for 30 seconds beyond flashover, 70 seconds beyond flashover and 180 seconds beyond flashover. To put these times in context, fire departments in big cities may have a three-minute response time. If they are not called until someone sees the fire venting out of a window (a sign of flashover), the chances of them extinguishing the fire with less than three minutes of post-flashover burning are practically zero. The results of the Oklahoma City experiment confirmed the data from Las Vegas obtained two years earlier. Further, it became clear that the longer the fire was allowed to burn after flashover, the less likely the fire investigators were to identify the quadrant of origin correctly. The results of the Oklahoma City experiment are shown in Table 3.2 .

Table 3.2 Post-flashover burning. Fires do not automatically stop burning once flashover is attained, so the assumption that burn patterns post-flashover are recognizable when compared with test fires ceased prior to flashover was suspect. The controlled burn was allowed to extend past flashover for different lengths of time in the Oklahoma City burn experiment to measure fire investigators' ability to determine the correct quadrant of origin

Time	# of responses	# correct	% correct
30 seconds	70	59	84
70 seconds	64	44	69
180 seconds	53	13	25

The 25% correct quadrant of origin selection at 180 seconds post-flashover, although better than the 6% obtained in the Las Vegas test, is actually only equal to random selection. The percentages obtained (84%, 69% and 25%) are weak if the results are used as the basis either to send people to prison or to deny them coverage under their insurance policy.

While these experiments demonstrate that the field is improving through research, the results also demonstrate that the infrastructure for training fire investigators must be improved. This is a strong argument to reduce the use of short-lived fires, extinguished before flashover, in training programs. Fire investigators and the people who employ them need to be prepared to accept the reality that sometimes the best answer that can be obtained is 'undetermined' if either an accidental or an incendiary call is not supported by conclusive evidence [52].

3.5.5 Professional qualifications and certification of fire scene investigators

The National Fire Protection Association (NFPA), through a document known as NFPA 1033, *Standard for Professional Requirements for Fire Investigator*, publishes minimum requirements for the knowledge skills and ability of a fire investigator. Principal among these is a 13-point list of areas in which a fire investigator is required to have education beyond high school level. These 13 topics are [53]:

1. Fire science
2. Fire chemistry
3. Thermodynamics
4. Thermometry
5. Fire dynamics
6. Explosion dynamics

7. Computer fire modeling
8. Fire investigation
9. Fire analysis
10. Fire investigation methodology
11. Fire investigation technology
12. Hazardous materials
13. Failure analysis and analytical tools.

Although obtaining certification is essential in a true profession, there remains a large number of fire investigators who are not yet certified. Fire scene investigators may become certified through the International Association of Arson Investigators (IAAI) or the National Association of Fire Investigators (NAFI). NFPA 1033 and NFPA 921 provide the basis of both certification programs. To enter into the certification process, applicants must provide information regarding their education, training and experience. If accepted into the process, the applicant is then required to complete successfully a written examination. Certificates are valid for a period of five years, at which time an investigator must demonstrate continued participation in the field and a minimum amount of continuing education for recertification.

In a very proactive move in the late 1980s, ATF required their agents to attend classes in fire behavior and, near the same timeframe, they initiated the ATF CFI program. An agent that is accepted in the program must take a series of college-level courses in fire science, complete a research paper and participate in a series of ATF training sessions, some taught by college professors or researchers, which include hands-on fire investigation and analysis exercises [54].

3.5.6 Fire debris analysis: history

In the early 1900s, when petroleum products were not commonly found inside structures, merely detecting a petroleum odor at a fire scene or on a debris sample would suffice for proof of the presence of an ignitable liquid residue (ILR). Separation of the ILR was carried out by steam distillation or by solvent extraction. Initially, that was the end of the process. An analyst (or anyone else) could simply sniff the extract (55, p. 20). One fire investigation text from 1945 recommended dipping a piece of bread in suspect fire debris and tasting it [56].

The first analytical techniques applied to ILR extracts were determinations of boiling point, refractive index and specific gravity, and also infrared spectrophotometry. None of these techniques are in use today for ILR analysis. Douglas Lucas (one of the leaders of the AAFS for most of its existence) first reported the use of gas chromatography for ILR analysis in 1960 [57]. This section covers the evolution of both separation techniques and analytical techniques since that time.

3.5.7 Evolution of ILR Isolation Techniques

Although there have been significant advances in the sensitivity of the analytical instruments, it is the advances in the efficiency of separation procedures that has allowed for the effective determination of miniscule quantities of ignitable liquid residues in fire debris samples.

Paul Kirk reported in 1969 that the normal manner of isolating a liquid accelerant from other materials is 'to distill the liquid from a solid residue in a current of steam' (58, p. 153).

Figure 3.13 shows a steam distillation apparatus. In this classical separation technique, the debris is covered with water and boiled, and the steam and other

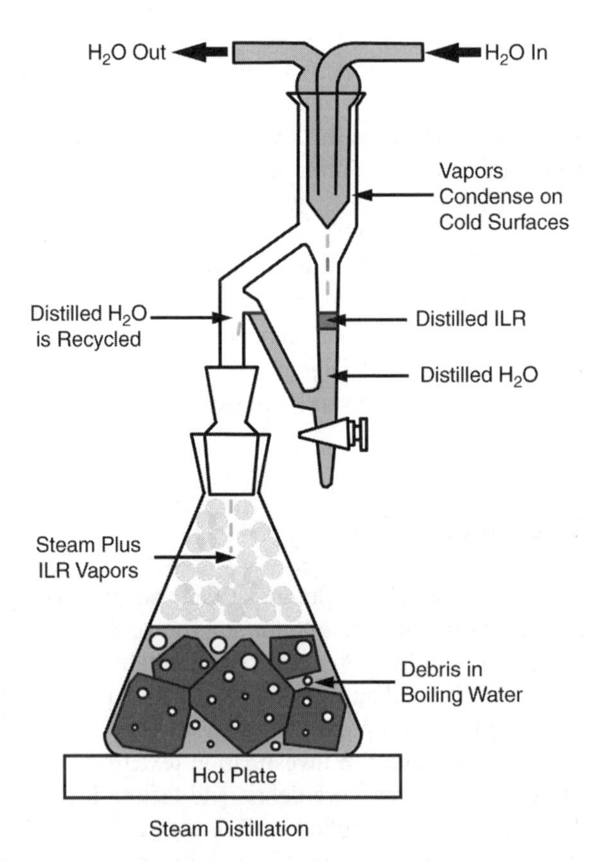

Figure 3.13 Steam distillation apparatus. The sample is boiled. Vapors condense on a 'cold finger' and fall into the trap, which allows the water to recycle, while the immiscible oil layer builds up on top of the water column.

vapors are condensed in a trap that recycles the water and allows any non-miscible oily liquids to float on top. There is actually a visible layer of liquid isolated from the sample. More often than not, this layer consisted of a drop or two, or simply a rainbow film on top of the column of water. This could be extracted with a solvent and analyzed but, even so, steam distillation was not a very sensitive technique. If the sample did not have detectable petroleum odor, steam distillation was almost always ineffective at isolating any ignitable liquid.

As with fire scene investigation techniques, this is an example of what was accepted at the time due to technology limitations. Therefore, because of its limitations, steam distillation, once an approved ASTM practice (E1385), has been relegated to 'historical' status.

Kirk also proposed that the debris could be heated in a closed container and the internal gaseous phase could be sampled and analyzed by gas chromatography, but he indicated that he was unaware of its use in routine analyses. By the mid-1970s, this technique, known as 'heated headspace', was routinely used, but it had sensitivity limitations similar to steam distillation and was ineffective at isolating the heavier components of common combustible liquids, such as diesel fuel. Heated headspace (Kirk called it a 'shortcut') remains in use as a screening tool in some laboratories today.

The first quantum jump in sensitivity was reported in 1979 in the *Arson Analysis Newsletter*, when Joseph Chrostowski and Ronald Holmes reported on the collection and determination of accelerant vapors [59]. These two chemists, from the Bureau of Alcohol, Tobacco and Firearms' Philadelphia laboratory, employed a dry nitrogen purge and a vacuum pump to draw ignitable liquid vapors from a heated sample through a Pasteur pipette filled with activated coconut charcoal. The vapors were eluted using carbon disulfide and analyzed by gas chromatography. Over the next decade, many research articles were devoted to the refinement and expansion of this process, now known as 'dynamic headspace concentration'. Dynamic headspace concentration remains a recognized analytical technique but, because it is both destructive and complicated, not many laboratories use it. Many laboratories, especially in Europe, use Tenax® in place of charcoal. The Tenax is thermally desorbed, rather than eluted with a solvent as with charcoal.

The case for 'passive headspace concentration', applied to fire debris, wherein an adsorbent package is placed in the sample container and heated, was made by John Juhala in 1982 (60, p. 32). Juhala used charcoal-coated copper wires and Plexiglas beads. He reported an increase in sensitivity of two orders of magnitude over distillation and heated headspace analysis, but many laboratories had just completed setting up their dynamic systems. Consequently, adoption of passive techniques took some time, but gradually its advantages made it the dominant method of separation.

Dietz reported on an improved package for the adsorbent, called C-bags, in 1991 (61, p. 111) but these quickly gave way to activated carbon strips (ACS), which required much less preparation. In 1993, Waters and Palmer (62, p. 165) reported on the essentially non-destructive nature of ACS analysis, performing up to five

Figure 3.14 Schematic drawing of passive headspace concentration using an activated carbon strip. Vapors are produced by heating the container with debris to 80°C. The ACS adsorbs the vapors for 16 hours, then is rinsed with an appropriate solvent. The resulting solution is then analyzed by GC-MS. An internal standard may be added to the sample to assess adsorption efficiency, and an additional internal standard may be added to the eluting solvent to assess injection efficiency.

consecutive analyses on the same sample with little discernable change, and no change in the ultimate classification of the residue. This separation technique is the method of choice in most laboratories today.

Figure 3.14 is a conceptual drawing of the procedure for passive headspace concentration. A typical adsorption device consists of a 10 × 10-mm square of finely divided activated charcoal, impregnated on a polytetrafluoroethylene (PTFE) strip. The technique was actually adapted from the industrial hygiene industry. Charcoal disk badges are worn by employees to determine their exposure to hazardous chemicals. The charcoal adsorbs a wide variety of organic compounds.

Solid-phase micro-extraction (SPME) represents yet another kind of passive headspace concentration technique. The SPME fiber is a more active adsorber of most ignitable liquid residues than is an activated carbon strip. Exposing an ACS to the headspace of a sample at elevated temperatures for 16 hours allows for the isolation of less than 0.1 μl of ignitable liquid residue if there is no competition from the substrate; an SPME fiber can accomplish the same feat in 20 minutes. A comparison of the advantages and disadvantages of the separation techniques is given in Table 3.3 .

Table 3.3 Advantages and disadvantages of ASTM methods. Although established methods have historical significance, their use should be evaluated against new technology, research findings and the type of evidence. This table provides an explanation as to the advantages and disadvantages of ASTM International standards in regard to the analysis of ignitable liquid residue

ASTM Method	Advantages	Disadvantages
E1385 Steam distillation	Produces a visible liquid, simple to explain	Labor-intensive; destructive; not sensitive; requires expensive glassware
E 1386 Solvent extraction	Useful for small samples and empty containers; does not cause significant fractionation; useful for distinguishing HPDs from each other	Labor-intensive; expensive; co-extracts non-volatile substances; increased risk of fire; solvent exposure, destructive
E 1388 Headspace sampling	Rapid, more sensitive to lower alcohols; nondestructive	No archivable sample; not sensitive to heavier compounds; poor reproducibility
E 1412 Passive headspace	Requires little analyst attention; sensitive; nondestructive; produces archivable sample; inexpensive	Requires overnight sampling time
E 1413 Dynamic headspace using activated charcoal	Rapid; sensitive; produces archivable sample; inexpensive	Labor-intensive; subject to breakthrough; destructive
E 1413 Dynamic headspace using Tenax	Rapid; sensitive	Labor-intensive;, requires thermal desorption; no archivable sample; destructive
E 2154 Solid phase microextraction (SPME)	Rapid; highly sensitive; useful for field sampling with portable GC/MS	Labor-intensive; expensive; requires special injection port; reuse of fibers; no archivable sample

3.5.8 Analysis of the extracted ILR

The gas chromatograph (GC) has been the primary instrument used in the identification of petroleum-based hydrocarbon liquids, which are the most commonly used accelerants.

As is the case with many forensic science disciplines, identification is based on pattern recognition and pattern matching. The 'pattern' arises because petroleum distillates are complex mixtures of up to five hundred different compounds. There is, for example, no such entity as a 'gasoline molecule'. Gasoline and most petroleum distillates are resolved by the chromatographic column into separate compounds, usually in ascending order by boiling point or molecular weight (lighter compounds pass through the chromatographic column more quickly than heavier compounds). The laboratory analyst comes to recognize the patterns produced by particular classes of petroleum products, and the 'match' is made by overlaying the chromatogram from the sample extract onto the chromatogram of a known standard. This comparison can be made on a computer screen, but the traditional method involves laying one paper chart over another on a light box.

There is some room for judgment in this pattern matching, and it is for this reason that coupling mass spectrometry with gas chromatography is the generally accepted methodology. While gas chromatography produces a pattern of peaks, the mass spectrometer is capable of identifying the compounds that produce the peaks and presenting displays showing only those peaks with a specified mass-to-charge (m/z) ratio. Such a display is known as an extracted ion *profile* (EIP) if more than one m/z ratio is presented, or an extracted ion chromatogram (EIC) for a single m/z ratio.

Sole use of gas chromatography may cause the analyst to confuse patterns produced by background materials with patterns of petroleum-based liquids. This is less likely to happen when gas chromatography-mass spectrometry (GC-MS) is used, because there is much less guesswork about the identity of the compounds responsible for the peaks on the chart. Caution is still required, however, because a piece of carpeting (or any combustible solid) pyrolyzing in the process of combustion may produce compounds that also are found in petroleum distillates. Thus analysts must be aware that merely detecting benzene, toluene and xylenes, or higher molecular weight aromatics, is not sufficient for identifying gasoline. The concentrations of all of the compounds of interest relative to one another are a critical step in pattern identification.

Proficiency tests, manufactured by Collaborative Testing Services (CTS) and sponsored by the American Society of Crime Laboratory Directors (ASCLD), revealed that the error rate for laboratories using gas chromatography alone is significantly higher (50–100% higher) than laboratories using GC-MS [64,65]. These same CTS studies show a decreasing reliance on GC alone.

3.5.9 Increasing sensitivity

Using the described techniques, it is possible to detect 0.1 μl of ignitable liquid residue in a 1 kg sample, assuming the sample substrate is not overly tenacious. In most arson fires, the fire setter uses an excess of accelerant, and it is usually not necessary to achieve this lower limit of detection.

There may be some issues regarding the wisdom of attempting to lower the detection limit any further. The background of a modern residence or office contains numerous petroleum-derived products, and finding ignitable liquid residue at this level may not be meaningful (66, p. 968). This is especially true when the total ion chromatogram (TIC), which is the top-level output of the gas chromatograph-mass spectrometer, does not resemble the total ion chromatogram of a petroleum product such as gasoline.

Using extracted ion profiles, it is possible to 'weed out' the compounds that obscure the ILR in the TIC [67]. However, when this occurs, it is wise to insist that the extracted ion profiles bear more than a passing resemblance to the EIPs from petroleum products. Toluene, xylenes and C_3 alkylbenzenes are commonly produced during the pyrolysis of polymers and adhesives. When the only evidence supporting an incendiary determination is a finding of exceedingly low concentrations of compounds common to gasoline, and that finding is solely based on an interpretation of the aromatic EIPs, the chemical analysis should be subjected to increased scrutiny.

3.5.10 Identity of source

Once an ignitable liquid has been exposed to a fire, its character changes to the extent that its source is very difficult to identify. Some work has indicated that source identification is possible if a sample has at least 70% of the original weight. There are times, however, when ignitable liquids are mixed, producing a unique pattern that can conceivably be identified with a source. There also exist occasions when it is possible unequivocally to eliminate a suspected source of an ignitable liquid residue.

There exist 'chemometric' methods for comparing sources. These generally involve quantifying the signals from compounds that are very similar in terms of volatility, so there is little effect from evaporation on the ratio between the two [68]. Another method to examine the identity of source involves quantifying the polynuclear aromatic hydrocarbons in the sample [69].

The exposure of a petroleum distillate to a fire results in its evaporation, with the lower boiling point compounds being preferentially evaporated over the higher boiling point compounds. This results in an irreversible increase in the average molecular weight of the mixture. Thus, when a suspected source of ignitable liquid exhibits a higher average molecular weight than the fire-exposed residue, the source can be unequivocally eliminated.

It is generally recognized that it is not possible to distinguish whether a sample has been exposed to a fire or to room temperature evaporation. A sample of petroleum distillate that has burned to 50% of its original volume or weight will give a gas chromatographic pattern that is currently indistinguishable from a sample that has naturally evaporated to that point.

3.5.11 Professional qualifications and certification of fire debris analysts

According to the Scientific Working Group for Fire and Explosions (SWGFEX), fire debris analysts should have 'a bachelor's degree in a natural science or equivalent, which shall include lecture and associated laboratory classes in general, organic, and analytical chemistry' [70]. Fire debris analysts with two years' experience can apply for certification from the American Board of Criminalistics (ABC).

Certification as a Diplomate of the ABC, denoted by the designation D-ABC, is awarded to individuals with a bachelor's degree in a natural science, two years of forensic laboratory or teaching experience, and upon successful completion of the ABC specialty examination in fire debris analysis. Certification as a fellow of the ABC, denoted by the designation F-ABC, is awarded to individuals with the qualifications of a Diplomate who successfully perform annual proficiency testing. ABC certification is valid for five years from the date of the award of the diplomate or fellowship certificate. Recertification can be accomplished by continued participation in the forensic sciences for training, casework, publishing or re-testing.

Affiliate status is available to an individual who meets all of the requirements for certification except for the two-year experience requirement. Upon successful completion of the fire debris analysis examination, applicants become 'certification eligible' until the experience requirement is met.

3.5.12 Standards creation

Because it is based on fundamental principles of chemical analysis, laboratory analysis of fire debris is one area of forensic science where there is a near consensus on methodology and terminology (71, p. 40). Because of this consensus, there has been a considerable amount of research published on the characterization of ignitable liquid residues recovered from fire debris. Much of the credit for this general consensus goes to ASTM Committee E30 on Forensic Sciences. The E30 committee has diligently worked with the community to write standards that could fall into one or more of ASTM's six categories [72]:

- *Standard Specification* – defines the requirements to be satisfied by subject of the standard.
- *Standard Test Method* – defines the way a test is performed.
- *Standard Practice* – defines a sequence of operations that, unlike a test, does not produce a result.
- *Standard Guide* – provides an organized collection of information or series of options that does not recommend a specific course of action.
- *Standard Classification* – provides an arrangement or division of materials, products, systems or services into groups based on similar characteristics such as origin, composition, properties or use.
- *Terminology Standard* – provides agreed definitions of terms used in the other standards.

Continued review of the standards has resulted in retiring some standards to 'historical status'. For example, the identification of limitations with steam distillation and solely using gas chromatography for analysis, both previously approved ASTM standards (E1385 and E1387 respectively), were relegated by the E30 committee to historical status. This responsive review activity continues to ensure the supported methodology is up to date.

The ASTM standards are useful educational tools that provide the analyst with nuances of the discipline. One example of this is shown in ASTM E1618, which advises analysts that merely detecting benzene, toluene and xylenes, or higher molecular weight aromatics, is not sufficient for identifying gasoline. The relative concentrations of all of the compounds of interest must be such that a recognizable pattern is produced.

The ASTM standards support both accreditation and certification. Laboratories seeking accreditation can adopt ASTM standards 'by reference', and auditors will know that the laboratory's procedures are valid. Certification bodies like the ABC can use the ASTM standards as the basis for their examinations.

3.5.13 Research opportunities

Fire debris analysis is about as 'settled' as any of the criminalistics disciplines. There still exist areas that could be improved, particularly in identifying the source of the residue. Comprehensive two-dimensional gas chromatography, followed of course by mass spectral analysis, holds the potential for making very precise identifications of ILR. Other analytical techniques that hold promise include GC-MS-MS and GC-isotope ratio mass spectrometry.

Automating fire debris analysis by the use of computer algorithms to classify the ILR is receiving significant attention as analysts attempt to produce quantitative data to make classifications more 'objective' and less subject to an analyst's judgment. It is known that some ILRs are subject to microbial degradation. Research into characterizing this degradation and preventing it is under way.

Recent studies by environmental chemists, attempting to measure the age or source of petroleum discharges, have identified several classes of higher molecular weight compounds that are more likely to survive a fire, which can be used for individualization. Considerable additional work is required before this technology will be applicable to routine fire debris analysis [73].

The annual meeting of the AAFS usually presents attendees with about a dozen oral papers and an equal number of posters on current research in fire debris analysis. In addition, the researcher interested in delving into the world of fire debris can obtain suggestions by reading the National Academies 2009 report, *Strengthening Forensic Science in the United States: A Path Forward*. The Committee on Identifying the Needs of the Forensic Science Community made the same recommendation that the LEAA had made 32 years earlier, i.e. more research is needed on the variability of post-fire effects [74].

3.6 Trace evidence

3.6.1 Introduction

Trace evidence consists of small (often microscopic and easily overlooked) quantities of matter of almost any type that may have significance in demonstrating contact between two sources. For example, fibers of the sweater worn by a murder victim may adhere to clothing of the perpetrator, the soil on a suspect's shoe may be compared to the site where a dead body was found, or spray paint spheres found on a corpse may correspond with a particular product used to paint only certain vehicles. These examples provide only a hint of the vast range of particles, stains and smears of almost any substance that may, under the right circumstances, become trace evidence. Such evidence from a victim, crime scene, vehicle, etc. can be compared to a known source – for example, a suspect, location, carpet, etc.

They can also be used to help further an investigation when there is no suspect, or where even basic facts, such as the location where a murder took place, are unknown. Thus, the dust on an item of clothing can be examined to determine the occupation of the owner, or the soil on the shoe of a murder victim can be studied to determine the location where they last stood. The techniques of detecting, isolating, analyzing, identifying and interpreting these minute quantities of unknown matter are the responsibility of the forensic trace evidence analyst. Because these dusts or smears are often minute, it is virtually impossible for the criminal to guard against picking them up or leaving them behind. Their small size and quantity can, however, make their detection, recovery, analysis and comparison challenging.

The scientific examination of these microscopic traces is conducted in forensically clean laboratories by specially trained scientists using sophisticated techniques and instruments. Cutting-edge, state-of-the art techniques, as well as older established methods of analysis, are employed to perform these analyses and comparisons. Trace evidence analysts must have academic credentials in chemistry, microscopy, biology and physics, as well as practical experience in dealing with these vanishingly small and easily overlooked specks of evidence.

Although a contemporary forensic trace evidence laboratory may look like the television set of *CSI* (albeit with brighter light), the work, which is often painstakingly slow and tedious, is performed according to the principles of the well-established scientific method and, more often than not, the results are only slowly developed as data are accumulated – unlike TV, where analytical results are delivered as a 'one liner' after a glance through a microscope or at a graph from an instrument.

3.6.2 Historical background

We live in a world that is disintegrating about us. The bright flecks we observe floating in a sunbeam, the debris that fills our vacuum cleaner or the black smear on our finger when we draw it across an open window sill are all composed of minute particles with their own identities that started out as something different from the dust they have become. Because of their small size and ubiquitous nature, it is almost impossible for animals (including people), objects or locations to avoid collecting these particles as they settle from the air or become deposited by contact.

Even though dust is ubiquitous, its composition is not. In other words, all dust is not the same. 'Inside' dust differs from 'outside' dust. Dust from a kitchen is different from dust in the bathroom and dust from a city differs from that on a farm. Furthermore, dusts from different farms are different due to the crops that are grown (and the minerals in the soil on which they are planted), the fertilizers different farmers apply to their fields, the plants that grow near the house or along the road and the animals, domestic and wild, that share the property. The proof of these statements is easy to demonstrate by examining such dusts under a microscope or analyzing them chemically or elementally with the sensitive analytical instruments of the modern chemical laboratory.

Due to these very characteristics, such microscopic traces can often provide compelling evidence of contact between two people, a person and a particular location, an object and a certain environment, etc., when properly collected, isolated, analyzed and identified. It was recognized by the last half of the 19th century that traces such as these could be exploited to establish facts in legal matters, particularly with regard to criminal investigations or to prove facts to support an assertion at trial in a court of law.

Hans Gross (1847–1915), an Austrian jurist (and arguably the founder of modern scientific criminal investigation) in his now classic textbook, *Handbuch für Untersuchungsrichter* (Handbook for Examining Magistrates), recommended that microscopists, chemists, botanists and mineralogists should be consulted to examine physical evidence that might arise in connection with criminal investigations. He went on to describe a case in which a coat of suspicious origin was placed in a strong paper bag and beaten with sticks. The dust recovered from the bag was recovered and subjected to a microscopical and microchemical analysis. Based on the results of this analysis, it was inferred that the coat belonged to a cabinet maker/joiner – a fact of significance in the case, which was subsequently confirmed by independent means. This was one of the earliest examples (but by no means the only one) of the use of dust in assisting the police in an investigation.

Early in the 20th century, a young doctor who had studied forensic medicine under Alexandre Lacassagne, then the dominant medical jurist in France, became strongly influenced by the ideas expressed in Gross's writings, but also the approach taken by Sherlock Holmes, the fictional creation of Sir Arthur Conan Doyle. Dr. Edmond Locard (1877–1966), the founder and director of the police laboratory in Lyon, applied the methods of the modern scientist to criminal investigation as advocated by both Gross and Holmes. There was practically no type of physical evidence that did not interest him. He studied in depth and conducted original research on any objects or materials that might potentially help him to solve a crime, including firearms and bullets, questioned documents, fingerprints (including the pattern of microscopic pores in fingerprint lines), cryptography (he was a crypt-analyst during WWI) and trace evidence, to name only a few.

Through his exhaustive research into and analysis of dust traces, Locard laid down the fundamental principle of trace evidence analysis that today bears his name. Locard's Exchange Principle states: 'Whenever two objects come into contact there is always a transfer of material between them.'

The importance and implications of this apparently simple statement are far-reaching and constitute the justification for studying trace evidence in the modern forensic laboratory. As it is an applied science that borrows heavily from, and is almost entirely based on, the principles of chemistry, physics, geology and biology, the Exchange Principle is one of the few precepts that forensic science can call its own.

Experiments and research (principally with respect to fiber transfers) conducted since Locard's time, and particularly in the last decades of the 20th century, have proven and validated Locard's thesis. Recognition of the truth of this principle has stimulated research into:

- methods for detecting, collecting and isolating these minute traces;
- developing rigorous scientific techniques for identifying and comparing minute amounts of unknown substances;
- developing and implementing methods for preventing cross-contamination, both at the crime scene and in the laboratory;
- interpreting the results of such analyses in ways that will aid in assisting investigators in active investigations or assist the court by providing facts or well supported interpretations at trial.

3.6.3 Trace evidence in modern forensic science

Locard's influence on the development and practice of trace evidence analysis continues to this day, particularly with regard to the Exchange Principle, which is cited whenever a forensic scientist gives testimony regarding the results of an association made by means of a trace evidence analysis. As utilized today, trace evidence may consist of almost anything (natural, man-made or man-altered) that can be imagined. The necessary characteristics of a substance to make it useful as trace evidence are that it be identifiable (even though present in small amount or size); suitable for comparison of its individual characteristics to a potential source; and that it is either uncommon or, if not, at least characteristic of a particular location (e.g. pollen grains), profession (e.g. welding spheres), object (e.g. carpet fiber), etc. The frequent use of *et cetera* when listing examples here is due to the fact that it is impossible to predict exactly what it is that might be important in any particular case until the evidence is subjected to a preliminary examination.

Table 3.4 Some types of trace evidence. Examples of substances that should be considered as trace material when examining evidence. This list is not complete, as any substance could potentially have importance to solving a case

1	Fibers: natural, regenerated and synthetic
2	Hairs: human and animal
3	Glass: including fiberglass, mineral wool and rock wool insulation
4	Paint: automotive, architectural, tool, appliance, etc.
5	Sand and soil
6	Wood: particles, splinters, sawdust
7	Pollen and spores
8	Plant parts: leaf fragments, seeds, roots, vegetables, fruits
9	Food particles: flour, condiments, food additives (non-dairy coffee creamer), sugar and non-nutritive sweetener
10	Powdered vegetable drugs, tobacco and spices
11	Starches
12	Feathers
13	Plastics and polymers
14	Metals
15	Fecal matter, vomit, stomach contents
16	Animal tissue: skin and epithelial cells, brain matter, liver, etc
17	Fats and oils: mineral, fuel, vegetable, etc.
18	Waxes, creams and ointments
19	Body fluids*
20	Bones and teeth
21	Plaster and gypsum drywall
22	Cement and concrete
23	Bricks, mortar and other ceramics
24	Fish scales
25	Diatoms, foraminifera, and other microfossils
26	Explosives
27	Cosmetics, lipstick
28	Glitter
29	Opal phytoliths
30	Combustion products from wood, oil and coal
31	Rubber

*Identification of species and comparison of these substances fall within the domain of forensic serology, but their identification as to type (e.g. blood, saliva, urine, etc.), which must proceed comparison, is normally performed by microscopical, microchemical or biochemical tests.

Table 3.4 lists some of the substances, of which the author is aware, that have been exploited at one time or another as trace evidence.

Some types of trace evidence examinations are so commonly performed that they are conducted, at least in larger laboratories, on a more or less routine basis. These include comparisons of hairs, fibers, glass and paint. Others are undertaken less frequently and depend upon the ability and expertise of the laboratory staff to perform them and interpret the results. These include analyses of explosives, wood, soil and pollen. Still others are yet more rarely performed, since they are only infrequently encountered as evidence, or become important only under special circumstances. Examples of these might include feathers, bricks and fecal or vomit stains. Proper analysis of these items requires specialized techniques and knowledge, which the average trace evidence examiner may not possess. If the evidence is important, outside subject experts are consulted in order to acquire the needed knowledge or to perform the analysis and interpret the results.

The important knowledge is of what to observe. Before any type of trace evidence is examined, it must be unequivocally identified by methods that are appropriate to the task. For unknown small particles and minute stains or residues, this is often the most difficult and taxing part of the examination. It is, however, essential to know what something *is* before one can begin to compare it with its suspected source. Therefore, before comparing a questioned hair to a known source, it is necessary to establish whether it originated from a human being or from an animal. If it is human, the somatic (i.e. body region) origin must be determined (e.g. head, pubic, eyebrow, limb, etc.), since it is necessary to compare like with like. Thus head hairs may only be compared to other head hairs under the microscope.

Similarly, before two fibers can be compared, they must not only be the same color but from the same generic class (e.g. cotton, polyester, nylon, etc.), since different types of fibers each have their own set of chemical and morphological properties and these cannot necessarily be determined for every fiber type. As an example, the melting point is an essential part of a nylon fiber comparison, because even fibers composed of the same type of nylon may be melted and extruded at different temperatures, depending on molecular weight, additives and impurities; however, acrylic fibers do not melt but only char when heated. It is a fault of principle to state that two objects cannot be excluded as having originated from the same source when one cannot state what the items are.

At times, trace evidence may carry still smaller trace evidence upon it, which can help to individualize the original evidence by making it more unusual or even unique. An example from a case in California illustrates this:

A foreign pubic hair, found in the pubic combings of a rape victim, was compared to the known pubic hair of a state patrol officer, whom the rape victim accused of attacking her. The questioned hair was found to be microscopically indistinguishable from that of the officer. Under the microscope, one could also observe an irregular deposit of an unidentified, colorless substance sticking to the surface of both the known and questioned pubic hairs. Minute specimens of the deposits were isolated from the hairs while observing through a stereomicroscope. Infrared microspectroscopy identified the same substance on both the known and questioned hairs. From these spectra, the deposits were identified as sodium carboxymethyl cellulose, the major ingredient in a prescription cream used by the patrolman to treat pubic lice. This evidence proved an intimate connection between the victim and the accused, and it supported the victim's version of event.

3.6.4 How does trace evidence help to solve crimes?

As noted earlier, microscopic particles and other residues are being deposited and transferred all the time. They only become trace evidence, however, when they occur in connection with a crime – then they can be exploited to answer a specific question or questions that arise in the course of an investigation or at trial. Trace evidence rarely solves crimes by itself, but helps by establishing facts. Since facts are, by their very nature, always true, proving something is indeed a fact helps us to get at the truth, which is the ultimate goal of an investigation or trial. Although trace evidence has been used for many different purposes throughout its long history, it is primarily employed today in two distinct roles.

1. **Comparative examinations**. This is the way in which trace evidence is most commonly utilized in the forensic laboratory today. An unknown sample (commonly referred to as the *questioned 'Q' sample*) is first identified (e.g. as a nylon fiber) and then compared to material (e.g. the carpet at a crime scene) from a suspected source. Samples from this are referred to as *known 'K' samples*. Since almost anything can become trace evidence in a real case, it is essential that, after a substance has been identified, its characteristics must be researched so that appropriate tests and observations can be performed that will truly demonstrate if the two substances are likely to have come from the same source, have only a superficial similarity or are distinctly different. For commonly occurring evidence, such as fibers, paint and glass, many methods of analysis are available. When the substance in question is more unusual, or rarely or never before encountered, literature research and consultation with specialists in the subject area or the manufacturer (for a man-made product) must be pursued. Comparative examinations of trace evidence are subject to intensive scrutiny, because the results will be used to help the court decide on guilt or innocence in criminal cases and liability in civil ones. Results must be documented, and the evidence, reports and notes are subject to review by the defense, which has the responsibility of determining if mistakes have been made or whether evidence or interpretations were overlooked or misunderstood. Because mistakes can result in a wrong verdict, both prosecution and defense have the responsibility to do their best to see that they do not occur.
2. **Developing investigative leads**. This activity is less often performed today than in the past, because of current emphasis on standard methods and legal procedures. Here, the laboratory scientist and detective work together to attempt to understand the source or purpose of evidence recovered from the crime scene, victim, etc. at an early stage in an investigation or even later, if the investigation has bogged down. Because the penalty for mistakes here is smaller than in court, scientists can utilize their imagination in the way that a research chemist does when trying to make a new discovery, and they can make full use of the scientific method. Questions such as 'Where could this soil have come from?', 'Who made

this paint?' or 'What could this fiber be used for?' have all been asked of the author by investigators over the years. Sometimes the question is simply 'What is this?', and many times the answer has opened a whole new line of inquiry. Using the scientific talent of the forensic laboratory to aid an investigation by answering questions and developing investigative leads is one of the most useful functions the trace evidence section can perform. Unfortunately, there are no standard methods for such undertakings, and the scientists who perform them must be highly skilled and experienced. The results can, however, be dramatic, as demonstrated in the examples that follow below.

Example 1: fiber comparison

The following hypothetical example describes the types of examination that might be performed in a typical fiber comparison case. Let us assume, for example, that fibers found on a suspect's shoe cannot be distinguished from those comprising the carpet in the dead victim's home. To a lay person, this conclusion may sound trivial and perhaps unimportant. If the analysis is conducted properly and the results carefully interpreted, however, it can actually mean that the shoes were either on the rug in the victim's home or that they had somehow picked up identical fibers from another source. If it can be shown that the shoes of the suspect never came into contact with another source of identical fibers, then the fibers in question must be from the victim's rug. The last point may be difficult or impossible to prove in practice, so another approach must be taken to narrow the possibilities of from where the fiber in question could have originated.

One of these is to determine just how common or uncommon are the fibers comprising the victim's rug. If they are common, it may be necessary to state that, while the fibers from the victim's rug and suspect's shoes are exactly the same in all of the properties compared, there are too many rugs with this same type of fiber in this case (perhaps, for example, the murder occurred in a new apartment building, in which all the units have same new, identical carpeting) to do more than state this fact and note that there are many possible sources of these fibers.

On the other hand, the fibers may have unusual or rare features. It is only possible to know if a feature is rare if one has examined many fibers of the same type over a long period of time and has become experienced in knowing just what is normal with respect to each type of fiber and what is not.

Any carpet fiber has many characteristics by which it can be distinguished from other fibers, although many of these would never even be noticed by an inexperienced observer. A fiber comparison is conducted by performing a series of tests and observations of independent characteristics on both the *questioned* and *known* fibers. If, at the conclusion of the testing, the known and questioned fibers cannot be distinguished from each other, then the laboratory would report that the fibers from the two sources are indistinguishable and, therefore, could have originated from the same source or from another source of identical fibers. If, however, any differences were detected between the 'K' and 'Q' fibers, the examination would be at an end, because the two fibers could be distinguished from each other and thus could not originate from the same source.

The types of tests and observations are not necessarily obvious to someone who does not look at fibers all the time. A trace evidence microscopist, on the other hand, spends a great deal of time learning about fibers and how to tell one from another, and in mastering the techniques and instruments used to analyze and compare them.

The depth of the analysis described below should make it clear that a positive association is not arrived at lightly, and certainly not with the ease depicted on television. It is, for the most part, painstaking and unglamorous work, but the rewards of arriving at the truth are extremely satisfying from the viewpoint of scientific inquiry.

In our example, we will assume that both the Q and K fibers appear to be the same color, and go on from there. Starting with fiber type, most carpets are made of nylon, polyolefin, polyester or wool fibers. If our hypothetical carpet is made from nylon fibers, they may be either nylon type 6 or type 6,6. Both types are nylon, but with slightly different chemical variations. In the laboratory, we can determine which type our fibers are made from by using infrared microspectroscopy. Table 3.5 lists and describes the instruments and methods that are available to the modern trace evidence analyst.

Table 3.5 The principal techniques and tools for trace evidence analysis. Instruments and methods available to the trace evidence analyst

1. Microscopy	
Stereomicroscope	Used to examine evidence *in situ*, isolate and prepare recovered particles and deposits, dissect fabrics and threads, and perform any type of manipulation that requires low magnification.
Polarizing microscope	This microscope permits the trace evidence analyst to study both the morphological and optical properties of particles, crystals, fibers and hairs to identify them and to observe features and quantitative data for comparison purposes.
Comparison microscope	This instrument consists of two microscopes placed side by side and connected by an optical bridge, which allows the images from the two microscopes to be viewed side by side. Hairs, multi-layered paint chips, fibers, etc. can all be compared directly and the image photographed.
Hot stage microscope	In this microscope, the stage is replaced by a small oven. Specimens can be watched as they melt and the temperature at which thermal events occur can be recorded. It is useful for identifying unknown crystals and, especially, for comparing thermoplastic fibers.

(*continued*)

Table 3.5 (*Continued*)

Fluorescence microscope	The presence of optical brighteners, fluorescent dyes and pigments, oils, grease and other fluorescent substances, which cannot be seen in ordinary light, can be observed and compared with this instrument.
Scanning electron microscope	Polarizing microscopes have a maximum useful magnification of about 1,500×. The SEM can magnify up to nearly 100,000×. When equipped with an x-ray analyzer (EDS), it can also be used to perform elemental analyses of particles smaller than can be done with the light microscope.
Microchemical analysis	Classical chemistry performed on microscope slides, in capillary tubes and on single fibers permits the microanalyst to identify minute amounts of chemical substances comprising stains, dust particles and other unknown substances.
2. Spectroscopy	
UV-visible microspectrophotometer	This instrument is used to obtain the spectrum of colored microscopic objects. It allows the microscopist objectively to compare the colors of fibers, particles of paint and lacquer, etc. It can also collect the fluorescence spectrum of microscopic specimens excited with the fluorescence microscope.
Infrared microspectrophotometer	The Fourier transform infrared microspectrophotometer (micro-FTIR) permits the chemical analysis of microscopic particles and fibers by identifying organic compounds, anions and complex polymers such as those comprising synthetic fibers, paints and plastics.
X-ray fluorescence microspectroscopy	This technique helps the trace evidence analyst to determine the trace element composition of microscopic evidence of diverse types. Although it is most commonly employed for the comparison of float glass, it can be used to compare the trace element composition of almost kind of evidence that might be encountered.
Raman microscpectroscopy	This is one of the newest scientific tools in the trace evidence laboratory. It can identify many chemical substances when they are present only as particles down to one micrometer in size. It is especially useful for identifying pigments, dyes and other organic and inorganic compounds.

Table 3.5 *(Continued)*

Mass spectrometry	The mass spectrometer is used to determine the mass spectrum of chemical species. It can be combined with a gas or liquid chromatograph, which separates mixtures of compounds and impurities before they pass into the spectrometer. They are part of an ICP (inductively coupled plasma) MS, which is used to determine the trace element composition down to the parts per billion (and sometimes less) range. MS instruments are also available which permit the direct insertion of unknown specimens for chemical identification.
3. Chromatography	
Gas chromatography -mass spectroscopy	The gas chromatograph is used to separate complex chemical mixtures, which are then identified by their retention times and mass spectra. For substances to be analyzed, it is necessary that they can be volatilized by raising the temperature of chemical derivitization. Polymers, drugs and post-explosion blast residues are all excellent candidates for analysis using this method.
Liquid chromatography -mass spectroscopy	Complex mixtures that do not volatilize or are thermally sensitive can be separated and identified using this instrument. Explosives, dyes and many other substances lend themselves to analysis by this method.
Thin layer chromatography	This form of chromatography is simple to perform. It is an excellent method of dye separation and analysis for comparing dyes extracted from fibers, because the jury can see the individual dyes that colored the Q and K fibers, rather than just peaks on a chart. The pure dyes can be recovered from the chromatogram and identified by micro-FTIR and Raman microspectroscopy.
4. X-ray diffraction	
X-ray diffraction	This technique is used to identify chemical compounds, not just elements, ions or functional groups, by their crystalline structure. It can be used to identify by name almost any crystalline compound and is thus useful for analyzing dyes, explosives, clays . . . in short, anything made of crystals. It can also be used to sub-classify substances by identifying the polymorphic form in which a particle is present. For example, it can show if the titanium delustrant in a synthetic fiber is the anatase or rutile polymorph.

It goes without saying that, to have originated from the same source, both the known and questioned fibers must be the same in all respects. Let us say that spectroscopy shows that the fibers in this case are both made from nylon 6. This is a good start, because about two-thirds of all nylon used in carpets is made from nylon 6,6, so we have fibers here that are a little rarer than normal, although there is still a lot of this nylon out there. We can go a step further with this, however, because nylon is a thermoplastic and its fibers are formed by melting and extrusion through a spinneret.

The temperature at which the nylon melts determines its viscosity, which in turn determines if, or how, it will go through the spinneret and become a fiber. The melting point, in turn, is determined by a number of factors, including molecular weight, additives, impurities, etc. that change the melting characteristics. A minute piece of even the smallest nylon fiber can be cut off and placed in a microscope hot stage, which permits the analyst to determine the exact point at which the nylon melts. This melting point not only confirms the nylon type but the temperature at which the fiber was extruded. Two nylon fibers from the same source must have the same melting point or melting range, because they were made at that temperature.

The cross-sectional shape of a carpet fiber is an important characteristic, since it is possible for a manufacturer to give a thermoplastic fiber such as nylon almost any shape one might wish. There are many reasons for imparting a particular cross-section. One of the earliest modified cross-sections had a triangular shape (referred to as trilobal), and it was developed by DuPont Fibers in the early 1960s. Nylon carpets made with these fibers got just as dirty as normal ones but, because of the way the cross-section interacted with light, it did not look dirty. Other manufacturers made their own modified cross-sections so that they could offer this feature and still circumvent DuPont's patents. For some fibers, it is still possible to determine the manufacturer simply by examining the cross-section of a single fiber. Because our known fiber is definitely from a carpet, we know beforehand that it will almost certainly exhibit a modified cross-section. If the fibers exhibit unexplained differences, our examination is at an end but, if the cross-sections of the Q and K fibers are the same, they provide yet another independent point of comparison.

There are so many features that can be determined on a single fiber that our list of properties to compare can be extended much further. Remember that the more properties the K and Q fibers have in common, the more likely it is that they came from the same source. This is particularly true when the properties being measured are easily varied. For example, the refractive indices and birefringence of a man-made fiber are subject to manufacturing parameters and can be used to distinguish two fibers that are may be identical in all other respects.

The fluorescence microscope can be used to look for fluorescence from dyes and optical brighteners. The microspectrophotometer is used to objectively determine if the color of the K and Q fibers is the same, or only appears to be the same by eye. The dyes from a single fiber can be extracted with an appropriate solvent in a capillary tube only slighter wider than a hair, and that extract can be separated into the individual dyestuffs used to color the fibers by chromatography. It is even possible to identify the separated dyes using infrared microspectrophotometry and Raman microspectroscopy. The spectra from the dyes are compared to those in the laboratory's reference database.

If, at the end of all these tests and observations, the Q and K fibers cannot be distinguished from one another, there is good reason for believing that they have originated from the same source (in our example the victim's carpet), or at least from another source of identical fibers. This is based on the factual information developed by analysis. The interpretation of this fact is made only after considering other possible sources of these fibers, as discussed earlier in this section.

Example 2: assisting in an investigation

This is a real-life example of the use of trace evidence to aid in an investigation. New high-end business computers were shipped from a manufacturer in Texas to a customer in Buenos Aires. Upon opening, it was discovered that the crates were filled with cement blocks and not the computers that had been ordered. After several months of investigation, pieces of the concrete were sent to a forensic laboratory in an attempt to determine where the switch had occurred.

Since the blocks had been stored outdoors immediately after being removed from the packing crates, there was little hope of trying to recover and examine any pollen for clues as to where the blocks were stored before being sealed in the crates. Instead, pieces of the cement blocks were gently dissolved in acid to remove the cement, and the washed and dried sandy aggregate was subjected to a petrographic examination. The majority of the sand was composed of grains of quartz and the particle size distribution was very uniform, which pointed to a beach sand source. Examination of the quartz in the scanning electron microscope (SEM) showed that marine diatom frustules were growing in crevices on some of the grain surfaces, which confirmed the ocean beach origin of the sand.

Next, the cleaned sand was poured into a glass tube and separated into 'heavy' and 'light' mineral fractions using bromoform, a liquid with a very high density of 2.9 g/cm^3. The majority of the sand (which consisted almost entirely of quartz) floated on the surface of the dense liquid, while the trace heavy minerals sank to the bottom of the tube, where they were recovered and identified using the polarizing microscope.

This was found to be composed of a mixture of metamorphic minerals, along with zircon and rutile grains that characterize the beach sands of the southeastern coast of the United States, and particularly the sands of southeastern Florida near Miami. The mineralogy was inconsistent with that of Argentina, in the region of Buenos Aires. Based on these results, the investigation was concentrated at Miami International Airport, where the baggage theft ring responsible for the thefts was discovered and brought to justice.

3.6.5 Case histories

Perhaps there is no better way to illustrate the importance of trace evidence analysis to modern criminal investigation than by demonstrating its role in real life cases. The following cases show where microscopic traces played a pivotal role in establishing facts and arriving at the truth.

Case 1. the green river murders

The Green River murders became, over the course of almost 20 years, the biggest serial murder case in US history. In July of 1983, the bodies of several women were found in and on the banks of Seattle's Green River. Over the ensuing months, and then years, more bodies of young women were discovered in the Seattle area. They had all been murdered in the same way, and a task force was formed to coordinate leads and organize the investigation. By the year 2000, the murders seemed to have stopped and the task force was reduced to a single detective.

In 2002, scientific advances led to the isolation and identification of DNA from four of the victims. The DNA from a single donor, who had been questioned and released in the late 1980s because he passed lie detector tests and did not match the psychological profile, was identified in each case. The DNA belonged to Gary Ridgway, a truck detailer at the Kenworth truck factory, who had lived and worked in the Seattle area for most of his life. Ridgway was arrested in November of 2002 and charged with four of the Green River murders. He admitted that he had hired each of the women, who worked as prostitutes, but maintained that he had not killed them.

Facing the deadline imposed by the demand for a speedy trial and having somewhere between 60 and 100 victims to deal with, the task force turned to trace evidence to determine whether there were any additional links that connected Ridgway to any of the other Green River victims. Microscopical examination of the dust vacuumed from the clothing that remained from some of the victims revealed the presence of microscopic spheres of glossy paint, in many colors and in large numbers. Spray paint particles are not normally found in the dust carried on a person's clothes, unless they have been spray painting or have been somewhere where spray painting had been done on, or have been around someone who has spray painted. One only carried the color they were spray painting, but spheres of multiple colors were isolated from the dust sweepings obtained from the victims' clothing.

Furthermore, the longer it has been since exposure to the spray paint, the fewer particles will remain on the clothing. In some cases, dozens and dozens of these paint spheres were recovered. Since there was no good reason for them to have multiple colors of microscopic spray paint spheres on their clothing given their occupation, l*et al*one in great numbers, it was almost certain that they had become exposed to these particles shortly before they died.

The isolated spheres were isolated from the dust with fine pointed tungsten needles while observing under a microscope, and they were analyzed microscopically by means of infrared microspectroscopy, to identify their chemistry, and energy dispersive x-ray spectroscopy, to obtain their elemental composition. Interpretation of the data from these experiments showed that the spray paint was not just any paint but a special high-end product recently introduced by DuPont under the name Imron. This paint was intended primarily for the repaint industry, but was used as the original finish on the trucks produced at the Kenworth factory where Ridgway worked his entire career. Microscopic spheres of Imron spray paint from remains of clothing eventually linked Ridgway and six of his victims.

Faced with this new information and the DNA results, Ridgway confessed to his lawyers that he was the Green River killer. They entered into a plea agreement with the state not to seek the death penalty if he cooperated in the investigation, and in 2003 Ridgway confessed to the murders of 48 victims. He will spend the remainder of his life in prison and believes he murdered as many as 70 women, but he would only confess to those he specifically recalled.

Case 2. serial rapist in Maryland

Police in Montgomery County, Maryland were baffled by a serial rapist who left few clues at the crime scenes other than his DNA, which was not in the CODIS database. The attacker covered his victims' heads with athletics shirts so that none of them could even provide a description. He had, however, left the shirts behind in two of the cases, and these were sent to a forensic microscopist for examination. The dust was vacuumed from the otherwise clean-looking shirts and analyzed microscopically and microchemically. The microscopist provided the following report:

> *'Both shirts were worn by the same person. The subject was employed working indoors as a drywall installer and finisher. He worked on large scale commercial projects and was, therefore, probably a professional. He wore one of the shirts in the early spring while he was working in the vicinity of oak trees.'*

Based on this report, the police turned to the media with the description provided by the microscopist, along with photographs of the two shirts. They asked the public to contact them with any information they might have that might help lead them to the rapist. About two months after submitting the report, the laboratory received a call from the detective. Early in the morning on the previous day, an elderly woman noticed a man leaving the apartment of a young woman next door. She did not recognize him and called the police with his description. He was pulled over by a patrol car a few blocks from the apartment. The police were interested in the fact that the suspect was driving a commercial drywall installation truck and began to question him about the rapes.

'Here's the part you won't believe,' said Detective Collins, who had submitted the original evidence for analysis. 'He confessed!' Collins went on to explain how the suspect explained that he knew: 'It was only a matter of time before you found me,' because he had seen the televised images of his shirts and listened to the description based on the dust recovered from them. He withdrew the confession shortly after making it to the police but it was too late, because his DNA was detected on every one of the victims. The suspect was tried and convicted.

3.6.6 Trace evidence analysis and the future

In spite of the fact that trace evidence has played a major role in forensic science since its earliest days, today it faces an uncertain future. The budget and space demands of DNA analysis have caused some forensic science laboratories to close their trace evidence sections. Some administrators feel they cannot justify the expense of the time and labor-intensive analysis a trace evidence examination may require, when compared to DNA analysis that can specifically identify an individual. The trace evidence analyst would reply that probative DNA does not occur in every case, and that trace evidence analysis has often provided the only lead in many difficult investigations and has led directly to the solution in many other cases.

However, increasing emphasis on standardizing methods of analysis between all laboratories has made it difficult for trace evidence analysts to assist in such investigations, since the types of evidence encountered are often unusual and no standard methods exist for their analysis and interpretation. More often than not, these specks of evidence must first be identified – another skill that does not lend itself to standardized methods. The danger to forensic science is that the fundamental skills that form the basis for all scientific investigations, whether academic, industrial or analytical, are in danger of being lost and being supplanted by the 'cookbook' procedures used by technicians.

The future of trace evidence is, to a certain extent, dependent on the recollection of the past and a return to serving the original role of the forensic scientist in helping to advance an investigation. This requires that the trace evidence analyst to be both a good technician and a good scientist. It requires that the trace evidence examiner be an analyst, not just a comparator. Most importantly, it means that the examination and analysis must be driven by the evidence itself, with each new step in the analysis being determined by the results of the test or observation that preceded it, and not by a flowchart or book of instructions. The analyst who helps to advance an investigation must have adopted the point of view described by Kirk in his classic book *Crime Investigation*, which is characterized by imagination, ingenuity and curiosity, balanced by skepticism, common sense and conservatism in interpretation.

Acknowledgments

Appreciation is expressed to the authors for their yeomen's work and contribution to this publication. The short timeline for the writings was mandated during a period when other obligations and the interruptions of life weighed heavily on their shoulders. The authors met this obstacle with dedication and diligence – a true testimony to the exactitude of the forensic science professional. Thank you to Dr. Max Houck for his succinct review of management in *Managing forensic science for quality performance*, to Dr. Jay A. Siegel, for his in-depth analysis of the history and current status of seized drug analysis in this country on *Illicit drugs*, to Dr. Cecelia A. Crouse for her illuminating discussion of the development and acceptance of DNA testing in *Forensic DNA analysis: a primer*, to Mr. John Lentini on his rendering of the profession and the distinction between two associated disciplines in *Fire scene investigation and laboratory analysis of fire debris* and to Mr. Skip Palenik for his eloquent description of the application of trace material to criminal investigations in the section titled *Trace evidence*.

References

1. Kirk PL. *Crime Investigation.* 2nd ed. New York: John Wiley & Sons, Inc., 1974.
2. *Webster's New World Dictionary of the American, Language*. 2nd ed. New York: World Publishing Company;1971.
3. Ksir C, Hart CL, Ray R. *Drugs, Society and Human Behavior.* 11th ed. New York: McGraw-Hill; 2006.
4. *Congressional Record* 40, No. 102 (Part 1), (December 4, 1905 to January 12, 1906).
5. *The Harrison Narcotics Tax Act* (Ch. 1, 38 Stat. 785).
6. U.S. Code, Title 21, Chapter 13, Subchapter 1, Part B.
7. Omnibus Drug Act 1988. http://thomas.loc.gov/cgi-bin/bdquery/z?d100:H.R.5210 (accessed January 8, 2012).
8. Scientific Working Group for the Analysis of Seized Drugs. Revision 5.1. 2011; 14–15. www.swgdrug.org (accessed November 15, 2011).
9. Wambaugh J. *The Blooding*. New York: Bantam Books;1989.
10. Public Law 103–322 (HR 3355). http://federalevidence.com/pdf/FRE_Amendments/1994Amendment/Public%20Law%20103-322%20(1994).pdf (accessed January 5, 2012).
11. Quality Assurance Standards for Forensic DNA Testing Laboratories effective 9-1-2011. http://www.fbi.gov/about-us/lab/codis/qas-standards-for-forensic-dna-testing-laboratories-effective-9-1-2011 (accessed January 5, 2012).
12. The Nobel Peace Prize. http://www.nobelprize.org/nobel_prizes/peace/ (accessed January 5, 2012).
13. Kleppe K, Ohtsuka E, Kleppe R, Molineux I, Khorana HG. *Studies on Polynucleotides. XCVI. Repair replications of short synthetic DNAs as catalyzed by DNA polymerases. Journal of Molecular Biology* 1971; 56:341–361.
14. Saiki RK, Arnheim N, Erlich H. A Novel Method for the Detection of Polymorphic Restriction Sites by Cleavage of Oligonucleotide Probes: Application to Sickle Cell Anemia. *BioTechnology* 1985; 3:1008–1012.

15. Walsh PS, Fildes N, Louie AS, Higuchi R. Report of the blind trial of the Cetus AmpliType HLA DQ alpha forensic deoxyribonucleic acid (DNA) amplification and typing kit. *Journal of Forensic Sciences* 1991; 36:1551–1556.
16. Budowle B, Moretti TR, Niezgoda SJ, Brown BL. *CODIS and PCR-Based Short Tandem Repeat Loci: Law Enforcement Tools*. Second European Symposium on Human Identification. Madison, WI: Promega Corporation, 1998:73–88.
17. Hares D. Expanding the CODIS core loci in the United States. *Forensic Science International: Genetics* 2011; 6(1):e52–e54.
18. Butler JM. *Fundamentals of Forensic DNA Typing*. Burlington, MA: Academic Press Elsevier; 2010; 99–110.
19. Walsh PS, Metzger DA, Higuchi R. Chelex 100 as a Medium for Simple Extraction of DNA for PCR-based Typing from Forensic Material. *BioTechniques* 1991; 10 (4):506–513.
20. Sambrook J, Fritz EF, *Maniatis T. Molecular Cloning: A Laboratory Manual*. 3rd ed. Cole Spring Harbor, NY: Cold Spring Harbor Laboratory Press, 1989; E.3–E.4.
21. Opel KL, Chung D, McCord BR. A Study of PCR Inhibition Mechanisms Using Real Time PCR. *Journal of Forensic Sciences* 2010; 55(1):1–9.
22. Fregeauu'x CJ, arc Lettu M, Elliottu KL. Adoption of Automated DNA Processing for High Volume DNA Casework: A Combined Approach Using Magnetic Beads and Real-time PCR. *International Congress Series* 2006; 1288:688–690.
23. Crouse CA. Making the Case: Automatic Forensic DNA Analysis. *Laboratory Medicine* 2008 Dec; 39(12):709–717.
24. Lee SB, Crouse CA, Kline MC. Optimizing Storage and Handling of DNA Extracts. *Forensic Science Review* 2010 July; 22(2):131–144.
25. Frippiat C, Zorbo S, Leonard D, Marcotte A, Chaput M, Aelbrecht C, *et al.* Evaluation of Novel Forensic DNA Storage Methodologies. *Forensic Science International: Genetics* 2010 Nov; 5(5):386–392.
26. Horsman KM, Hickey JA, Cotton RW, Landers JP, Maddox LO. Development of Human-Specific Real-Time PCR Assay for Simultaneous Quantitation of Total Genomic and Male DNA. *Journal of Forensic Sciences* 2006; 51:758–765.
27. Nicklas JA, Buel E. Simultaneous Determination of Total Human and Male DNA Using Duplex Realtime PCR Assay. *Journal of Forensic Sciences* 2006; 51: 1005–1015.
28. Butler JM. Forensic DNA Typing. Burlington, MA: Academic Press Elsevier; 2005; 325–372.
29. Butler JM, Shen Y, McCord BR. The Development of Reduced Size STR amplicons as tools for analysis of degraded DNA. *Journal of Forensic Sciences* 2003; 48(5):1054–1064.
30. Fatolois L, Ballentyne J. The US Y-STR Database. *Profiles in DNA* 2008; 11(1): 13–14.
31. Tiselius AW. The Moving-Boundary Method of Studying the Electrophoresis of Proteins [dissertation]. *Nova Acta Regiae Societatis Scientiarum Upsaliensis* 1930 Ser. IV; 7.(4)
32. Cohen AS, Najarian DR, Paulus A, Guttman A, Smith JA, Karger BL. Rapid separation and purification of oligonucleotides by high-performance capillary gel electrophoresis. Proceedings of The National Academy Of Sciences Of The United States Of America 1988 Dec; 85(24):9660–9663.

33. Watts D. Genotyping STR loci using an automated DNA sequencer. In: Lincoln PJ, Thomson J, editors. *Methods in Molecular Biology*, Vol 98: Forensic DNA Profiling Protocols. Totowa, NJ: Humana Press Inc, 1998;193–208.
34. The Federal Bureau of Investigation, Laboratory Services. *SWGDAM Interpretation Guidelines for Autosomal STR Typing by Forensic DNA Testing Laboratories.* http://www.fbi.gov/about-us/lab/codis/swgdam.pdf (assessed January 5, 2012).
35. The Federal Bureau of Investigation, Laboratory Services. *Combined DNA Index System (CODIS).* http://www.fbi.gov/about-us/lab/codis (accessed January 5, 2012).
36. Butler JM. *Advanced Topics in Forensic DNA Typing: Methodology.* Burlington, MA: Academic Press Elsevier; 2012;497–514.
37. Boudreau JF. *Arson and Arson Investigation: Survey and Assessment.* National Institute of Law Enforcement and Criminal Justice, LEAA, USDOJ; 1977.
38. Brannigan F, Bright R, Jason N. *Fire Investigation Handbook.* Washington, DC: US Government Printing Office; 1980.
39. Technical Committee on Fire Investigations. *NFPA 921, Guide for Fire and Explosion Investigations.* Quincy, MA: National Fire Protection Agency (NFPA); 1992.
40. *Millers Mutual Insurance Company v. Janelle R. Benfield*, 140 F. 3d 915 (11th Cir. 1998). http://caselaw.findlaw.com/us-11th-circuit/1396573.html.
41. *Kumho Tire v. Carmichael*, 526 U.S. 137 (1999). http://www.law.cornell.edu/supct/html/97-1709.ZO.html.
42. USDOJ. *Fire and Arson Scene Evidence: A Guide for Public Safety Personnel*; 2000. http://www.ncjrs.gov/pdffiles1/nij/181584.pdf (accessed November 15, 2011).
43. Beyler CL. *Analysis of the Fire Investigation Methods and Procedures used in the Criminal Arson Cases Against Ernest Ray Willis and Cameron Todd Willingham, Report to the Texas Forensic Science Commission, August 17, 2009.* http://www.acslaw.org/taxonomy/term/751 (accessed November 15, 2011).
44. Mann D. Fire and Innocence. *Texas Observer.* 2009; Dec 2. http://www.texasobserver.org/cover-story/fire-and-innocence (accessed January 12, 2012).
45. Carman S. *Improving the Understanding of Post-Flashover Fire Behavior, Proceedings of the 3rd International Symposium on Fire Investigations Science and Technology (ISFI).* Cincinnati, OH, May 19–21, 2008. http://www.carmanfireinvestigations.com (accessed November 15, 2011).
46. Official Documentation. http://fire.nist.gov/fds/documentation.html (accessed November 15, 2011).
47. McGrattan K, Hostikka R, McDermott S, Floyd J. *Fire Dynamics Simulator (Version 5) User's Guide.* Gaithersburg (MD): National Institute of Standards and Technology; 2010 Oct. NIST Special Publication 1019-5.
48. Forney GP. *Smokeview (Version 5) – A Tool for Visualizing Fire Dynamics Simulation Data Volume I: User's Guide.* Gaithersburg (MD): National Institute of Standards and Technology; 2010 Dec. NIST Special Publication 1017-1.
49. Post-Flashover fires. http://www.cfitrainer.net (accessed November 15, 2011).
50. *A Ventilation-Focused Approach to the Impact of Building Structures and Systems on Fire Development.* http://www.cfitrainer.net (accessed November 15, 2011).
51. Heenan D. *History of the Post-Flashover Ventilation Study.* Presentation to the California Conference of Arson Investigators; San Luis Obispo, CA, 2010, Nov 9.

52. Marquardt M. *Understanding Post-Flashover Fires: Recognizing the Importance of Ventilation*. Presentation to the California Conference of Arson Investigators; San Luis Obispo, CA, 2010, Nov 9.
53. Technical Committee on Fire Investigator Professional Qualifications. *NFPA 1033 Standard for Professional Qualifications for Fire Investigator*. Quincy, MA: National Fire Protection Agency (NFPA), 2009.
54. The Bureau of Alcohol, Tobacco, Firearms and Explosives, Public Affairs Division, ATF Fact Sheet, Washington, DC, 2010. http://www.atf.gov/publications/factsheets/factsheet-certified-fire-investigators.html (accessed November 15, 2011).
55. Stauffer E, Dolan J, Newman R. *Fire Debris Analysis*. Boston, MA: Academic Press; 2008.
56. Rethoret H. *Fire Investigations*. Toronto, Canada: Recording & Statistical Corporation, Limited, 1945.
57. Lucas D. The Identification of Petroleum Products in Forensic Science by Gas Chromatography. *Journal of Forensic Sciences* 1960; 5(2):236–247.
58. Kirk P. Fire Investigation; including fire-related phenomena: arson, explosion, asphyxiation. New York: Wiley, 1969.
59. Chrostowski J, Holmes R. Collection and Determination of Accelerant Vapors. *Arson Analysis Newsletter* 3 (5). Columbus, OH: Systems Engineering Associates, 1979;1–17.
60. Juhala JA. A Method for Adsorption of Flammable Vapors by Direct Insertion of Activated Charcoal into the Debris Samples. *Arson Analysis Newsletter* 6 (2). Columbus, OH: Systems Engineering Associates, 1982; 32.
61. Dietz WR. Improved Charcoal Packaging for Accelerant Recovery by Passive Diffusion. *Journal of Forensic Sciences* 1991; 36(1):111–121.
62. Waters L, Palmer L. Multiple Analysis of Fire Debris using Passive Headspace Concentration. *Journal of Forensic Sciences* 1993; 38(1):165–183.
63. Lentini J. *Scientific Protocols for Fire Investigation*. Boca Raton, FL: CRC Press; 2006.
64. Collaborative Testing Services. 1998. Flammables analysis Report No. 9716.
65. Collaborative Testing Services. 2000. Flammables Analysis Report No. 99–536.
66. Lentini J, Dolan J, Cherry C. The Petroleum-laced Background. *Journal of Forensic Sciences* 2000; 45(5):968–989.
67. Keto RO. GC/MS Data Interpretation for Petroleum Distillate Identification in Contaminated Arson Debris. *Journal of Forensic Sciences* 1995; 40(3):412–423.
68. Dolan JA, Ritacco CJ. Gasoline Comparisons by Gas Chromatography-Mass Spectrometry Utilizing an Automated Approach to Data Analysis. *Proceedings of the American Academy of Forensic Sciences*2002 Feb 11–16; Atlanta, GA. Colorado Springs, CO: American Academy of Forensic Sciences; 2002;62.
69. Sandercock M, Du Pasquier E. Chemical Fingerprinting of Gasoline: 3. Comparison on Unevaporated Automotive Gasoline Samples from Australia and New Zealand. *Forensic Science International* 2004; 140:71–77.
70. Scientific Working Group for Fire and Explosions (SWGFEX). Laboratory Fire Standards and Protocols Committee, *Quality Assurance Guide for the Forensic Analysis of Ignitable Liquids, Version 2.0*. Dec 2003.
71. U.S. DOJ. *Forensic Sciences: Review of Status and Needs*. 1999 Feb. http://www.ncjrs.org/pdffiles1/173412.pdf (accessed November 15, 2011).

72. ASTM International. *ASTM International, Standards Worldwide*. www.astm.org (accessed November 15, 2011).
73. Stout SA, Uhler AD. Chemical "Fingerprinting" of Highly Weathered Petroleum Products. *Proceedings of the American Academy of Forensic Sciences*; 2000 Feb 21–26; Reno, NV. Colorado Springs, CO: American Academy of Forensic Sciences; 2000;82–83.
74. National Research, Council. *Strengthening Forensic Science in the United States: A Path Forward*. Washington, DC: The National Academies Press; 2009.

Further reading

Collins J. *Good To Great*. New York: Harper Business; 2001.

Collins J. *Good to Great and the Social Sectors*. New York: Harper Collins; 2005.

Deming WE. *Out of the Crisis*. Boston: MIT Center for Advanced Engineering Study; 1986.

Drucker P. *Management Challenges for the 21st Century*. New York: Harper Business; 1999.

Drucker P. *The Effective Executive: The Definitive Guide to Getting the Right Things Done*. New York: Collins; 2006.

Houck MM, Riley R, Speaker P, Witt T. FORESIGHT: A Business Approach to Improving Forensic Science Services. *Forensic Science Policy and Management* 2009; 1(2):85–95.

Kobus H, Houck MM, Speaker P, Riley R, Witt T. Managing Performance in the Forensic Sciences: Expectations in Light of Limited Budgets. *Forensic Science Policy and Management* 2011; 2(1):36–43.

Mintzberg H. *Managers not MBAs*. San Francisco: Berrett-Koehler; 2005.

Sclar E. *You Don't Always Get What You Pay For: The Economics of Privatization*. Ithaca, NY: Cornell University Press; 2001.

Shewhart W. *Statistical Method From The Viewpoint of Quality Control*. New York: Dover; 1939.

Speaker PJ. Key Performance Indicators and Managerial Analysis for Forensic Laboratories. *Forensic Science Policy and Analysis* 2009a; 1(1):32–42.

Speaker PJ. The Decomposition of Return on Investment for Forensic Laboratories. *Forensic Science Policy and Management* 2009b; 1(2):96–102.

Speaker PJ, Fleming AS. Monitoring Financial Performance: An Approach for Forensic Crime Labs. *The CPA Journal* 2009; **79**(8):60–65.

4

Forensic pathology – the roles of molecular diagnostics and radiology at autopsy

James R. Gill[1], Yingying Tang[2], Gregory G. Davis[3], H. Theodore Harcke[4] and Edward L. Mazuchowski[5]

[1]*Department of Forensic Medicine, New York University School of Medicine, New York, New York, USA*
[2]*Molecular Genetics Laboratory, New York City Office of the Chief Medical Examiner, New York, New York, USA*
[3]*University of Alabama, Birmingham, Alabama, USA*
[4]*Department of Radiologic Pathology, Armed Forces Institute of Pathology, Washington D.C., USA*
[5]*Armed Forces Institute of Pathology, Washington DC, USA*

For much of the 20th century, the practice of forensic medicine and pathology hardly differed from the practice of the late 19th century. Autopsy work then, as now, depended upon trained physicians with sharp eyes and a desire to determine the underlying cause of death. The tools available to physicians in the morgue were limited to rulers, scales, scalpels, cameras and, for the fortunate, plain film radiography. Microscopes and laboratory support for culture of microorganisms and chemical and toxicological analysis completed the pathologist's tool kit.

During the course of the 20th century, the analytical methods for toxicological analysis became far more sophisticated with the introduction of enzyme-mediated screening procedures, gas chromatography, mass spectrometry, high performance liquid chromatography and the computers linked to these devices. The unprecedented abilities of these analytical devices transformed the accuracy of death certification for deaths due to intoxication or poisoning.

Other causes of death have resisted all but the most elementary determination of cause – for example, pulmonary thromboemboli caused by thrombophilias, or determination of the specific virus responsible for initiating a case of lymphocytic

Forensic Science: Current Issues, Future Directions, First Edition. Edited by Douglas H. Ubelaker.

myocarditis. As scientists around the world probe the genomes of humans, bacteria and viruses, they are finding genetic alterations that can lead to death – alterations that can be detected even after death.

Meanwhile, another change has affected the practice of forensic medicine. The concept of a mass disaster is not new, but the regularity with which mass disasters occur has been revealed by means of telecommunication systems that bring images of disaster in one part of the world to the rest of the world within minutes of the event. Disasters may arise naturally or can be man-made. Man-made disasters may be unintentional or a deliberate violation of human rights.

The rise of the global community has made new demands upon the practice of forensic medicine, because forensic medicine can identify those who have died in such disasters and provide evidence of what caused those deaths – evidence that can be used in court in the event of a human rights violation. Because forensic medicine exists, it allows identification and determination of cause of death in ways that were not possible before, and now the global community expects that expertise to be brought to the evaluation of a mass disaster.

The deployment of forensic scientists and specialized equipment from around the world to the regions affected by the tsunami of December 26, 2004 would not, and could not, have happened a century earlier, but the number dead (over 220,000) [1] revealed that some mass disasters require more than an earnest desire to help. Advances in radiology and the incorporation of this technology into forensic practice should make the practice of autopsy more efficient both in day-to-day work and, with mobile scanners, even in mass disasters.

This chapter will therefore focus on how developments in DNA analysis and genetic diagnostics, and in the use of advanced radiologic techniques such as CT, will transform the practice of autopsy in the 21st Century. This is appropriate, as the sister fields of anatomical and clinical pathology will continue to alter practice in the morgue and in the laboratory, refining autopsy practice together.

4.1 Molecular testing and the forensic autopsy

Clinical medicine and the basic sciences routinely use genetic techniques in their work. Until recently, a forensic pathologist's involvement with DNA was the collection of specimens for profile typing for identification [2–4]. Currently, however, the use and potential use of molecular testing in autopsy practice is much more advanced. Today, molecular studies that are available to the autopsy pathologist include analysis for genetic cardiac diseases, inheritable thrombophilias, pharmacogenetics and detection of infectious agents. The valuable information obtained from these molecular tests may have considerable medicolegal ramifications, and is often important to the deceased's immediate family and to the community as a whole.

4.1.1 Cardiac diseases [5–11]

Due to its common role in many sudden, unexpected deaths that frequently come under the jurisdiction of the medical examiner/coroner (ME/C), heart disease is of particular interest to the forensic pathologist. Atherosclerosis, cardiomyopathies, infections and congenital anomalies are usually evident, grossly or microscopically. Molecular studies have detected mutations for structural cardiomyopathies (myosin, troponin) and electrolyte channels (sodium, chloride, calcium) which have further elucidated these diseases.

The 'structural' cardiomyopathies typically have diagnostic gross and microscopic features. For example, hypertrophic cardiomyopathy (HCM) is an autosomal-dominant disease of myocytes with mutations that primarily involve genes encoding for sarcomeric proteins [12–18]. Molecular analysis is not required for the diagnosis when the autopsy finding is clear. Deaths due to genes known to cause myocardial structural changes, however, may present as sudden unexplained death before the structural defect becomes grossly visible [19,20]. This is particularly true in children whose hearts are still under development and where the structural changes may be less apparent.

Channelopathies are caused by mutations in genes encoding ion channel proteins that affect the cardiac conduction system and cause sudden cardiac death [7,9]. Channelopathies, however, cause sudden cardiac death with a grossly and microscopically normal-appearing heart [21,22]. Since the 'channelopathies' are functional alterations of cellular ion channels, molecular techniques are required to make post-mortem diagnosis in these hearts that appear 'too good to die'. Several studies from Europe and the United States on a large number of sudden, unexplained, autopsy-negative deaths have found putative cardiac channel mutations in 10–15% of infants [23–26] and in up to 35% of children and adults [27].

For channelopathies, molecular diagnostics have allowed the autopsy diagnosis of diseases that had previously been diagnosed only by clinical and electrocardiographic features (e.g. Brugada syndrome, long QT syndrome) [6,7,9,27–30]. Channelopathies also have been implicated in some instances of sudden infant death [9,23,31,32]. Cardiac 'channelopathies' should be considered when no cardiac abnormality or extra-cardiac cause has been identified in a sudden death.

Molecular analysis for cardiac channelopathies typically consists of all or selected exons in six major cardiac ion channel genes (KCNQ1, KCNH2, SCN5A, KCNE1, KCNE2 and RyR2) in which disease-causing sequence variants have been previously reported [7,28]. DNA may be extracted from dried blood samples or from post-mortem tissue samples (e.g. heart, spleen or liver) that have been preserved in special tissue storage solutions. DNA is amplified by the polymerase chain reaction and then undergoes bi-directional (or repeat uni-directional) sequencing with BigDye terminator chemistry, followed by automated capillary electrophoretic separation. The sequence variants are evaluated according to the American College of Medical Genetics Recommendations for Standards for Interpretation and Reporting of Sequence Variations [33].

The unique challenges of post-mortem molecular investigation include issues of sample collection, cost and test indication and interpretation [8]. The traditional method to preserve autopsy tissues in formalin for microscopic examination makes the tissues unusable for current PCR and Sanger sequencing-based molecular analysis. A non-formalin based storage solution, such as RNA*later*®, should be used. Non-formalin fixatives allow rapid stabilization and protection of the cellular nucleic acids without creating damage (e.g. cross-linking) in the DNA double strands.

The samples can be stored in the long term at –80°C. A dry blood card sample can be analyzed, but this is not ideal because the concentration of DNA is usually low. Since post-mortem samples may need to be retained for years, storage capacity – particularly freezer space – is an issue. Therefore, other means of long-term storage of post-mortem tissue samples are being investigated. In addition, with the arrival of the next-generation sequencing technology that provides the capacity of massive parallel sequencing analysis of much shorter DNA fragments (100 bp), our ability to analyze archived formalin-fixed tissues may improve.

In the clinical setting, when the diagnosis is evident by symptoms (e.g. syncope), characteristic ECG findings (e.g. prolonged QT interval) or a specific clinical response to certain medications, genetic testing may help confirm the disease and help to identify genetic carriers in the family [7]. In the forensic setting, however, aside from the sudden death, there usually is no clear clinical phenotype to support a channelopathy.

Some believe that genetic testing in these instances is a 'fishing expedition' that is not cost-effective [7]. This is understandable, given that the current test panel consists of a limited number of genes for inherited arrhythmia syndromes that will yield a low rate of positive results at a high cost. In addition, molecular testing and interpretation of any genetic disease is complicated by genetic heterogeneity, penetrance and variable expressivity, as well as genetic and environmental alterations.

Even though forensic investigations may involve deaths without specific channelopathy phenotypes, with proper case selection, testing is still useful. In certain cohorts (sudden death, young age, with a negative autopsy including toxicology testing), routine genetic testing can yield positive results in a considerable percentage (e.g. 10 of 49 cases evaluated by Tester and Ackerman) [27]. The interpretation of some of these results, however, may be a challenge.

Interpretation of the relevance of a previously unreported sequence variant can be particularly problematic in post-mortem forensic investigations. Since there usually are no other clinical or laboratory functional pre-mortem studies in these instances, there is little other information available to discern the importance of the variant. Functional characterization of these variants (such as *in vitro* electrophysiology functional study of a mutant ion channel), or correlation with living family members with clinical signs of disease, may potentially help to further elucidate the disease-significance of these variants.

As the molecular autopsy is technology-dependent, current techniques and cost are major obstacles for widespread use. Since over 50 genes are thought to contribute to sudden unexplained death, current capillary electrophoresis sequencing is limited by low throughput, long sequencing times, and high costs (in 2011, the cost of reagents and supplies was approximately $550 for a full panel). Next-generation or third-generation sequencing technology, however, has increased throughput and maintained accuracy, while also reducing costs. Hopefully, this trend will continue.

4.1.2 Hereditary thrombophilia [34–42]

Forensic pathologists frequently encounter pulmonary thromboemboli at autopsy. This is due to the sudden and often unexpected nature of the death and that pulmonary thromboemboli are often associated with injury and surgical procedures. Venous thromboembolism (VTE) is often multi-factorial and accounts for 100,000 deaths annually. As such, recognition of common hereditary defects in persons dying from a pulmonary thromboembolism (PE) has potentially life-saving importance to surviving family members. It also may be the proximate or contributing cause of death.

Virchow described the three risk factors (Virchow's Triad) of stasis, hypercoagulability and vascular injury. The hypercoagulability risk factor includes thrombophilias. These are diseases or conditions associated with an increased thrombotic risk, and inheritable thrombophilias include mutations in factor V Leiden and prothrombin and deficiencies of antithrombin (AT) and proteins C and S. Abnormalities involving antithrombin III, proteins C and S, plasminogen, dysfibrinogenemia and antiphospholipid antibodies use functional and serologic diagnostic assays that are ill-suited for post-mortem blood.

There are, however, molecular studies that can detect some thrombophilias even in post-mortem blood. There is a relatively high prevalence of certain thrombophilic heterozygous mutations, the most common ones ranging from 1–7% in the Caucasian population. Molecular analysis can be performed for two common hereditary thrombophilias: mutations in factor V Leiden (FVL); and prothrombin G20210A (PT) using various standard single nucleotide polymorphism detection methods. A recent molecular study of 578 out-of-hospital PE deaths revealed unique epidemiological differences among ethnic groups [42]. Future research is warranted to understand these aspects of fatal PE.

4.1.3 Pharmacogenetics [4,43–51]

Pharmacogenetics is the study of genetic variations that cause individuals to have different responses to the same dose of a drug. Drug concentrations can vary substantially for two people with the same weight, age, and dose. Genetic variations can affect a drug's efficacy or toxicity. A person's response to a drug is a complex trait that is influenced by different genes. A drug's absorption, distribution, metabolism and interactions with targets are affected by genetic factors. Differences in these genetic factors have been estimated to account for 15–30% of differences in drug metabolism and response [52].

In the 1950s, the first observations of genetic variation in drug response were noted with the muscle relaxant suxamethonium. Some people have a less efficient variant of butyrylcholinesterase, the enzyme that metabolizes suxamethonium, and this results in slower recovery from surgical paralysis. These variations in the N-acetyltransferase gene resulted in the classification of patients as 'slow and fast acetylators' and affect the half-lives and blood concentrations of other medications, including isoniazid and procainamide.

These genetic variations may occur at the stage of metabolizing enzymes (e.g. P450 system), drug transporters (e.g. P-glycoprotein efflux pump) and drug targets (e.g. insulin receptors) [52]. The cytochrome P450 oxidases (CYPs) are examples of genetic variations of metabolizing enzymes. The isoenzyme CYP2D6 has over 75 known allelic variations, of which some cause enhanced activity and others have no activity. For example, since codeine is activated by this enzyme, some people will not be adequately treated with standard doses if they have an inactive allelic form.

Due to the genetic polymorphisms of the CYP P450 enzymes, a phenotype classification has been developed (e.g. ultra-rapid, extensive, intermediate and poor metabolizers). Interestingly, poor metabolizers may be at risk for a drug overdose due to poor metabolism of some drugs, but they also can be at risk for therapeutic failure in instances of prodrugs that need to be metabolized to make the active metabolite. In addition to these polymorphisms, some drugs also may act as inhibitors or inducers of the P450 system, thereby further affecting its function.

Since interpretation of toxicology results for the determination of the cause and manner of death may depend on qualitative and quantitative factors, there is potential for these genetic alterations to result in an incorrect determination of suicide. For example, a person who dies with an elevated amitriptyline concentration may raise the suspicion of suicide. For forensic pathologists, CYP phenotyping may potentially reveal that an elevated post-mortem concentration is physiologic and not due to an intentional overdose [53].

How useful this additional information is to a practicing forensic pathologist, however, is not clear, and some believe that routine genotyping in suspected intoxications is probably not worthwhile at this time [53]. This additional information may not be that useful, because death certification is not based solely on a drug concentration from a toxicology report. Forensic pathologists already contend with other factors (e.g. decomposition and post-mortem redistribution) that may alter post-mortem drug concentrations.

For a forensic pathologist, it is important to recognize factors that may result in a potentially high (or low) drug concentration. This knowledge may prevent the pathologist from incorrectly certifying a death which has an elevated drug concentration, but no other support for an intentional self-destructive act, as a suicide. Regardless, the entire case investigation always must be considered when interpreting toxicology results and certifying a death.

4.1.4 Infectious disease [54–62]

There are numerous infectious agents that can be detected and typed with molecular analyses or probes. This largely remains the realm of the public health laboratory and clinical pathology, but forensic pathologists need to be aware of these techniques and to ensure that specimens are properly and promptly collected, stored and submitted.

Forensic pathologists frequently encounter tuberculosis, HIV, bacterial meningitis, H1N1 influenza, myocarditis, etc. Molecular analysis offers rapid, reliable and specific tests for a variety of microorganisms and viruses. The extent of molecular tests available for the detection of microorganisms is rapidly increasing and is too large to describe here, but there are several detailed review articles available [56,57,60,61,63]. As evidenced by the recent H1N1 pandemic, molecular techniques can diagnose new and changing microorganisms rapidly and accurately [64–66].

4.1.5 Other

There are other rare or unusual instances where the ability to perform post-mortem genetic testing may be beneficial for the next of kin. These include instances of questioned paternity and unusual genetic disease. For example, Leber optic neuropathy is a rare hereditary disease that eventually causes blindness. The family of a man who committed suicide by hanging contacted the medical examiner's office several weeks after the autopsy. They requested that a sample of his blood be released so that it could be analyzed for Leber optic neuropathy, since the decedent had been going blind and his physician suspected the inherited disease. PCR/RFLP analysis of post-mortem blood was performed at a local university, and the most common primary mitochondrial point mutation associated with the disease was detected. Since the disease is hereditary, this was important information for the family to know.

4.1.6 Conclusion

Molecular analysis, in conjunction with the forensic autopsy, requires proper collection, storage abilities and availability of cost-effective testing. Blood is routinely collected and stored in forensic autopsies because, even if there is no obvious need for molecular testing, a need may arise later. One may not predict when a DNA sample may be of use. If there is a specific concern for sudden unexplained death in infants, children and adults, additional collection of tissue in special solutions is prudent. Funded regional or national molecular testing centers would allow smaller offices the ability to investigate these deaths further. Such facilities would also provide data for large research studies that could help ascertain the phenotypes of new variant sequences. All this sounds extraordinary, but what seems extraordinary today soon becomes the standard of care.

4.2 Incorporating radiologic imaging into medicolegal death investigations

4.2.1 Introduction

For over 100 years, radiologic imaging has been used in medicolegal death investigations. The radiograph, developed by William Roentgen in 1895, was the first x-ray imaging modality employed in these investigations. These images were used to document both injuries and the presence of either projectiles or foreign bodies. The processing of these images was labor-intensive, and the exposure time to develop these plates in the late 1800s was approximately 60 minutes. Within a year of Roentgen's discovery, fluoroscopes were developed, which allowed the user to view moving images in real time. However, the diagnostic quality of the fluoroscope was inferior to that of the still radiographs. In addition, the fluoroscope exposed the user to the deleterious effect of the x-rays – hard radiation [67].

Throughout the last century, there have been numerous technological advances in the field of radiology, which have led to the development of radiographic machines with near-instant image availability and markedly decreased radiation exposure. Despite these advances, the incorporation of radiologic imaging into medicolegal death investigation has been inconsistent. The imaging modalities used in a medicolegal death investigation vary widely from medicolegal system to medicolegal system. In addition, there is variation between medicolegal systems in regards both to the types of cases that are imaged and to how much of the body will be imaged.

The National Association of Medical Examiners (NAME) has established minimum standards for the use of scientific tests, procedures and support services. In regards to radiography, NAME outlines the following standards:

1. Radiographs of infants are required to detect occult fractures which may be the only physical evidence of abuse.
2. Radiographs detect and locate foreign bodies and projectiles.
3. Charred remains have lost external evidence of penetrating injury and identifying features.

This guideline further states that the forensic pathologist or representative shall x-ray all infants, explosion victims, gunshot victims and charred remains. This section of the guideline does not explicitly state whether or not it is necessary to perform whole body radiographs in each case. Radiography is also addressed in the section 'Procedures Prior to Disposition of Unidentified Bodies'. In this section, it is recommended that whole body radiographs be taken to document skeletal characteristics and radio-opaque foreign bodies such as bullets, pacemakers and artificial joints [68].

This guideline uses the general term 'radiograph' and does not address any other x-ray modality. In addition, it does not further delineate the extent of radiographs that should be performed on infants, explosion victims, gunshot victims and charred remains. For example, is it acceptable to radiograph an infant with only one image, or should there be multiple images of each of the body parts, with both anterior-posterior and lateral views? Similarly, does one have to perform radiographs of all body parts, when there is only a gunshot wound of the head? Furthermore, no standard currently exists on whether or not imaging should be completed on cases of blunt force injury, sharp force injury, asphyxia, sudden death, and bodies found in water.

4.2.2 Imaging modalities

At the present time, there are three radiologic imaging modalities that can be used in medicolegal death investigations: fluoroscopy, radiography and computed tomography (CT). Each modality has its unique advantages and disadvantages in comparison to the other modalities.

Fluoroscopy displays a continuous x-ray on a monitor. Fluoroscopes have been available since the discovery of x-rays and, in their simplest form, they consist of an x-ray source and a fluorescent screen, with the body part being scanned placed in between. Improvements of this basic set-up couple the screen to an x-ray image intensifier and charged-coupled device (CCD) video camera, allowing the images to be viewed on a monitor and recorded. Most of the currently available fluoroscopes have the ability to print a hard copy of the image or to capture the image on a digital storage device.

The type of fluoroscopy most commonly used in medicolegal investigations is the C-arm fluoroscope (Figure 4.1). Although possible, it is difficult to conduct whole body imaging using a fluoroscope. The field of view is limited and there is significant artifact when viewing the images real-time as the body is moved. This motion artifact may also allow one to overlook a radio-opaque object. Selected orthogonal views can be obtained by rotating the C-arm. In addition, there are radiation concerns for the operators of the fluoroscope, and each individual must don personal radiation safety equipment such as lead aprons, lead gloves and lead thyroid protectors.

Imaging using a C-arm fluoroscope works best with at least two individuals: one individual to operate the machine and one to manipulate the body into the correct field of view. If a radiologic technologist is not used to perform the scans, the individuals performing the scan must be trained on the system and there must be a mechanism in place to measure the radiation exposure, regardless of who is

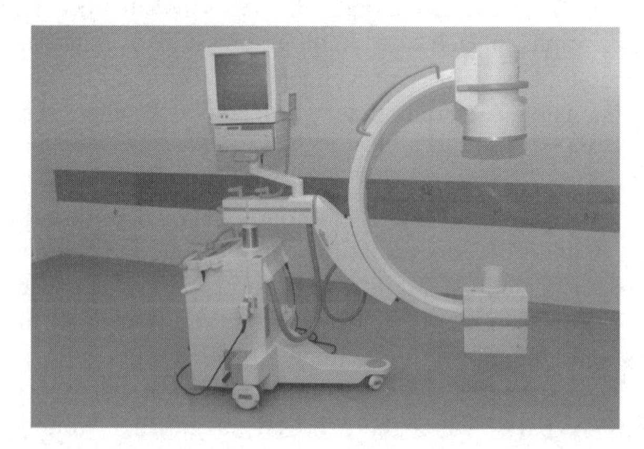

Figure 4.1 C-Arm Fluoroscopy Unit. GE XR 7700 Compact (GE Medical Systems, Milwaukee, WI).

performing the scans. The life expectancy of a C-arm is between 5–8 years and the cost of a complete system is roughly between $20,000 and $50,000.

The main use of the fluoroscope in medicolegal investigations is to facilitate the localization and recovery of either projectiles or metallic foreign bodies. This is achieved by locating the object with the fluoroscope and noting its approximate location in relation to skeletal landmarks. The object is then recovered after diligent dissection. Occasionally, the object may move during the autopsy search procedure and it becomes necessary to re-image. The fluoroscope has the advantage of allowing the user to dissect and re-image in near real time.

Fluoroscopes can also be used to detect vascular injury on selected cases by injecting the vessel of concern with contrast material and looking for extravasation of the contrast material at the suspected area of injury. The advantage of this technique is that the location of the extravasation of the contrast material can be observed as it is being injected. The disadvantage is that a three-dimensional reconstruction of the image cannot be generated.

Radiography uses x-rays to generate and record a static image after termination of exposure. This was the type of imaging modality first described by Roentgen. In this modality, an x-ray beam is passed through the body, a portion of the x-rays are absorbed or scattered, and the remaining x-ray pattern is transmitted to a detector and the image recorded. The image is produced by differential attenuation of the x-ray beam by body structures and it may be recorded either on film or electronically.

If film is used as the medium for viewing the image, one can wait up to 45 minutes for a complete study to be processed [67]. Computed radiography (CR) substitutes a receptor plate for the film. CR makes use of an existing X-ray tube and table, so it is less expensive to implement than a direct digital radiography (DDR) system. The CR plate is scanned by laser after exposure and a digital image is produced. DDR systems, popular in the last decade, link the X-ray tube and receptor and make it possible to view radiographs almost immediately after the images are taken, to ensure that the quality of the image is sufficient. Both CR and DDR systems permit some digital processing to correct an image that has not been optimally exposed. This avoids the need to repeat views for over- or under-exposure, but it is still necessary to repeat images for incorrect position.

In order to be able to optimally store and view digital images without making a hard copy, it is necessary to network acquisition, viewing and storage. This is done commercially by a picture archiving and communication system, or 'PACS' network. A PACS system allows the images to be viewed at multiple locations within a facility, including at the autopsy station, and archived for future retrieval. PACS networks also have the capability of remote access, allowing individuals to view images remotely for diagnosis, collaboration and educational activities.

In order to produce a 'whole body' image, individual radiographs of the head, neck, torso, pelvis and extremities must be taken separately and then viewed in succession, making sure that every area of the body is covered without gaps. Using film, this process can take several hours. Even if DDR is used, one must still take multiple images of the body in order to produce 'whole body' coverage. This requires constant manipulation of the body, either by an autopsy assistant or by a radiologic technologist. Acquiring a 'whole body' collage with DDR takes approximately 10–15 minutes per body.

Radiographic studies require a radiologic technologist or an individual that has been trained and certified to use an x-ray machine. In addition, a person trained to be a PACS administrator is desirable for upkeep of a PACS system. There is minimal radiation exposure to the technologist when standard shielding and precautions are observed. The life expectancy of a DDR system is 6–8 years, and the cost of a complete system is currently roughly between $80,000 and $150,000.

In medicolegal death investigations, radiography is the most widely used radiologic imaging modality. Radiography can be used to determine if there are any radio-opaque fragments in the body. In cases where there has been a ballistic injury, the radiographs may depict bullets, bullet fragments, pellets, wads or blast fragments. Where there has been a suspected sharp force injury, the radiographs may depict retained fragments of the weapon. If only anterior-posterior images are obtained, precise localization of the object is limited to two dimensions. Orthogonal or lateral images must be obtained to locate the object precisely in three dimensions.

With radiographs, there is excellent edge detail of the borders of radio-opaque objects, but it must be noted that precise measurement of the dimensions of radio-opaque objects on radiographs is limited by geometric and physical factors. Magnification always occurs, due to the distance between the x-ray source, the object and the x-ray detector. The true shape and length of the object will be distorted on the image when the object is not positioned perpendicular to the x-ray beam.

Radiography is useful in documenting skeletal injuries. Although it is possible to determine a fracture of one of the long bones by manual manipulation at autopsy, a radiograph can further delineate the exact location of the fracture, the presence of multiple fractures in the same bone, and in some cases may help with the determination of the direction of force as it was applied to the bone. Some fractures of the vertebrae, especially spinous process fractures, are difficult to detect at autopsy unless a posterior approach is used. In order to detect these fractures using radiography, it is necessary to use orthogonal views.

As stated previously, NAME mandates radiographs on all infant deaths. Taking orthogonal images and separate images of the major regions of the body, as opposed to a single view 'babygram', enables greater resolution and can aid in the detection and delineation of acute and older injuries.

Similar to fluoroscopy, vascular injury can be documented by radiographs obtained after injecting the vessel of concern with contrast material and showing extravasation of the contrast material at the suspected area of injury [69]. Depending on the amount of extravasation of contrast material, it may be difficult to definitively depict the exact location of the injury to the vessel.

A recent development in the field of radiography has been the introduction of planar whole body digital scanning such as the Statscan™ system by Lodox (South Africa) [70]. This system was developed in the late 1980s for South African mine owners trying to fight widespread diamond theft by their workers, and it permits rapid whole body imaging with low dose radiation. In the past decade, multiple trauma centers throughout the world have used the technology to triage patients, as a whole body scan can be completed in as little as 13 seconds without moving the individual. These images, however, have a slightly distorted aspect ratio, due to the manner in which the x-rays are detected and processed for viewing.

The technology has recently been implemented by some medicolegal systems. Like conventional radiographs, the operation of the machine requires a trained individual and benefits from linkage to a PACS system. By scanning in orthogonal planes (anterior-posterior and lateral), rapid localization of projectiles or foreign materials through the body is obtained. The approximate cost of this system, with a five-year maintenance, contract is currently $400,000 [70].

Computed tomography (CT) uses multiple x-ray exposures acquired from circumferential positions around the body to produce computer-generated cross-sectional images in a plane or 'slice' through the body. Whole body CT scanners have been commercially available for the past 30 years (Figure 4.2).

The x-ray source and detector rotate around the individual as they are moved through a circular opening on a motorized table. The detector records the x-rays passing through the individual. This data is sent to a computer that maps the attenuation measurements to picture elements using a grayscale. The reconstruction of all of the individual data sets produces a series of 'slices' or multiple cross-sectional images.

Originally, selected images were printed onto film and viewed on a light board. Due to advances in computer technology and storage capability, one can now view and efficiently store all of the cross-sectional images on a computer network using a PACS. Post-processing workstations can be used to create multiplanar two-dimensional (2-D) three-dimensional (3-D) images that can be manipulated and reformed.

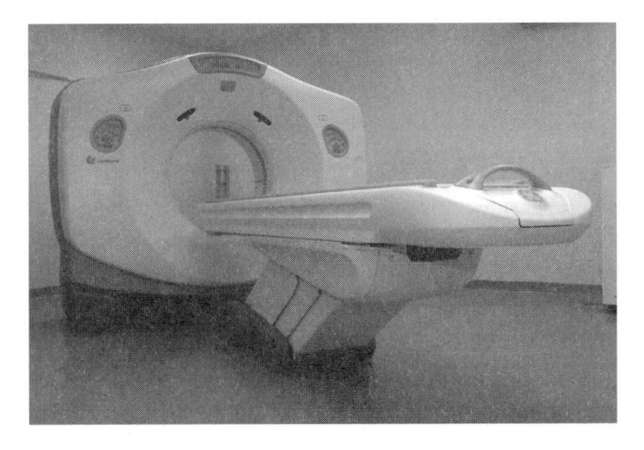

Figure 4.2 Forensic CT Scanner, with expanded bore and extended table travel (200 cm), enables a single-pass whole body scan to be acquired.
GE Lightspeed 16 Xtra (GE Medical Systems, Milwaukee, WI).

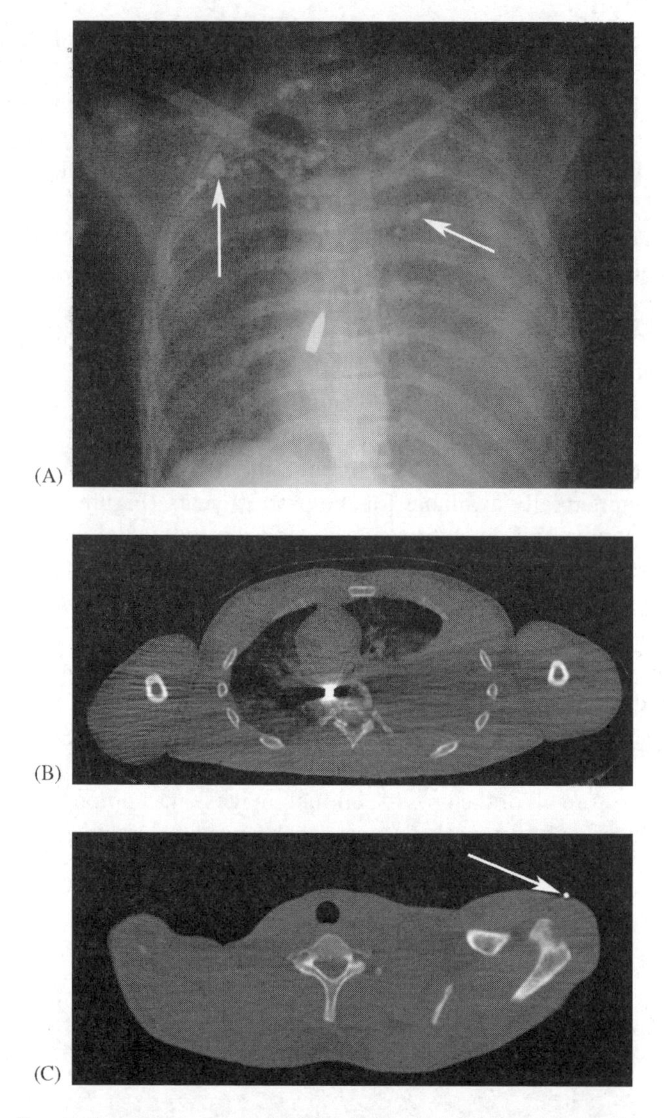

Figure 4.3 Penetrating gunshot wound of the thorax. (A) Digital chest radiograph showing bullet. Scattered opaque densities (arrows) are rocks between the clothing and body. (B) Axial CT image shows precise location of the bullet. (C) Axial section at level of the entry wound. Note metal surface marker at entrance wound anterior to the acromion process of the left shoulder (arrow). (D) Oblique coronal reconstruction in the plane of the wound path. Note entrance wound marker (arrow), fracture of the anterior margin of the acromion (black arrow), and fracture of the left first rib (arrow head). The opaque left hemothorax reflects the large hemothorax. The track, anterior to posterior, left to right and downward crosses the posterior mediastinum. A lacerated aorta was found at autopsy.

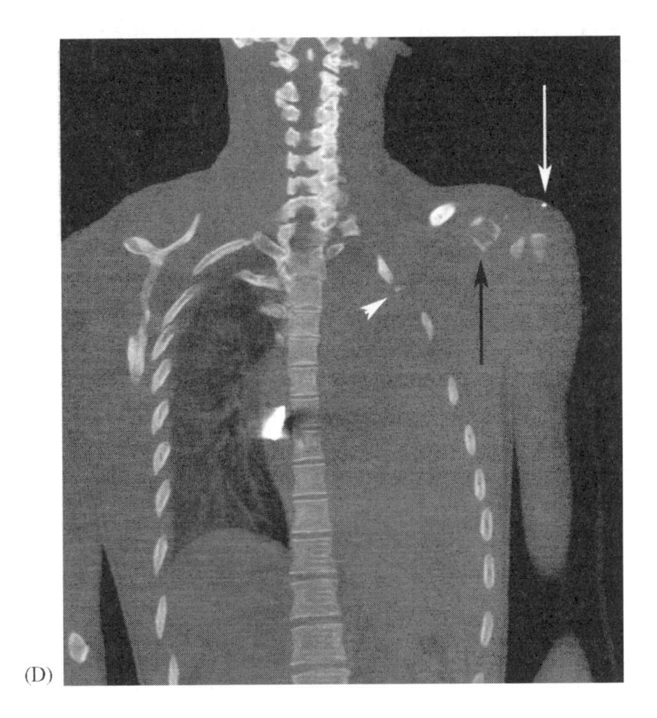

Figure 4.3 *(Continued)*

The software developed for image processing in clinical (ante-mortem) radiology is also suited for post-mortem imaging, with no special adaptations required. Whole body CT requires a radiologic technologist, and a PACS administrator is also desirable. With the use of multi-detector CT scanners, it takes approximately 5–10 minutes (transfer to/from the table, positioning and scanning) to complete a whole body scan. As with radiographs, there should be no appreciable radiation exposure to the technologists, as they are shielded from the equipment.

The present cost of an entry level CT scanner and necessary software is approximately $500,000, and the life expectancy is between 7–10 years. The CT scanner and software require regular preventative maintenance and upgrades that vary widely in cost, and this should be incorporated into the original purchase contact.

In cases where there are ballistic injuries, CT imaging allows for the precise location of the object by looking at the cross-sectional, axial images. Using this modality obviates the need for orthogonal images to be taken with radiographs (Figure 4.3). One disadvantage of CT imaging in cases of ballistic injuries is that metallic objects create a streak artifact, and the edge detail of the borders is blurred (Figure 4.3B, 4.3D). This accounts for continued use of radiographs after CT equipment is acquired. This streak artifact also occurs near teeth that have amalgam restorations.

CT imaging can also aid in the trajectory of the projectile through the body. For example, in gunshot wounds to the head, the beveling of the skull that has been described for decades in the forensic literature can be readily seen. In addition, the distribution of bone fragments through which a projectile passed may also aid in the determination of the trajectory (Figure 4.3D).

One of the major limitations of CT imaging in the evaluation of ballistic injuries is that it may be difficult to determine the location of the entrance and exit wounds on the skin surface. Although the presence of gas in the soft tissue and disruption of tissue surfaces may indicate a wound on the skin surface, the collapse of the temporary cavities, compression of the soft tissue defects and position of the body on the CT table limit their detection. Embolization of projectiles is detected using whole body CT imaging by finding a fragment in an area where there is no suspected wound path.

Some sharp force injuries can be documented with CT imaging. As with ballistic wounds, air in the subcutaneous tissue may aid in the determination of the wound on the skin surface. Internal hemorrhage, if present, is often readily apparent. When the images are viewed prior to autopsy, this gives the prosecutor an indication of where the wound path may be. If the weapon has injured bone, the images may show either a fracture or a defect that has similar characteristics to the weapon used (Figure 4.4). CT imaging can document whether a fragment of the weapon has broken off and can demonstrate its precise location. Air in the vascular system and heart can also be

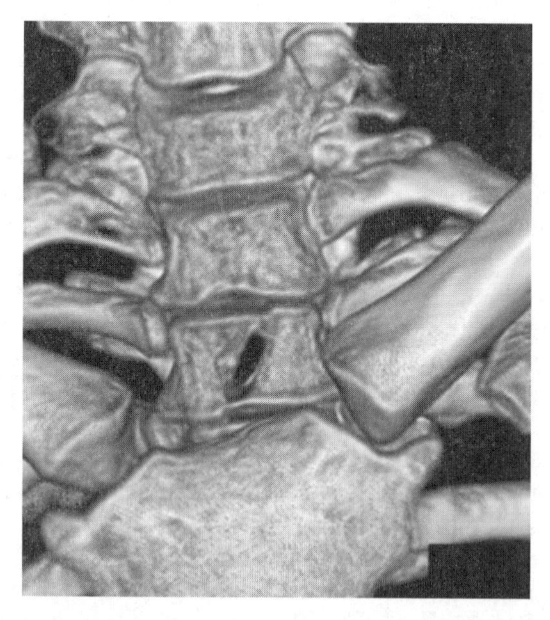

Figure 4.4 One of multiple stab wounds of the back injured the second thoracic vertebral body. A 3-D reconstruction of the spine shows where the knife injured the spine. Note the defect shows one sharp end and one blunt end, which can indicate the characteristics of the weapon.

visualized using CT imaging in cases where the sharp force injury is to the neck. However, caution must be used, as decomposition can also lead to air in these structures.

As with radiographs, CT imaging is an effective modality in documenting skeletal injuries. The main advantages that CT have over radiographs are that orthogonal views are not necessary, fractures of the spine are easily detected and occult fractures can be detected. Post-processing of the data, using 2-D and 3-D reconstructions, aids in fracture documentation (Figure 4.5). These three-dimensional displays may facilitate the understanding of the mechanism of injury.

In contrast to radiographs, CT imaging can also document some soft tissue injury. Some examples are hematomas underlying the skin surface, ruptures of the diaphragm with organ displacement, hematomas of the brain and transection of the spinal cord. Hemothoraces, hemopericardium and hemoperitoneum can also be documented. However, without dissection, it is difficult to define where the injury that caused the hemorrhage is located.

Similar to fluoroscopy and radiography, vascular injury can be detected by injecting the vessels with contrast material. Some systems have employed injection of the entire vascular system with contrast material [71], and other systems have employed directed injections with contrast material [72]. Regardless of the method employed, CT imaging can document the extravasation, or lack thereof, of the contrast material. After post-processing with three-dimensional software, the structures surrounding the location of vessel injury can be visualized (Figure 4.6).

Natural disease processes can also be depicted using CT imaging. Coronary artery calcifications can be detected and documented by CT (Figure 4.7). Intracranial hemorrhage and subarachnoid hemorrhage can be visualized with non-contrast CT imaging, but precise location and mechanism are not able to be determined (Figure 4.8). In these cases, it is necessary to know the circumstances surrounding the death, as hemorrhage from trauma will be similar. CT imaging may also depict alveolar consolidation, suggesting either the presence of severe bronchopneumonia or pulmonary emboli. Cystic disease of the liver and kidney can also be visualized [72].

The cross-sectional imaging produced by CT can add objective and reproducible anatomic data to the medicolegal death investigation. Such data can be used both to document findings and as a demonstrative aid. When used in conjunction with autopsy, CT imaging provides an anatomic overview of the body prior to dissection. Injuries and radio-opaque objects are localized more efficiently, complex fractures are better documented and visualized, and occult injuries may be detected. Although CT imaging has the potential to augment a limited autopsy – or, in some instances, replace autopsy – prospective studies must be performed that validate the accuracy of using only CT imaging in the determination of the cause of death. Future studies and research that compare the accuracy of CT imaging to autopsy are necessary to establish the effectiveness of CT alone in the determination of the cause of death.

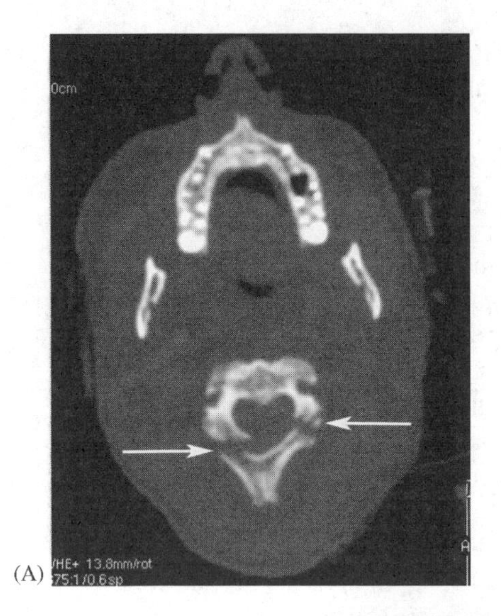
(A)

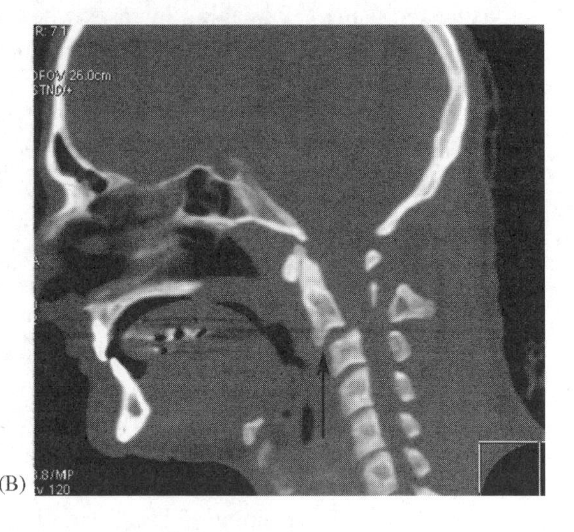
(B)

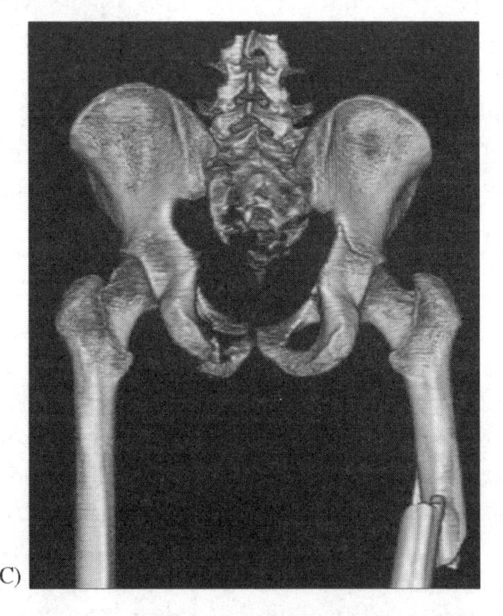
(C)

Figure 4.5 Spine and pelvic blunt force injury. (A and B) Axial and sagittal CT shows fracture of the posterior arch of second cervical vertebra (arrows) with anterior subluxation of the body (black arrow). (C) 3-D CT reconstruction of a pelvis viewed from posterior. Note fractures of the sacrum, right iliac and pubis, left ischium and pubis, and the shaft of the left femur.

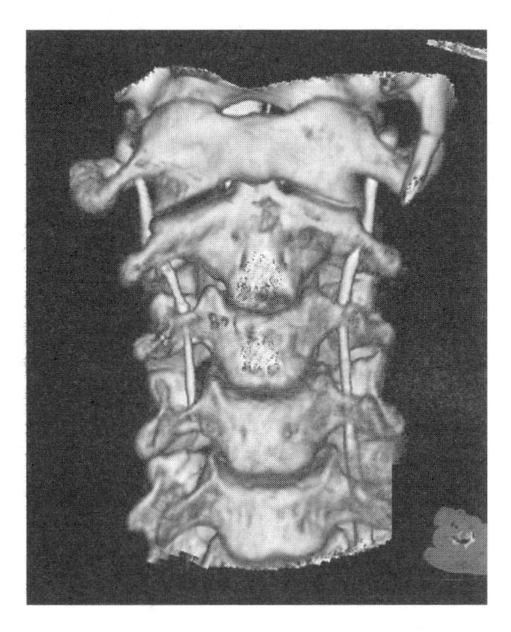

Figure 4.6 Post-mortem vertebral angiography in a trauma case with head injury. 3-D CT reconstruction shows intact vertebral vessels bilaterally as they pass through the upper cervical area.

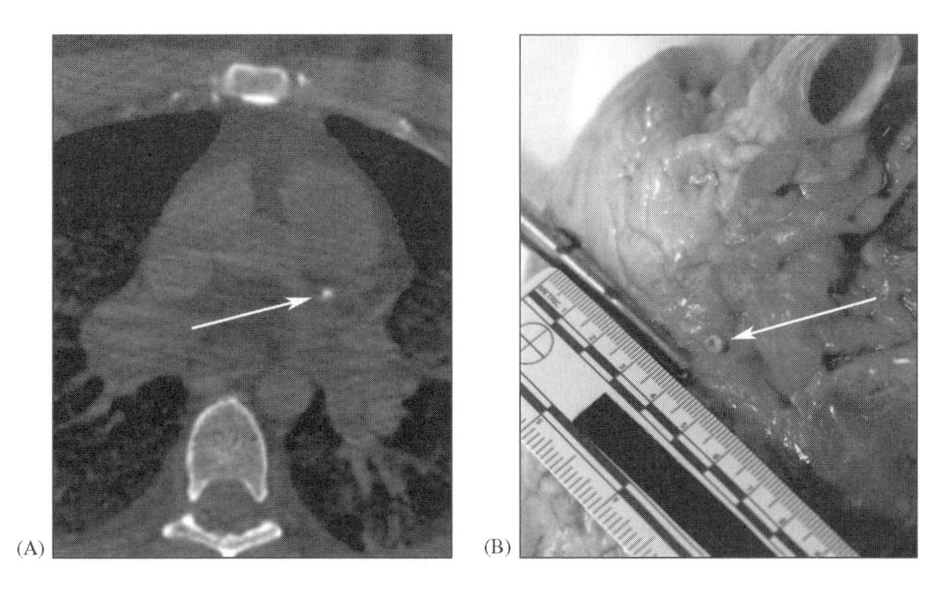

Figure 4.7 Coronary occlusion. (A) CT shows calcification on the left coronary artery (arrow). (B) Definitive assessment of the coronary artery lumen (arrow) at autopsy.

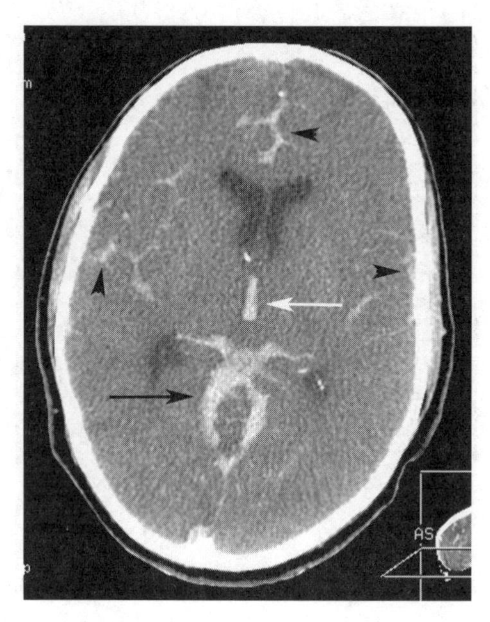

Figure 4.8 Diffuse subarachnoid hemorrhage. Post-mortem brain CT shows blood in the subarachnoid space (black arrow heads) as well as the third ventricle (arrow) and the base of the brain (black arrow). Only at autopsy was it established that this was due to a cerebral artery aneurysm rupture.

4.2.3 Research and future directions

In recent years, a few medicolegal death investigation systems throughout the world have incorporated CT imaging into their autopsy protocols. This has taken different directions, depending on the needs and structure of the system. Some systems have used CT imaging, alone or in conjunction with radiography, as an extension of their autopsy protocol, to supplement the findings from dissection and autopsy. This could be termed 'radiology assisted autopsy' [72–74]. Other systems have used CT imaging as a tool to aid in the determination of whether or not a conventional autopsy is necessary. This could be termed 'triage autopsy' [73,75,76]. The complete replacement of a forensic autopsy by CT constitutes what can be termed a 'virtual autopsy'. In areas where there is cultural opposition to dissection and autopsy, there is great interest in supplanting dissection and autopsy with CT imaging, and this is already occurring [73].

Fiscal, personnel and space requirements have limited the rapid and widespread incorporation of CT into death investigation. Some medicolegal death investigation systems are considering the feasibility of incorporating CT imaging into their system and are asking for specifics on cost of equipment, operating expenses and the return value. Specifically, the cost-effectiveness of CT in relation to forensic pathologist staffing is of particular interest. As discussed above, future studies and research are necessary to validate the effectiveness of incorporating CT imaging in determination of the cause of death. In addition to these studies, there are numerous related topics that can be examined scientifically to verify the strengths and limitations of CT. Below are some current areas of research and some areas that warrant further study. Hopefully, future investigations will include these issues.

The effect of post-mortem changes and decomposition on CT findings is a major area where there is a lack of studies and future research needs to be conducted. Due to hemoconcentration from livor mortis, there is increased attenuation of the dependent aspects of organs, cardiac chambers, great vessels and dural sinuses. This finding must not be mistaken for thrombosis or injury unless there are additional supportive findings. In addition, decomposition introduces air into the organs and tissues, and limits the ability to detect air emboli and pneumothoraces. The effects on a body of time since death under differing environmental conditions has been extensively studied by forensic scientists [77], but not using CT imaging.

One of the main limitations of utilizing CT in the evaluation of gunshot wounds is the inability of CT to locate the surface entry and exit wounds precisely. Although the presence of gas in the soft tissue and disruption of tissue surfaces may be indicators of the general location of these wounds, the collapse of the temporary cavities, compression of soft tissue defects and the position of the of the body on the scanning table can obscure the exact location of the entry and exit wounds.

In order to overcome this limitation, radio-opaque markers can be placed on the surface entry and exit wounds after external examination and before imaging of the body with CT (Figure 4.3C, D). The marked images can be processed with imaging software to produce a three-dimensional image of the body with the precise location of the entry and exit wounds on the skin surface. It must be noted, however, that this technique does not overcome the limitation of CT in distinguishing entrance gunshot wounds from exit gunshot wounds. This distinction is best made by combining the post-mortem forensic imaging with the findings from the external inspection and internal dissection of the body.

Since reconstructed images can be manipulated to obtain any desired orientation of the body and wound pathway, CT images have the potential to demonstrate gunshot wound pathways clearly in medicolegal proceedings. Since the data produced by CT imaging is in a coordinate system, the data can be translated into a variety of new and existing computer software applications that deal with modeling of the scene [78].

The documentation of intravascular lines, chest tubes, endotracheal tubes and airway devices, as well as less common devices such as intraosseous infusion catheters, is part of the external examination of the autopsy protocol. CT imaging can aid in the assessment of the internal positioning of these devices. This information may be helpful to first responders and other emergency personnel who are using a new device or who have a question regarding success of a particular technique.

It must be noted that device assessment at the time of the post-mortem CT imaging may not be accurate, since devices can shift position during transport and the handling of bodies. Furthermore, the clinical effectiveness cannot be assessed, due to the complex physiologic factors involved in injury and resuscitation. However, by providing this feedback with regard to anatomic position and physical characteristics, inferences can be made about the specific device or intervention. As an example, the precise tip location of intraosseous catheters cannot be assessed during routine autopsy. In contrast, CT imaging can easily assess the axial or three-dimensional location of the needle tip (Figure 4.9). In a study of 61 intraosseous tibial catheters, 95% were successfully positioned in medullary bone [79].

In mass fatality incidents and natural disasters, CT imaging, with or without digital radiography, has been proposed as a triage tool to determine which individuals need a complete autopsy. The objectives of the triage are to make sure that there is no evidence of foul play (i.e. projectiles in the body) and to confirm the presence of CT imaging findings which are consistent with the recovery circumstances. For example, if the circumstances surrounding the death are that the individual died in an earthquake, and the CT imaging depicts multiple fractures consistent with blunt force injuries, then CT may allow determination of the cause and manner of death to a reasonable degree of medical certainty without the need for a complete autopsy. As an example, 28 deaths from the 2010 Haiti earthquake were triaged using an algorithm incorporating CT imaging and digital radiographs with external examination. The CT and visual findings were adequate for issue of a death certificate in 22 (79%) of the 28 cases. Only six (21%) required a complete autopsy to determine the cause and manner of death [80]. Further studies of this nature have the potential to alter remains processing in disaster mortuaries.

In medicolegal investigations of deaths resulting from motor vehicle collisions, CT images – with or without digital radiography – can be used to further delineate and accurately document skeletal trauma in more exact detail. CT has greater accuracy than radiography and can show fractures not recognized on autopsy. Currently, some medicolegal death investigation systems are involved in providing data to crash injury databases such as the Crash Injury Research Network (CIREN). The goal of these networks is to understand what injuries are occurring, how they are occurring and what can be done either to prevent or to mitigate these injuries. Accurate and complete data gathered from actual events will increase the fidelity of and refine the modeling systems that these networks have developed and are developing, and should lead to improved automobile design.

This review has purposely omitted a detailed discussion of magnetic resonance imaging (MRI). While it is under discussion as a future direction for post-mortem imaging, very few medicolegal systems have the resources to consider MRI. Research in post-mortem MRI is being conducted [73], however, and it is expected that this will continue at a few centers. Much of the future of post-mortem MRI will depend on the acceptance and successful incorporation of CT into medicolegal systems. Should CT, which is less costly to acquire and operate than MRI, prove to be a successful addition to a medicolegal system, then interest in MRI is likely to follow.

4.2.4 Training and education

Post-mortem forensic imaging draws upon two bodies of knowledge: forensic pathology and radiologic imaging. A logical assumption is that individuals

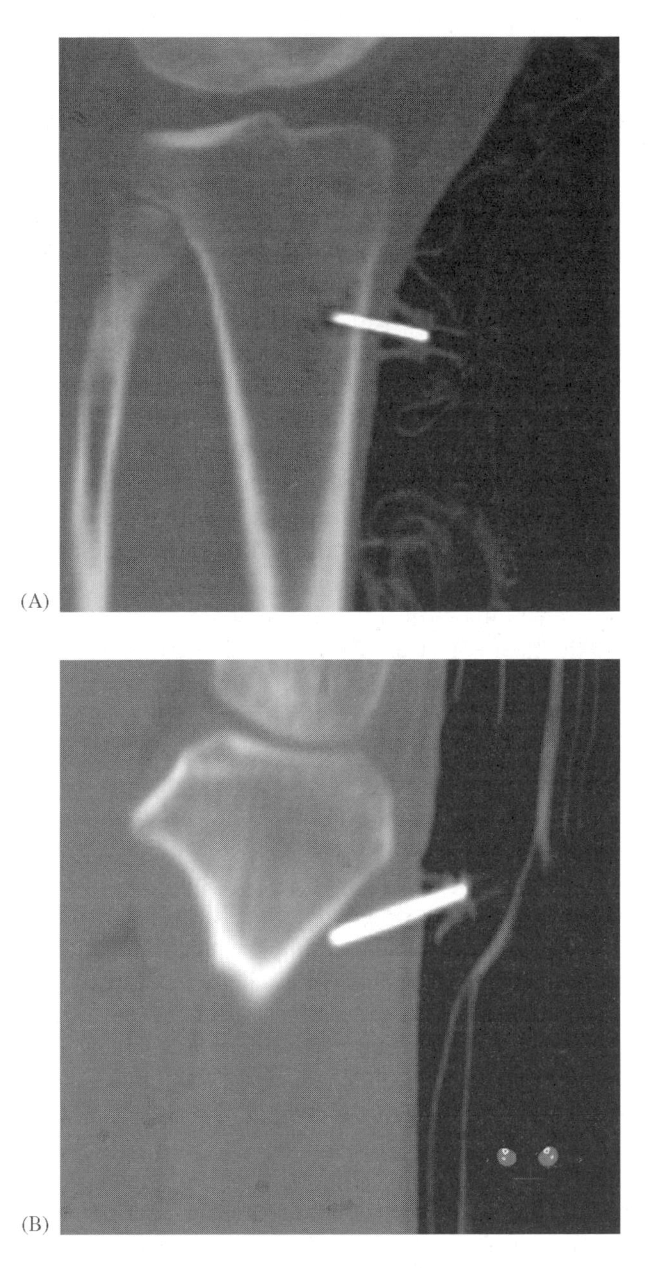

Figure 4.9 Coronal CT images of tibial intraosseous infusion needles. (A) The needle tip is in the medullary cavity of the proximal tibia. (B) The needle tip is in soft tissue adjacent to the tibial cortex.

interested in advanced forensic imaging will come from one of these two disciplines. Educationally, these have been separate pathways, culminating in fellowships in forensic pathology or a subspecialty of radiology, such as cross-sectional imaging with CT. Since forensic pathology currently does not include training in cross-sectional imaging, and radiology does not address post-mortem imaging and principles of forensic pathology, any plan for training in forensic imaging should be tailored to the needs of individuals drawn from one of these backgrounds.

In defining the content of a training program, the forensic pathology community will need to define necessary principles of forensics and post-mortem anatomic concepts that are not familiar to radiologists trained in ante-mortem radiology and pathology. These have already been defined in training programs for personnel working in forensics as investigators, criminalists and law enforcement officers, and should be readily adapted to training radiologists. After didactic education, radiologists will need to be reacquainted with autopsy procedures and taught about forensic autopsy, including autopsy's strengths and limitations.

The radiology community will need to teach forensic pathologists the basics of interpreting cross-sectional anatomic images and the use of multiplanar two-dimensional and three-dimensional reconstructions to supplement axial images. In a case where the pathologist will be supervising the individual (technician or radiologic technologist) acquiring CT images, the pathologist will need to understand some basic technical principles affecting image acquisition and quality. Viewing images is one aspect of forensic radiology, but manipulating the images is equally critical. This means that the pathologist must become skilled in the application of the computer workstation software used to manipulate contrast, reconstruct multiplanar views and produce 3-D images.

The required training time for a pathologist or radiologist could be different for the two disciplines; however, for simplicity, we can assume it will be about equal. It now becomes necessary to estimate this time or its equivalent in experience. The number of cases done seems preferable to a temporal measure only. Our experience suggests that 100–120 cases of the correct spectrum of pathology would be adequate. It is important to remember that each physician trainee brings considerable skill and basic knowledge from their respective discipline (e.g. human anatomy, knowledge of disease), and that this is readily applied to the new material and facilitates learning.

Using case number as the criterion allows time to be discounted, giving the trainee the flexibility to do all these cases within a single time period or to train for short periods over an extended time interval. It will be important to define the required case spectrum, e.g. the number of ballistic, blunt force, drowning, burn, and natural cause cases necessary for a full experience. Experience should also include cases with varying amounts of decomposition, predation and skeletalization. There are now textbooks which combine advanced imaging and basic forensics, and these can serve as a resource for developing case criteria for training programs.

Training will be most efficient when the faculty and student are at a site where the cross-sectional imaging equipment and the autopsy area are co-located. This, however, would limit the opportunity for forensic imaging at present to a small number of medical examiner's offices in the United States (Armed Forces, Maryland and New Mexico). Overseas, there are additional sites where training is offered (Australia, Switzerland, and Japan), and it is expected that further sites will be added, both in the US and abroad, in the future.

An alternative to a co-located site is the situation where the autopsy facility is a short distance from the imaging facility and post-mortem CT is done by transporting the body between the locations. In this scenario, it should be no problem for the trainee to be able to make radiologic-pathologic correlations. These situations expand the total number of sites where effective training should be possible. Since imaging today is digital, the transmission of imaging data to the autopsy site can be timely; the trainee could interpret the images at the autopsy site under the guidance of a radiologist located elsewhere, who is simultaneously viewing the case.

The cost of training is likely to be an '*ad hoc*' situation which will vary from place to place and differ with individuals. It is unlikely that government and forensic organizations will have the resources to support a full-scale training program in forensic imaging. Perhaps a few individual or named 'fellowships' might be created. It is more likely, however, that funding will be pruned from existing budgets, where a pathologist or radiologist already on salary is given some time away from the schedule to pursue training (e.g. sabbatical leave).

For current and future budgeted fellows in forensic pathology, it becomes a matter of allowing time for training in imaging during the fellowship experience. Radiologic-pathologic correlation has always been required for radiology residents, but this has not included forensics. At fellowship level, it may be possible for individuals doing cross-sectional or other fellowships to use elective time for forensic instruction. This would be negotiated on an individual basis, initiated by the interest of the individual fellow.

Oversight responsibility for training will be an evolving issue. No one can predict at this point whether forensic imaging will evolve to a recognized subspecialty such as 'radiopathology' or 'forensic radiopathology'. From the practical and fiscal points of view, it is most likely that requirements will be generated from the pathology specialty rather than radiology, because jobs will most likely be tied to the medical examiner system. It is logical to have existing programs of forensic training and certification incorporate post-mortem imaging into their curriculum over a period of time. Accredited programs without proximity to CT scanners will need to consider sending fellows to a training site. Ultimately, training might include web-based components and established regional/national training sites.

References

1. U.S. Geological Survey. *Earthquakes with 50,000 or More Deaths.* [updated 2011 Apr 14; cited 2011 Oct 28]. Available from: http://earthquake.usgs.gov/earthquakes/world/most_destructive.php.

2. Pinckard JK. Memorial Eckert paper for 2007 forensic DNA analysis for the medical examiner. *American Journal of Forensic Medicine and Pathology* 2008; 29(4):375–381.
3. Karch SB. Changing times: DNA resequencing and the "nearly normal autopsy". *Journal of Forensic and Legal Medicine* 2007; 14(7):389–397.
4. Madea B, Saukko P, Oliva A, Musshoff F. Molecular pathology in forensic medicine – Introduction. *Forensic Science International* 2010 Dec 15; 203(1–3):3–14.
5. Tang Y, Bieschke ET, Labitzke E, D'Andrea J, Sainte-Marie S, Wang D, *et al.* Comprehensive molecular genetic testing for the cardiac channelopathy genes in 42 cases of sudden infant death and sudden unexplained death in the city of New York revealed high mutation rate. *Proceedings of the American Academy of Forensic Sciences* 2008 Feb 18-23; Washington (DC). 2008:265(G34).
6. Basso C, Carturan E, Pilichou K, Rizzo S, Corrado D, Thiene G. Sudden cardiac death with normal heart: molecular autopsy. *Cardiovascular Pathology* 2010; 19(6):321–325.
7. Campuzano O, Beltrán-Alvarez P, Iglesias A, Scornik F, Pérez G, Brugada R. Genetics and cardiac channelopathies. *Genetics in Medicine* 2010; 12(5):260–267.
8. Bai R, Napolitano C, Bloise R, Monteforte N, Priori SG. Yield of genetic screening in inherited cardiac channelopathies: how to prioritize access to genetic testing. *Circulation: Arrhythmia and Electrophysiology* 2009; 2(1):6–15.
9. Tester DJ, Ackerman MJ. Cardiomyopathic and channelopathic causes of sudden unexplained death in infants and children. *Annual Review of Medicine* 2009; 60: 69–84.
10. Thiene G, Corrado D, Basso C. Revisiting definition and classification of cardiomyopathies in the era of molecular medicine. *European Heart Journal* 2008; 29(2):144–146.
11. Basso C, Burke M, Fornes P, Gallagher PJ, deGouveia RH, Sheppard M, *et al.* Guidelines for autopsy investigation of sudden cardiac death. *Virchows Archiv* 2008; 452(1):11–18.
12. Pigozzi F, Rizzo M. Sudden death in competitive athletes. *Clinics In Sports Medicine* 2008; 27(1):153–181, ix.
13. Thiene G, Basso C, Corrado D. Cardiovascular causes of sudden death. In: Silver M, Gotlieb AI, Shoen FR, editors. *Cardiovascular pathology.* New York: Churchill Livingstone, 2001; 326–374.
14. Nimura H, Bachinski LL, Sangwatanaroj S, Watkins H, Chudley AE, McKenna W, *et al.* Mutations in the gene for cardiac myosin-binding protein c and late-onset familial hypertrophic cardiomyopathy. *New England Journal of Medicine* 1998; 338(18):1248–1257.
15. St John Sutton M, Epstein J. Hypertrophic cardiomyopathy – beyond the sarcomere. *New England Journal of Medicine* 1998; 338(18):1303–1304.
16. Litovsky SH, Rose AG. Clinicopathologic heterogeneity in hypertrophic cardiomyopathy with regard to age, asymmetric septal hypertrophy, and concentric hypertrophy beyond the pediatric age group. *Archives of Pathology and Laboratory Medicine* 1998; 122(5):434–441.
17. Basso C, Thiene G, Corrado D, Buja G, Melacini P, Nava A. Hypertrophic cardiomyopathy and sudden death in the young: pathologic evidence of myocardial ischemia [see comments]. *Human Pathology* 2000; 31(8):988–998.

18. Cedrone AJ, Makaryus JN, Catanzaro JN, Ruisi P, Romich TJ, Horan P, *et al.* Sudden cardiac death in a 20-year-old male swimmer. *Journal Southern Medical* 2010; 103(5):464–466.
19. Koenig X, Dysek S, Kimbacher S, Mike AK, Cervenka R, Lukacs P, *et al.* Voltage-gated ion channel dysfunction precedes cardiomyopathy development in the dystrophic heart. *PLoS One* 2011; 6(5):e20300.
20. McNair WP, Sinagra G, Taylor MR, DiLenarda A, Ferguson DA, Salcedo EE, *et al.* SCN5A mutations associate with arrhythmic dilated cardiomyopathy and commonly localize to the voltage-sensing mechanism. *Journal of The American College Of Cardiology* 2011; 57(21):2160–2168.
21. Davis JH, Wright RK. The very sudden cardiac death syndrome – a conceptual model for pathologists. *Human Pathology* 1980; 11(2):117–121.
22. Chugh SS, Kelly KL, Titus JL. Sudden cardiac death with apparently normal heart. *Circulation* 2000; 102(6):649–654.
23. Arnestad M, Crotti L, Rognum TO, Insolia R, Pedrazzini M, Ferrandi C, *et al.* Prevalence of long-QT syndrome gene variants in sudden infant death syndrome. *Circulation* 2007; 115(3):361–367.
24. Ackerman MJ, Siu BL, Sturner WQ, Tester DJ, Valdivia CR, Makielski JC, *et al.* Postmortem molecular analysis of SCN5A defects in sudden infant death syndrome. *JAMA* 2001; 286(18):2264–2269.
25. Millat G, Kugener B, Chevalier P, Chahine M, Huang H, Malicier D, *et al.* Contribution of long-QT syndrome genetic variants in sudden infant death syndrome. *Pediatric Cardiology* 2009; 30(4):502–509.
26. Chung SK, MacCormick JM, McCulley CH, Crawford J, Eddy CA, Mitchell EA, *et al.* Long QT and Brugada syndrome gene mutations in New Zealand. *Heart Rhythm* 2007; 4(10):1306–1314.
27. Tester DJ, Ackerman MJ. Postmortem long QT syndrome genetic testing for sudden unexplained death in the young. *Journal of The American College Of Cardiology* 2007; 49(2):240–246.
28. Chopra N, Knollmann BC. Genetics of sudden cardiac death syndromes. *Current Opinion In Cardiology* 2011; 26(3):196–203.
29. Carturan E, Tester DJ, Brost BC, Basso C, Thiene G, Ackerman MJ. Postmortem genetic testing for conventional autopsy-negative sudden unexplained death: an evaluation of different DNA extraction protocols and the feasibility of mutational analysis from archival paraffin-embedded heart tissue. *American Journal of Clinical Pathology* 2008; 129(3):391–397.
30. Towbin JA. Molecular genetic basis of sudden cardiac death. *Cardiovascular Pathology* 2001; 10(6):283–295.
31. Otagiri T, Kijima K, Osawa M, Ishii K, Makita N, Matoba R, *et al.* Cardiac ion channel gene mutations in sudden infant death syndrome. *Pediatric Research* 2008; 64(5):482–487.
32. Schwartz PJ, Stramba-Badiale M, Segantini A, Austoni P, Bosi G, Giorgetti R, *et al.* Prolongation of the QT interval and the sudden infant death syndrome. *New England Journal of Medicine* 1998; 338(24):1709–1714.
33. Richards CS, Bale S, Bellissimo DB, Das S, Grody WW, Hegde MR, *et al.* ACMG recommendations for standards for interpretation and reporting of sequence variations: Revisions 2007. *Genetics in Medicine* 2008; 10(4):294–300.

34. Ely SF, Gill Jr., Fatal pulmonary thromboembolism and hereditary thrombophilias. *Journal of Forensic Sciences* 2005; 50(2):411–418.
35. Seligsohn U, Lubetsky A. Genetic susceptibility to venous thrombosis. *New England Journal of Medicine* 2001; 344(16):1222–1231.
36. McGlennen RC, Key NS. Clinical and laboratory management of the prothrombin G20210A mutation. *Archives of Pathology and Laboratory Medicine* 2002; 126(11):1319–1325.
37. Copeland AR. Sudden natural death due to pulmonary thromboembolism in the medical examiner's jurisdiction. *Medicine, Science and The Law* 1987; 27(4): 288–293.
38. Goldhaber SZ. Pulmonary embolism. *New England Journal of Medicine* 1998; 339(2):93–104.
39. Bauer KA. The thrombophilias: well-defined risk factors with uncertain therapeutic implications. *Annals of Internal Medicine* 2001; 135(5):367–373.
40. Dalen JE. Pulmonary embolism: what have we learned since Virchow? Natural history, pathophysiology, and diagnosis. *Chest* 2002; 122(4):1440–1456.
41. Goldhaber SZ, Elliott CG. Acute pulmonary embolism: part I: epidemiology, pathophysiology, and diagnosis. *Circulation* 2003; 108(22):2726–2729.
42. Tang Y, Sampson B, Pack S, Shah K, Yon Um S, Wang D, *et al.* Ethnic differences in out-of-hospital fatal pulmonary embolism. *Circulation* 2011; 123(20):2219–2225.
43. Sajantila A, Palo JU, Ojanperä I, Davis C, Budowle B. Pharmacogenetics in medicolegal context. *Forensic Science International* 2010; *203(1-3)*:44–52.
44. Pilgrim JL, Gerostamoulos D, Drummer OH. Review: Pharmacogenetic aspects of the effect of cytochrome P450 polymorphisms on serotonergic drug metabolism, response, interactions, and adverse effects. *Forensic Science, Medicine, and Pathology* 2011; 7(2):162–184.
45. Hutson S. Pharmacogenetics raises new legal questions. *Nature Medicine* 2010; 16(7):729.
46. Kaeferstein H. Forensic relevance of glucuronidation in phase-II-metabolism of alcohols and drugs. *Legal medicine (Tokyo, Japan)* 2009; 11 Suppl 1:S22–S26.
47. Gatzidou ET, Zira AN, Theocharis SE. Toxicogenomics: a pivotal piece in the puzzle of toxicological research. *Journal of Applied Toxicology* 2007; 27(4):302–309.
48. Langman LJ, Kapur BM. Toxicology: then and now. *Clinical Biochemistry* 2006; 39(5):498–510.
49. Kupiec TC, Raj V, Vu N. Pharmacogenomics for the forensic toxicologist. *Journal of Analytical Toxicology* 2006; 30(2):65–72.
50. Jin M, Gock SB, Jannetto PJ, Jentzen JM, Wong SH. Pharmacogenomics as molecular autopsy for forensic toxicology: genotyping cytochrome P450 3A4*1B and 3A5*3 for 25 fentanyl cases. *Journal of Analytical Toxicology* 2005; 29(7):590–598.
51. White R.M.Sr., Wong SH. Pharmacogenomics and its applications. *Medical Laboratory Observer (MLO)* 2005; 37(3):20–27.
52. Musshoff F, Stamer UM, Madea B. Pharmacogenetics and forensic toxicology. *Forensic Science International* 2010; 203(1–3):53–62.
53. Koski A, Sistonen J, Ojanperä I, Gergov M, Vuori E, Sajantila A. CYP2D6 and CYP2C19 genotypes and amitriptyline metabolite ratios in a series of medicolegal autopsies. *Forensic Science International* 2006; 158(2–3): 177–183.

54. Olano JP, Walker DH. Diagnosing emerging and reemerging infectious diseases: the pivotal role of the pathologist. *Archives of Pathology and Laboratory Medicine* 2011; 135(1):83–91.
55. Leggieri N, Rida A, François P, Schrenzel J. Molecular diagnosis of bloodstream infections: planning to (physically) reach the bedside. *Current Opinion In Infectious Diseases* 2010; 23(4):311–319.
56. Cathomas G. Molecular diagnostic in infectious disease pathology: an update. *Annales de Pathologie* 2009; 29 Spec No 1:S19–S21.
57. Bravo LT, Procop GW. Recent advances in diagnostic microbiology. *Seminars In Hematology* 2009; 46(3):248–258.
58. Sintchenko V, Gallego B. Laboratory-guided detection of disease outbreaks: three generations of surveillance systems. *Archives of Pathology and Laboratory Medicine* 2009; 133(6):916–925.
59. Muldrew KL. Molecular diagnostics of infectious diseases. *Current Opinion In Pediatrics* 2009; 21(1):102–111.
60. Ratcliff RM, Chang G, Kok T, Sloots TP. Molecular diagnosis of medical viruses. *Current Issues in Molecular Biology* 2007; 9(2):87–102.
61. Millar BC, Xu J, Moore JE. Molecular diagnostics of medically important bacterial infections. *Current Issues in Molecular Biology* 2007; 9(1):21–39.
62. Craven RR, Hamilton LR, Deetz CO, Dunne WM. Postmortem analysis and the role of the clinical microbiology laboratory. *Acad Forensic Pathol* 2011; 1(1):32–38.
63. Wagar EA, Mitchell MJ, Carroll KC, Beavis KG, Petti CA, Schlaberg R, *et al.* A review of sentinel laboratory performance: identification and notification of bioterrorism agents. *Archives of Pathology and Laboratory Medicine* 2010; 134(10):1490–1503.
64. Ginocchio CC, Zhang F, Manji R, Arora S, Bornfreund M, Falk L, *et al.* Evaluation of multiple test methods for the detection of the novel 2009 influenza A (H1N1) during the New York City outbreak. *Journal of Clinical Virology* 2009; 45(3): 191–195.
65. Bucher D, Tumpey T, Lowen A, Gill J, Shaw M, Matthews J, *et al.* 2009 H1N1 swine flu: the 2010 perspective. *Annals of The New York Academy Of Sciences* 2010; 1205 Suppl 1:E10–E20.
66. Faix DJ, Sherman SS, Waterman SH. Rapid-test sensitivity for novel swine-origin influenza A (H1N1) virus in humans. *New England Journal of Medicine* 2009; 361(7):728–729.
67. Brogdon BG. *Forensic radiology.* Boca Raton: CRC Press, 1998.
68. National Association of Medical, Examiners. *Forensic Autopsy Performance Standards.* [updated 2006 Aug; cited 2011 Oct 28. Available from: http://thename.org/index.php?option=com_docman&task=cat_view&gid=45&Itemid=26.
69. Dolinak D, Matshes E, Lew E. *Forensic pathology: principles and practice.* San Diego: Elsevier Academic Press: 2005.
70. http://www.lodox.com, cited 2011 Oct 28.
71. Grabherr S, Djonov V, Yen K, Thali MJ, Dirnhofer R. Postmortem angiography: review of former and current methods. *American Journal of Roentgenology* 2007; 188(3):832–838.
72. Levy AD, Harcke HT. *Essentials of forensic imaging: a text-atlas.* Boca Raton: CRC Press; 2011.

73. Thali MJ, Dirnhofer R, Vock P. *The virtopsy approach: 3D optical and radiological scanning and reconstruction in forensic medicine*. Boca Raton: CRC Press; 2009.
74. Nolte KB, Mlady G, Zumwalt RE, Cushnyr B, Paul ID, Wiest PW. Postmortem x-ray computed tomography and forensic autopsy: a review of the utility, the challenges and the future implications. *Acad Forensic Pathol* 2011; 1(1):40–51.
75. O'Donnell C. An image of sudden death: utility of routine post-mortem computed tomography scanning in medicolegal autopsy practice. *Diagnostic Histopathology* 2010; 16(12):552–555.
76. O'Donnell C, Rotman A, Collett S, Woodford N. Current status of routine post-mortem CT in Melbourne, Australia. *Forensic Science, Medicine, and Pathology* 2007; 3(3):226–232.
77. DiMaio VJ, DiMaio D. *Forensic pathology*. 2nd ed. Boca Raton: CRC Press;2001.
78. Mazuchowski EL, Berran PJ, Harcke HT. Use of multidetector computed tomography in the evaluation of gunshot wounds. *Proceedings of the American Academy of Forensic Sciences* 2011 Feb 21–26; Chicago (IL). 2011:302 (G95).
79. Harcke HT, Crawley G, Mabry R, Mazuchowski E. Placement of tibial intraosseous infusion devices. *Military Medicine* 2011; 176(7):824–827.
80. Harcke HT, Mazuchowski EL, Berran PJ. Use of multidetector computed tomography in the medicolegal investigation of human remains in a natural disaster. *Proceedings of the American Academy of Forensic Sciences*, 2011 Feb 21–26; Chicago (IL). 2011:309(G106).

5

The places we will go: paths forward in forensic anthropology

Dawnie Wolfe Steadman

Forensic Anthropology Center, University of Tennessee, Knoxville, Tennessee, USA

5.1 Introduction

Anthropology is the scientific study of the origin and physical and cultural variation of humans across space and time. Since this is a broad topic, anthropology is traditionally divided into four sub-disciplines that examine different facets of human evolution and cultural change:

- Socio-cultural anthropology studies cultural variation among living populations.
- Archaeology seeks to understand cultural variation of the past through the physical remnants that people have left behind.
- Linguistic anthropology examines the variation of languages and language transmission in the past and present.
- Biological (physical) anthropology is the study of the evolution of humans and non-human primates and the cultural, biological and genetic variation among past and present human populations.

Forensic anthropology is one component of biological anthropology, in that it applies our understanding of modern human biological variation to address questions of medicolegal significance. Traditionally, these questions concern personal identity – scientifically matching decomposed, burnt or fragmentary remains to a missing person. In addition, forensic anthropologists are asked to interpret skeletal injuries, estimate the post-mortem interval and recover human remains. New technologies and global events have expanded the roles of forensic

Forensic Science: Current Issues, Future Directions, First Edition. Edited by Douglas H. Ubelaker.

anthropologists, as they are increasingly called to address medicolegal questions concerning living humans, to assist in small and large-scale international investigations of human rights abuses and to contribute to mass fatality mitigation.

The strong presence of forensic anthropology in forensic scientific contexts is somewhat startling, given the short history of the professional discipline, but its rapid growth also poses many challenges for the future. The purpose of this chapter is to provide a brief history of the discipline, to present a short overview of the methods used for identification, to illustrate some of the key methodological, theoretical and training challenges that forensic anthropology currently faces, and to discuss international perspectives related to these issues. It will also highlight some of the current research efforts that will not only address these problems but define the future directions of the discipline.

5.2 The history of forensic anthropology

Like many of the forensic science disciplines, the historical roots of forensic anthropology in America reach back in time to isolated cases near the turn of the 20th century. Biological anthropology was not well defined at that time, but anatomists and anthropologists who were interested in skeletal and dental tissues and human variation were key figures in this emerging field.

One of the most famous historical cases of skeletal and dental identification involved the demise of Dr. Parkman, a Harvard professor, at the hands of an institutional colleague in 1849. Despite the best efforts of the perpetrator to eliminate Dr. Parkman's remains, including dismemberment, burning and chemical processing, anatomists Oliver Wendell Holmes and Jeffries Wyman were able to determine that the remains were of a single human individual, while a dentist recognized the recovered dentures as those he had personally made for Dr. Parkman.

Nearly 50 years later, the death of Louisa Luetgert at the hands of her sausage-maker husband in Chicago ushered in the first involvement of a physical anthropologist in a criminal trial. Louisa's remains had been dismembered and placed in a sausage vat to decompose. Anthropologist George Dorsey identified the remains and testified as to his findings. By this time, biological anthropology was establishing itself as the primary field of research into morphological (physical) variation among humans (compared to biochemistry or blood types). Further, the availability of skeletal remains in museums enabled researchers to compare skeletal variation of past and present peoples. The stage was set for physical anthropologists to expand into the forensic arena.

A key element to the early development of the discipline was an intimate link between research and practice. At the turn of the century, biological anthropologist Thomas Dwight exemplified this relationship in his studies of sex, age and stature variation of several skeletal elements [1–4]. During this nascent period, law enforcement would, from time to time, bring bones to biological anthropologists working at museums or universities. In response, Wilton Krogman published the landmark *A Guide to the Identification of Human Skeletal Material*, in 1939, which detailed the scope of services that a well-trained biological anthropologist could provide when presented with skeletonized remains. This article helped lay the foundation for what medicolegal authorities could expect from forensic anthropologists. The typical question was whether the remains were human; however, as the anthropologists began to better understand skeletal variation, they could provide more information to the medicolegal community than just species identification.

The 1970s ushered in the initial professionalization of the discipline, marked by the founding of the Physical Anthropology Section in the AAFS in 1977. The title of the section related to the principal way the early founders defined themselves – most were university professors of physical anthropology, who studied many aspects of skeletal biology, while few practitioners focused exclusively on forensic anthropology.

Today, there are more professionals whose careers are dedicated entirely to forensic anthropology, yet the great breadth and depth of the field remains strengthened by wide-ranging perspectives in skeletal biology. Many of the AAFS Physical Anthropology members conduct research in bioarchaeology, primatology, human biology, histology, bone chemistry, pathology, genetics and a number of other rich areas of inquiry. In addition, the academic roots of the discipline allow forensic anthropologists to contribute actively to the understanding of skeletal variation among contemporary populations around the world.

5.3 Forensic anthropological practice

Critical to contextualizing the current challenges facing forensic anthropology and the future directions of the discipline is a general understanding of the breadth of the field. The general practice of forensic anthropology consists of four broad categories of analyses: biological profile, forensic archaeology, trauma analysis and taphonomy.

5.3.1 Biological profile

The essential foundation of forensic anthropology is a type of gap analysis – scientific means of personal identification where other modalities fail. Thus, forensic anthropologists are called in to cases in which facial, fingerprint and DNA identification cannot be rendered efficiently or at all. The biological profile is a tool to narrow the list of potential missing persons such that dental, radiographic and other ante-mortem sources of a few missing people can be compared to the post-mortem evidence. A traditional biological profile consists of age, sex, ancestry, stature and pathology. The methods used to estimate each parameter are developed from studies of reference samples containing individuals of known age, sex, ancestry and stature.

Age, sex, stature and ancestry are estimated using two approaches – observations of morphological characteristics of particular features of the skeleton and measurements of skeletal elements. For instance, the preferred method of sex estimation of the skeleton is observation of the *os coxae* (hip bones). The female pelvis broadens during sexual maturity, relative to the straight, narrow pelvis of males, as an adaptation to childbirth. Muscle attachments of the skull can also be useful, since the neck muscles that attach to the head tend to have greater mass in males than those of females. Discriminant function analysis using measurements of the skull, femur and other bones is also informative for sex estimation [5,6].

The methods utilized to assess skeletal age depend upon whether the individual is still growing or is skeletally an adult, in which case all of the bones and teeth have completed their development. The estimation of juvenile age involves comparing the growth pattern of the dentition and skeleton to that of appropriate reference samples of children of known age and sex. Dental development, dental eruption, bone measurements and fusion of the ends of the long bones to the shaft (epiphyseal fusion) are employed, depending on the relative age of the individual. For instance, deciduous dental development and bone measurements are the best age indicators for infants, while permanent tooth development and eruption are key indicators of age in adolescence, and epiphyseal fusion begins in late adolescence after dental development is complete. Fusion of all of the epiphyses to their shafts signifies the end of the growth period, so forensic anthropologists then turn to adult skeletal age indicators.

While juvenile aging techniques are based on skeletal and dental development, adult aging techniques focus on skeletal indicators that degenerate in a regular, age-progressive manner. There is little scientific consensus as to which aging technique is best, so forensic anthropologists are encouraged to use as many standard indicators as possible and to determine appropriate error rates. Because skeletal and dental development is tightly controlled genetically, juvenile age estimates have errors of only a few weeks (fetuses), months (infants) or years (adolescents), but adult aging techniques carry errors of at least a decade. This is because there is great variability in how an adult skeleton ages, and this is further compounded by sex and population variation.

The majority of adult aging methods are based on morphological observations, including the pubic symphysis [7–10], the auricular surface of the ilium [11], ectocranial cranial suture closure [12] and the sternal end of the ribs (particularly the right fourth rib) [13–15]. Quantitative aging methods are also gaining scientific favor, including the Lamendin method for anterior teeth [16,17], histological methods of aging [18,19], and tooth cementum annulation, which involves counting the layers of cementum deposited annually on a tooth root [20,21]. One problem with all these methods is that the errors may not be known, or are too narrow or broad to be realistic or useful.

Another method, Transition Analysis [22], combines morphological data from the pubic symphysis, auricular surface and cranial sutures but applies a Bayesian approach to calculate a maximum likelihood estimate, as well as 95% confidence intervals that other methods lack. However, Transition Analysis was originally developed for evaluating large cemetery samples rather than individuals, and it requires more validation tests to evaluate accuracy for individual skeletons [23].

'Ancestry' refers to the population history of an individual, which is a biological parameter, yet forensic anthropologists are often mistakenly asked to assess race – a construct denoting how a person aligns oneself socially. Thus, one may have African American ancestry, but self-identify as European American (white). Forensic anthropologists can assess ancestry by using morphological and metric analyses of the skull and certain long bones. A statistical software program, *Fordisc* 3.0 [5], employs discriminant function analysis to classify unknown individuals into groups of individuals of known sex and ancestry. *Fordisc* contains samples from the Forensic Data Bank (FDB), a collection of skeletal measurements and other data from contemporary skeletons, Howell's worldwide database and documented data from early 20th century medical school donations. An advantage of *Fordisc* over morphological approaches is that it generates a number of error values for the classifications. However, whether ancestry estimation should be included in the biological profile at all is passionately debated in the United States, as discussed below.

Stature is traditionally estimated by measuring the lengths of long bones and applying a regression formula derived from individuals of known stature from the same sex and population, if possible. The regression method of stature estimation dates back to the late 19th century [24], but more recent methods, or the revival of earlier methods, take a slightly different tact. The anatomical method, first described by Fully and Pineau in 1960 [25], combines measurements of the skull, vertebrae, leg and ankle bones with a correction factor to take into account height contributed by soft tissue in the living person. While forensic anthropologists were initially wary of this method (it is more time-consuming, some of the measurements lack clear definition, the soft tissue correction factor is ill-defined and it cannot be used for incomplete skeletal remains), a recent resurgence in attention to the method has proved productive [26,27]. Other statistical approaches also have merit and deserve greater attention, despite the computational workload [28]. Post-mortem stature estimates are compared to ante-mortem documentation of living stature, such as driver's licenses or medical records, though the accuracy of these sources can be questionable [29].

It is unlikely that the estimation of age, sex, ancestry and stature is sufficient to eliminate all but one individual from a pool of missing persons. Therefore, forensic anthropologists examine the skeleton for features that may be unique. They compare dental and skeletal pathological conditions and anomalies to the dental or medical records of missing persons. Some conditions are non-specific, because they are fairly common (such as osteoarthritis), but certain types of healed fractures, infectious diseases, cancers or congenital malformations may leave unique skeletal changes [30,31]. Orthopedic devises and other surgical hardware are very useful for identification [32,33]. In addition, the location and type of dental modifications (e.g. a filling, crown or root canal) can be compared to dental records of missing individuals [34,35].

Forensic anthropologists work closely with forensic radiologists and odontologists to assess the uniqueness of pathological or anomalous conditions, yet questions still arise as to how to competently determine if a 'positive' identification has been made. These challenges are addressed below.

5.3.2 Forensic archaeology

One of the first tasks in an investigation may be to locate the remains of missing persons. In the USA, most forensic anthropologists have extensive training in archaeological techniques and crime scene processing, so that they can work side by side with law enforcement and criminal science investigators to locate and recover buried or scattered remains. Searches of large areas benefit from advanced technological tools such as ground penetrating radar (GPR), and other remote sensing devices that detect subsurface disturbances which may indicate a burial. Once remains are found, either under the ground, on the surface or both, standard excavation and mapping techniques are utilized to document the location of the burial and the spatial relationships between bones and any related evidence (e.g. bullets, clothing, ligatures or other bodies).

Many forensic anthropologists feel that archaeological search and recovery is one of the most under-utilized services [36,37]. This is partly a historical problem, in that early forensic anthropologists preferred to stay in the laboratory and have remains brought to them. However, contemporary forensic anthropologists recognize that the burial context is crucial to understanding the post-mortem (taphonomic) processes that may have altered the remains, how the grave was dug (e.g. hand tools or machinery), the age of the grave (and hence the post-mortem interval) and associated evidence [38,39]. Therefore, most forensic anthropologists today prefer to attend the scene and assist in the retrieval of the remains.

5.3.3 Trauma analysis

Forensic anthropologists are increasingly asked to consult on cases involving skeletal trauma, even when the identity of the individual is known. There are five categories of skeletal trauma – blunt force, gunshot, explosive, sharp force and thermal (burning). Each type of trauma leaves specific markers, though fractures are an underlying characteristic of most all types. While forensic anthropologists have made inroads in developing standardized terminology for trauma [40–42], there is tremendous variation in how bone responds to force. Contributing variables include the amount of force, the distribution of force applied to the body/bone, the velocity of the force, the size and density of the bone, and the nature of the surrounding soft tissue.

Trauma research remains an exciting part of forensic anthropology, and the current trend is to move from actualistic studies (attempts to simulate the conditions that caused a particular fracture) to experimental studies that are developed to understand the underlying biomechanical principles of skeletal responses to trauma. The most common type of study is on fresh, de-fleshed bone using human remains or non-human models [43–45]. However, new research that utilizes whole, fleshed limbs or bodies is needed to better understand the relationship between soft and osseous tissue responses to force in controlled (experimental) situations. Collaboration between anthropologists and biomechanical engineers are proving especially productive in these endeavors.

5.3.4 Forensic taphonomy and the post-mortem interval (PMI)

Taphonomy is the study of what happens to the body after death. While paleontologists are interested in how remains become fossilized over millennia, forensic anthropologists are tasked with understanding what variables (e.g. water, insects, and animals) affect decomposition and distribution of the body during the first few years of decomposition. The post-mortem interval is very difficult to estimate, even when death has occurred only a few hours or days before the remains were found. Forensic anthropologists are particularly challenged, because they are asked to assess PMI when remains are in advanced stages of decomposition, buried, scattered or completely skeletonized. Thus, forensic anthropologists provide a *post hoc* assessment of what taphonomic factors have been at work and how quickly these are likely to have produced the observed changes in the body and the surrounding environment.

Recognizing the need for a better understanding of the taphonomic processes of decomposition, Dr. William Bass developed the Anthropological Research Facility at the University of Tennessee, Knoxville in the 1980s. For the past three decades, scores of researchers have tracked decomposition of donated human cadavers in a number of different microenvironments. A generalized progression of decomposition has been documented (fresh, putrefaction, bloat, decay and skeletal), but each stage contains a number of events (e.g. skin slippage, marbling) [46]. Temperature and insects are the interrelated prime movers of decomposition. Low temperatures, or other factors that make the body inaccessible to insects (e.g. burial), will slow decomposition [47].

Decomposition rates developed in the temperate environment of Knoxville are inapplicable to other environments, such as the arid Southwest United States or colder northern climes [48]. New human decomposition research facilities in Texas, Boston and North Carolina will allow critical cross-environmental research of the post-mortem interval. Decomposition studies have historically relied principally on observations rather than on quantitative methods. However, studies that quantify the exchange of materials at the body/soil interface have proven useful [49–54]. It is important to note that taphonomic research not only helps assess the post-mortem interval, but also provides empirical evidence to assist in discriminating perimortem and post-mortem damage to the body and skeleton.

5.4 Current challenges and research in forensic anthropology

Forensic anthropologists currently face many key methodological, procedural and theoretical issues. Some relate to long-standing tensions within anthropology, while others are more recent and reflect the rapid growth and expansion of the discipline. These 'growing pains' are perhaps the healthiest aspects of the field, for careful navigation and consideration of these problems lead to new avenues of research.

Most of these issues are inextricably linked, so the sections below should not be viewed as mutually exclusive. In addition, research is presented that demonstrates some of the important inroads that forensic anthropologists are making to address these challenges.

5.4.1 Validation studies and best practices

One of the most important ramifications of the recent changes in the rules of evidence is greater attention to the scientific merit of traditional anthropological methods. This, in turn, has led to a tremendous amount of research during the past decade, especially in the development of quantitative methods. Prior to *Daubert* and the Federal Rules of Evidence, a forensic anthropologist was able to base an opinion on his or her experience and informal agreement within the discipline concerning which methods were acceptable. However, experiential testimony is being replaced by standardized methods and realistic error rates that scientifically justify the opinion.

The intent of the *Daubert* guidelines is to ensure that experts follow the scientific method – whether the techniques have been tested, if error rates associated with the technique are known, if the technique has been subject to peer review, and if the technique is generally accepted within the discipline. Some forensic anthropological methods lack error rates due either to the nature of the technique (observations rather than quantified data) or because the original publication of a technique lacked confidence or prediction intervals.

Validation studies are used in all areas of science to quantify the performance of a particular technique by assessing objectively the accuracy (obtaining the correct answer) and reliability (obtaining the same answer across multiple trials or observers) and quantifying the error of the technique. In forensic anthropology, validation studies strive to utilize large samples of contemporary skeletons of known age, sex and population affiliation.

The process of validation often leads to refinements that improve the accuracy and reliability of the original method. For instance, Raxter and colleagues [26,27] found that the standard errors of the original Fully stature method (anatomical method) were too narrow, and that some measurement techniques, particularly of the vertebral bodies, were difficult to replicate. Further, the regression formulae proposed by Fully did not incorporate an age component that recognizes the loss of adult stature with advancing age. Raxter et al. revised the anatomical method and clarified some of the methodological and theoretical issues utilizing a large component of the Terry collection. A recent validation study demonstrates that the errors associated with the revised method are relatively small [55].

Validation studies can also help practitioners to answer a crucial question – which methods available in the literature are the best? Given the breadth of tasks that forensic anthropologists may be asked to carry out in a single case (e.g. recovery, age, sex, stature, trauma, pathology, PMI), developing best practices within the field is daunting.

In 2008, forensic anthropologists joined the network of Scientific Working Groups (SWG) and developed SWGANTH. Co-sponsored by the Federal Bureau of Investigation and the Central Identification Laboratory in Hawaii, SWGANTH is currently comprised of 15 committees tasked with establishing best practices guidelines that range from methods (such as age and ancestry estimation) to training, ethics and quality control. While these are true guidelines, in that they are reached by consensus and lack any means of enforcing their implementation among individual practitioners or labs, it is the first time that forensic anthropologists have had protocols for their practice that can be referenced in court. For instance, the SWGANTH age estimation document provides the scientific criteria by which to judge an aging method, and it provides biological and/or statistical justification for the selection of methods to be applied.

While age, sex, stature and ancestry estimations are amenable to validation studies, computing error rates or standardizing methods are more difficult for other important aspects of forensic anthropological casework. Trauma analysis, for instance, utilizes the principles of biomechanics to reconstruct how a bone was insulted, yet there is no statistical approach available to assess interpretive errors. For example, as a bullet impacts a cranial bone, it creates a cone-shaped defect that is beveled in the direction the bullet was traveling. Thus, a forensic anthropologist can distinguish an entrance wound from an exit wound and help to reconstruct the trajectory of the bullet through the skull. However, the caliber of the bullet, the angle of entry, the thickness and preservation of the bone, and a number of other factors, can significantly affect this seemingly straightforward interpretation. Rather than risk over-interpretation, the current SWGANTH best practices advocate that forensic anthropologists describe the timing (ante-mortem, perimortem or post-mortem) and category (e.g. gunshot wound) of the injury. The emphasis on description and basic categorization is also recommended for other, non- or semi-standardized components of forensic anthropological casework, including pathology and the post-mortem interval.

5.4.2 The certainty of 'positive' identification

Personal identification may be ascertained using a variety of techniques that depend upon the amount of remains recovered and available and the quality and quantity of the ante-mortem records. For instance, forensic anthropologists may use radiographic evidence to compare bony trabecular patterns, surgical hardware, anatomical variants, pathological conditions and dentition to establish identity [32,33,56]. While SWGANTH best practices state that surgical implants (especially those with unique serial numbers) and radiographic comparisons of trabecular patterns, pathological conditions and dental features can lead directly to positive identification, a fundamental understanding of the probative strength of different types of anthropological evidence for identification is lacking. Is ancestry estimation as contributory to identification as age? Does stature hold any probative value if the individual was of average height? Is a healed leg fracture just as useful for individualizing as a dental filling? What about two dental fillings? Are multiple rib fractures more or less unique than a single leg fracture? Just how many points of similarity are necessary to document before identification is rendered?

The last question is not unique to forensic anthropology. Latent print analysis is undergoing a similar debate concerning whether the number of minutiae required to certify a 'match' can, or should, be standardized. Yet how can forensic anthropologists justify a 'positive' identification, which implies zero uncertainty, if the level of confidence that the characteristics observed are individualizing is not quantified? Should forensic anthropology adopt the more conservative approach, whereby their analytical outcomes are limited to 'exclusion,' 'inclusion' or 'undetermined', and rely fully on DNA or other modalities for positive identification?

Many forensic anthropological identifications are based on non-technical sources of information, such as healed fractures. When bone fractures heal, they leave a callus that can be seen on radiographs or grossly for months or years. If the location and degree of healing is consistent with ante-mortem records of a missing person, then an identification can be made. However, forensic anthropologists are now asking whether this evidence is sufficient. Consider the following simplified example. A biological profile of human remains found in an abandoned building indicates that the remains are of a 25–35 year old African American female who stood between 62 and 68 inches tall. This biological profile reduces the number of missing persons in the local pool from 100 down to ten. The forensic anthropologist also observes a healed fracture on the left humerus (upper arm bone). Only one of these ten missing individual exhibits a healed humeral fracture. The location and degree of healing of the fracture is consistent, so the forensic anthropologist reports to the medical examiner that a positive identification can been made.

Such a process is very straightforward and well accepted in the discipline. The forensic anthropologist assessed a high probative value to the humeral fracture because it was unique in the 100 individuals available for comparison. However, the critical question should be addressed is how often a humeral fracture occurs in the 'population at large' – a contemporary cohort of individuals. In 2006, Komar and Lathrop published a study to assess the frequency of fractures among contemporary collections of human skeletons in Tennessee and New Mexico [57]. The authors found that not only were fractures a relatively common occurrence, many individuals exhibited two or more fractures. Moreover, they established that three individuals who shared the same biological profile also exhibited the same number and skeletal distribution of healed fractures and orthopedic repairs. Thus, the probability that forensic anthropologists might find multiple individuals with similar fractures may be higher than initially thought.

Adopting statistical models from molecular genetics, forensic anthropologists are beginning to assess the probative value of different lines of skeletal evidence quantitatively rather than qualitatively. Nuclear DNA is a highly effective identification tool, because of the genetic differences between targeted alleles, the independence of inheritance of those alleles and the large databases of contemporary human DNA that can be consulted to determine how often a particular sequence occurs in the 'population at large'. Likelihood ratios can then be calculated to quantify this frequency.

Most osteological data can be treated independently but, until recently, forensic anthropologists lacked any large databases to represent the 'population at large' for these data. No single database covers the diversity of anthropological evidence (age, sex, ancestry, pathology and stature), but *post hoc* databases can be compiled from a variety of data, including national health and crime databases, large skeletal collections, missing persons databases and the census [58,59]. For instance, *Odontosearch* is a powerful software database designed to assess the likelihood of finding a dental match made between a missing person and unknown remains among over 40,000 contemporary individuals in the database, the "population at large" [60,61]. A likelihood ratio is calculated that determines how often one would expect to find this particular dental pattern in the population at large. The strength of an identification increases with the size of the likelihood ratio.

While some forensic anthropologists have attempted to quantify the probative value of the biological profile, some researchers have focused on quantifying the likelihood of a correct identification using a single modality. For instance, sinuses are lobulated air cavities above the orbits in the frontal bone of the skull and in the nasal, maxillary and mastoid areas. The structure of the sinuses varies between individuals and they have been utilized for personal identifications [62], yet the assumption that they are unique had not been formally tested.

Christensen [63] developed a standardized method for superimposing tracings of frontal sinus outlines of nearly 600 individuals. She found that the likelihood ratios are very large, especially among complex sinuses, indicating that frontal sinuses are indeed useful for identification. While Christensen used complex geometric modeling in her study, a simpler method of aligning the frontal sinuses on post-mortem and ante-mortem radiographs has achieved high success [64]. A similar likelihood approach to quantifying the certainty of ante-mortem and post-mortem matches of anatomical structures has also been applied to the temporal bone [65,66] and may be practical for other anatomical structures.

5.4.3 To estimate or not to estimate ancestry

Ancestry estimation is perhaps the most controversial of the biological profile parameters due to a number of competing perspectives from within and outside anthropology. One is a historical perspective that sees ancestry estimation as a typological endeavor and hence an unwanted atavistic element from our anthropological past. Another perspective perceives an inconsistency between modern anthropology's mantra that races do not exist, and forensic anthropological use of ancestry in a biological profile [67]. Yet another view is that categories of modern humans by racial type/ancestral population are irrelevant given the immense amount of population variation seen today [68]. Others, however, assert that human populations do assort by geographic groups, and they advocate that such divisions should be used to identify unknown individuals [69–73].

There are other perspectives as well, both in favor of and against ancestry estimation in forensic cases, but the tension among US anthropologists is very high on this subject. Perhaps this is because anthropologists often talk past each other, or each side claims fundamental misunderstandings. For instance, several anthropologists have presented or published papers showing that *Fordisc* 3.0 is unable to correctly classify certain individuals of known ancestry, and they therefore conclude that race/ancestry classification is untenable and destructive [74,75]. In response, various authors have demonstrated that *Fordisc* is, indeed, incorrect when the measurements are taken improperly, or the appropriate reference sample is not represented in *Fordisc* [70,76]. The problem, they argue, is not that population classification is inherently faulty, but rather that the user of *Fordisc* did not understand the capabilities and limitations of this tool.

What is sadly lacking in this debate is a healthy discussion of what modern human variation means, and how forensic anthropologists can more meaningfully contribute to its understanding. Precious few articles [70] have contributed new frameworks in which to address modern variation, and none further seek to integrate genetic and morphological models of variation that can help in understanding some fundamentally different perspectives. Thus, while anthropological geneticists see that approximately 90% of human genetic material is the same across populations, forensic anthropologists are trained to tease out where that 10% of variation occurs and utilize it for the purposes of individualization. Each perspective makes a meaningful, but not exclusive, contribution to our understanding of modern human variation.

The skull provides the most information regarding ancestry, and forensic anthropologists apply both metric and non-metric (morphological) methods to assess ancestry. Of these, morphological techniques are under greatest scrutiny given their subjectivity (e.g. what constitutes a 'wide' nasal aperture) and their inability to assess statistical errors using a simple trait list approach. Some forensic anthropologists rely predominantly upon their extensive experience in observing morphological variation of the skeleton, rather than metric methods [77].

However, recent research has demonstrated that a commonly used trait list [78] based on very small samples is not nearly as discriminating as previously thought when applied to different samples [79]. Traits scored as binary variables (e.g. presence or absence, wide or narrow) are being replaced by ordinal scaling methods that encompass the range of expression of each trait [79,80]. In addition, the use of logistic regression of ordinal variables permits error estimates that have been previously lacking in morphological methods. This new statistical approach has also been introduced for morphological sex estimation, and it has tremendous potential to give morphological analyses greater scientific weight in the eyes of the courts [81].

5.4.4 Technology in forensic anthropology

Within the past decade, forensic anthropologists have embraced new or adopted existing technologies that assist with locating and identifying human remains. For instance, while subsurface detection techniques such as ground penetrating radar (GPR) technology has been used by gas, electric and construction companies for years, forensic archaeologists have initiated novel research to examine the limitations of the method for different forensic situations (e.g. burials under concrete, variations of burial depth, collapsed buildings) [82,83].

Other technologies can help determine the forensic significance of recovered remains. For instance, scanning electron microscope/energy dispersive spectroscopy (SEM/EDS) identifies the type and quantity of specific elements in a material and helps to determine whether remains are biological (this is especially useful for burned materials). Protein radioimmunological assay (pRIA) use species-specific proteins to determine if recovered remains are human. In addition, radiocarbon dating can assess whether an individual was alive before or after artificial carbon was introduced into the atmosphere with atomic bomb testing and usage in the mid-20th century [84,85].

Non-DNA identification-related technology has expanded in recent years, but its utility is not widespread, given the specialized training and equipment required. For instance, histological aging methods tend to have smaller errors than morphological aging techniques, yet require training and experience to perform competently. Similarly, computer imaging and facial superimposition techniques are useful to compare skeletal features with photographs of missing persons, but they involve expensive equipment and extensive training to ensure that the comparative features are scaled, measured and superimposed correctly. Three-dimensional morphometric techniques capture shape variables better than two-dimensional measurements with calipers and can be accomplished using CT scans or a 3-D digitizer [86,87].

While these technologies are often found in forensic anthropology laboratories, other equipment that proves useful in forensic anthropology is rarely a component of a forensic anthropology laboratory. Equipment such as mass spectrometry, scanning electron microscopes or CT scans are more likely to be found in chemistry, physics or biology departments, or in hospitals. Professional collaborations with other specialists can help to make the technology available to forensic anthropologists.

5.4.5 Training, accreditation and certification

Forensic anthropology is very popular in US colleges and graduate programs. Judicious coverage by television series, movies, books and televised criminal proceedings has made the field a well-recognized entity in popular culture. While the TV shows are disheartening, in that they lack realistic methods and promote pseudoscience, one positive aspect is that students enroll in classes where they can learn the real science and issues of the field. University administrators have capitalized on this interest, and forensic anthropology graduate programs have exploded onto college campuses in the past two decades. An unfortunate outcome of this interest is a job market that is flooded by newly minted forensic anthropologists, who compete for very few university positions and even fewer non-academic job opportunities at medical examiner offices or crime labs.

While some anthropologists practice forensic anthropology with a master's degree, a PhD is highly encouraged and may be required by upcoming legislation. A SWGANTH group on education and training is currently formalizing guidelines for graduate and undergraduate courses and training experiences to prepare for a career in forensic anthropology. Undergraduate students who wish to attend a graduate program in forensic anthropology should have a strong background in the natural sciences. Given the technological expansions of the field, courses in genetics, chemistry, biology and anatomy are crucial. A good foundation in statistics is also required in order to critically evaluate existing studies and conduct independent research. Forensic anthropology is first and foremost an anthropological pursuit, and classes in all four fields of anthropology provide both practical and theoretical skills. Archaeology is especially important, given that forensic anthropologists routinely assist in the search and recovery of human remains in outdoor settings.

A burgeoning pedagogical dichotomy exists within US forensic anthropology graduate programs. Some focus exclusively on forensic anthropology, particularly at the master's level, while others train students more broadly in anthropology but emphasize forensic anthropology. Specialized programs aim to provide students with intense, hands-on opportunities, yet may de-emphasize theoretical issues. Alternatively, broad training can provide students with a well-rounded education in anthropology, critical thinking and theory but, depending upon the program, may not provide students with the crucial practical experience (casework) they need to work independently upon graduation and attain professional certification.

Many fields of forensic science certify individuals who demonstrate high proficiency in the field. The American Board of Forensic Anthropology (ABFA) is currently the only certifying body for forensic anthropology in the world. Through rigorous proficiency exams and extensive reviews of training and casework, the ABFA certifies that its diplomates have achieved the highest levels of competency in the field. Certified diplomates must also undergo re-certification every three years. As of 2011, the ABFA has certified a total of 88 diplomates, though about a quarter of these no longer participate in casework.

The application requirements for ABFA certification include a PhD degree, three years of post-graduate practice and the submission of case reports for review. However, the current requirement of a PhD excludes individuals practicing with a master's degree to apply for certification. While other organizations promote standards for master's-level practitioners (e.g. Society of Forensic Anthropologists, International Association of Identification), certification is not offered. To address this issue, the ABFA is currently developing a certification process for 'specialists' in forensic anthropology who have not (yet) attained a PhD, but who can demonstrate that they can achieve the high practical standards expected for certification. Non-U.S. anthropologists can now also apply for ABFA certification.

Although it is extremely important for individual practitioners to demonstrate that they have achieved the highest standards of proficiency in the field, it is equally crucial to ensure that the laboratories in which forensic anthropologists work achieve the highest standards for receiving, storing, analyzing and handling forensic evidence, including osteological evidence. Laboratory accreditation is expected for crime labs, including DNA laboratories, which handle physical evidence. Forensic anthropology laboratories handle a variety of evidence, including clothing, bones, soft tissue, insects, hair and blood, so proficiency training on evidence handling and storage is especially important.

However, the two major certifying bodies in forensic science, the American Society of Crime Laboratory Directors (ASCLAD-LAB) and the National Association of Medical Examiners (NAME) are not currently prepared to evaluate forensic anthropology laboratories. Accreditation is daunting, because most forensic anthropology laboratories are on university property and have limited funding, infrastructure and space, all of which are crucial for accreditation. The only forensic anthropology laboratory to achieve ASCLAD accreditation to date is the Joint POW-Accounting Command – Central Identification Laboratory (JPAC-CIL) in Hawaii, a government-funded laboratory. Accredited in 2004 for trace evidence analysis, their process has highlighted some issues specific to forensic anthropology and has paved the way for a handful of university-based laboratories to engage in the preliminary stages of accreditation. The accreditation process requires not only a significant amount of dedicated time for the laboratory director, but also long-term commitments from the institution to achieve and maintain accreditation standards. Despite the institutional costs, forensic anthropology laboratory accreditation will be invaluable for demonstrating the scientific integrity of the field.

5.4.6 Going global: disaster victim identification and human rights investigations

Search and recovery, trauma assessment and estimation of the biological profile and PMI are central to 'local' cases – that is, cases in which forensic anthropologists are consulted by local medical examiner, coroner or law enforcement authorities. However, forensic anthropologists take on additional roles in extra-local situations, such as mass disasters and human rights investigations.

A mass fatality incident is typically defined as one more fatality than local authorities can handle. This is a relative number, and reflects the infrastructure capabilities of local disaster personnel more than the actual scale of the event. Recent mass fatality events include all of the September 11, 2001 sites, mass transportation accidents and natural disasters. Bombings and shootings involving multiple casualties may also be included. Mass fatality events always involve multiple agencies, including various governmental bodies, and local search and rescue, so forensic anthropologists must quickly learn how to navigate within this large response framework.

Medical examiners recognize the diverse roles that forensic anthropologists have in disaster mitigation, and nearly every medical examiner's office consults with in-house or local forensic anthropologists to assist with city, county or state-wide disaster planning and policy. Forensic anthropologists also assist as members of multidisciplinary disaster teams, such as the federal Disaster Mortuary Operational Response Team (DMORT), under the Department of Health and Human Services, and Kenyon International Emergency Services, a private organization. In addition, three forensic anthropologists are currently employed at the National Transportation Safety Board.

Because DNA is the primary identification modality in mass fatalities, assessing the anthropological biological profile is less imperative in most situations. Mass fatality incidents typically involve the rapid processing of multiple bodies, so triage, body part identification and re-association and quality control measures have become primary forensic anthropological roles.

Triage involves the separation of body parts that are not directly linked by hard or soft tissues, and the removal of nonhuman and non-biological materials. Mitigating commingling is especially important, because every dissociated part receives a distinct case number and undergoes DNA testing. If the remains of two or more individuals are commingled and are not tested, then someone may never be identified. Depending upon the scale of the event, anthropologists assist in re-associating dismembered body parts (e.g. upper and lower legs) by matching fracture edges or joint articulations [88]. This can decrease the number of DNA samples that are necessary for laboratory analysis.

Medical examiners have also asked forensic anthropologists to design and implement quality control measures – systematic checks to ensure that all body parts have been identified properly, and that all available parts of the correct individual are together when a body is released to the family [89].

Human rights investigations involve the location and exhumation of clandestine graves, as well as trauma analysis and personal identification of individuals who were victims of political violence. Thus far, all human rights investigations involving forensic anthropologists have occurred outside of the United States. Like mass fatalities, human rights investigations can vary significantly in scale but involve the processing of multiple bodies.

Anthropological involvement in human rights investigations began in the early 1980s, as the Argentine Forensic Anthropology Team (EAAF) began to systematically investigate mass graves of *desaparecidos* – individuals who were kidnapped, murdered and 'disappeared' at the hands of the Argentine military Junta. Their success in Argentina demonstrated that meaningful information could be gained from the victims, and investigations began across Latin America and in Africa [90–92]. In the 1990s, anthropology-based investigations were an integral part of the prosecution effort of the International Criminal Tribunals in Rwanda (ICTR) and in the former Yugoslavia (ICTY). Forensic anthropologists continue to be active in human rights investigations around the world [93,94].

The methodological deficiencies of forensic anthropology, as practiced in the USA, are magnified when forensic anthropologists work in international contexts. Most standards for age, sex, ancestry and stature are entirely based on US reference samples, and their application to other populations carries unknown error. In one study, aging methods based on US skeletal samples in the former Yugoslavia performed poorly on Bosnian war dead [95]. Similarly, Eastern Europeans are, on average, taller than most Americans, so new population-specific regression formulae have been created from Croatian and Bosnian war victims [96]. Many other studies of war victims have increased the reliability and accuracy of identification [97–99]. The study of large reference samples of victims provides the ability not only to modify techniques appropriately or produce new methods, but to quantify statistical error for use in court [97].

Forensic anthropologists find themselves in the proverbial catch-22, because methodological accuracy will only improve with the study of local population variability, yet research on victims of conflict or disaster is fraught with ethical issues and only rarely permitted. One principal issue is whether informed consent is required and, if so, who is eligible to provide consent. It is not uncommon for entire villages and family lineages to be eliminated in a conflict, such that identifications may be impossible and/or there are no surviving family members to give consent for research.

One solution is to study only identified victims for whom family consent has been given. The sample size may be small, but still informative. Another possibility is to find non-victim data to establish standards (e.g. census, health and growth records) [98]. Developing ethical guidelines for research in mass disaster and human rights contexts will continue to involve communication between anthropologists, the investigative agencies and family communities, to ensure that major stakeholders are consulted.

5.5 International perspectives on forensic anthropology

Forensic anthropology formally began in the United States, so professionalization, theory and method development in the US have outpaced that of other countries. Nonetheless, international forensic anthropologists, particularly in Europe and in Latin America, have gained tremendous practical experience based on local histories of human rights atrocities and mass disasters [100,101]. What is often lacking, however, is a formal infrastructure for training and practice. This section briefly examines international perspectives on some of the challenges presented above, and highlights advances occurring abroad.

European anthropologists have developed novel techniques for aging, juvenile sexing and identification. For instance, techniques utilizing the acetabulum (hip socket) for aging are gaining popularity in Europe [102–105]. Like their US counterparts, European researchers have also recently emphasized metric analyses and general method standardization. In addition, because many international colleagues routinely work in mass disaster and human rights contexts, emphasis is often placed on the metric analysis of single bones or features to accommodate fragmentary remains. For instance, some methods favored by European practitioners utilize one or two measurements of the petrous, mastoid and mandible for sex estimation [106–108]. Emphasis on population-specific standards for sex, age and stature estimation is well entrenched in the international literature [98,107,109,110], yet colleagues abroad are less likely to focus on ancestry, or at least debate the ancestry issue, than their US colleagues. In other words, 'race' remains primarily an American theoretical concern [111–113].

Forensic anthropological casework involving living (or fleshed) individuals has recently increased in the United States [114], but greater attention has been given to forensic-related questions of living individuals abroad. Most issues relate to whether an individual (victim, defendant or immigrant) is above or below a particular national legal adult age, usually 18 years. Because the bones cannot be directly observed in living individuals, attention has focused on increasing plain film and CT resolution of specific age-related features [115]. Other techniques are being developed to estimate stature from security cameras or closed-circuit television, to help identify perpetrators caught on tape or fleshed victims from CT scans [116,117].

Theoretical and methodological differences aside, two of the most difficult challenges for non-US forensic anthropologists are the lack of infrastructure to accommodate this new field into daily medicolegal casework and opportunities for formal academic training. Forensic anthropology is very new to most countries where the coroners and/or medical examiner systems are well entrenched and unaccustomed to consulting forensic anthropologists. Thus, forensic pathologists often complete all skeletal examinations, including trauma and PMI analyses, without the benefit of anthropological expertise [100,118–120].

Unlike the United States, few foreign universities outside of the UK, Canada and Australia offer degree programs in forensic anthropology. Training is brief and is concentrated on practical skills in either forensic archaeology or anthropology at the master of science (MS) level. In other countries, such as Guatemala and Columbia, forensic anthropologists gain tremendous practical experience of exhuming mass graves and identifying victims, yet few university training programs are available, and they are rarely asked to participate in non-humanitarian death investigations by medicolegal agencies. All forensic anthropologists face a unified challenge to help build training and laboratory infrastructures that are safe and productive.

5.6 Future goals in forensic anthropology

While this chapter has largely concentrated on identification techniques, the number of services that forensic anthropologists provide to the medicolegal community is quite extensive. This breadth poses some significant challenges, in that some methods have a strong scientific basis, while others are based more on experience and historically lack the statistical means to demonstrate associated errors. Spurred by changes in the rules of admissibility, some of the most exciting research in forensic anthropology today provides better quantification and hence more methodological transparency. Testimony based on years of experience, or untested methods handed down from professors to students, are actively being challenged, strengthened, or simply put aside if they fail to reach higher levels of scientific scrutiny. In some cases, novel research provides new statistical or theoretical frameworks to address standard methods. In other cases, the very foundations of our discipline are questioned, such as 'what is a positive identification?'

Given the international engagement of forensic anthropologists from around the world, one of the key issues for future research involves exploring the extent and nature of local and global modern human variation. Continued emphasis on data collection from groups around the world is integral to address the current and future questions that forensic anthropologists will face. For example, what standards should be used in a plane crash in which an international set of victims are represented? This will require the concentrated efforts of forensic anthropologists with medical schools, medical examiner offices, cemeteries or university administrators, in order to have access to or initiate skeletal collections in a number of different countries [121]. It is these skeletal collections that will also improve the rigor of the standards and error rates for estimating the biological profile. This will be of particular value for adult aging techniques and for discriminating among older individuals.

Despite the theoretical debates concerning ancestry and race, forensic anthropologists have, directly or indirectly, made tangible contributions to the study of modern human variation that have not only proven useful for forensic anthropological casework, but also have farther reaching implications. Until the 1980s, forensic anthropologists relied on skeletal measurements obtained from reference samples of individuals born at, or before, the turn of the 20th century, for sex, stature and ancestry estimation. However, these measurements did not appear adequately to describe the variation among contemporary individuals. The Forensic Data Bank (FDB) was created by Richard Jantz and colleagues to address directly our lack of knowledge of contemporary human variation by collecting demographic, morphological and metric data from nearly 3,000 modern forensic cases.

Another important effort to understand contemporary human variation stems from one of the largest body donation programs in the country. The William M. Bass Donated Skeletal Collection at the University of Tennessee consists of over 900 individuals of known demographic, health, occupation and life history data. While the donation program began as a means of studying taphonomy, the wealth of data has provided insight into human variation and, more broadly, modern human evolution. For instance, extensive research has demonstrated secular changes in height and sexual dimorphism over the past century [122,123]. Other body donation programs have begun at other anthropological institutions, and these will provide more data concerning contemporary geographic and population variation.

Forensic anthropologists continue to increase empirical knowledge through well-designed experimental research. This avenue of research is particularly important in trauma analysis, such as differentiating between perimortem and post-mortem fractures. Historically, animal models have provided the basis for such studies [44,45,124] but, given different bone composition and density, it is imperative to study humans as well. Facilities that maintain long-term research of human cadavers, such as the University of Tennessee, Texas State San Marcos and Western Carolina University, provide the best opportunities for understanding how human bones fracture in a variety of post-mortem environments.

Another exciting change in forensic anthropology is greater attention to soft tissue anatomy and its relation to osseous structures. This trend is largely due to the incorporation of forensic anthropologists into medical examiner settings and medical schools, in which daily observations of whole cadavers are possible. For instance, a better understanding of the functional anatomy of the ventral arc helped to explain its variation between the sexes [125]. In another study, Adams and Herrmann [114] developed regression formulae for anthropometric (soft tissue) measurements of body and body segments, to allow stature estimations when bodies are fleshed or dismembered.

Studies of soft tissues are perhaps most crucial for trauma analysis, since the expansion and tearing of soft tissues has direct implications for certain bony injuries [41,46]. More extensive biomechanical experiments of fleshed remains are required to assess the nature of fracture initiation and propagation of complex bone fractures that are commonly seen in forensic cases. In addition, the behavior of burned bone is tightly linked to the amount and thickness of the overlying soft tissue, how quickly the bone is exposed and the duration of exposure [126]. Efforts to achieve a preliminary standardization of bone trauma and burning patterns are exciting areas of research, and they promote collaborations with medical examiners, bioengineers and medical schools.

Another important trend that will certainly continue is expansion into more diverse professional positions, working directly with families as well in government and policy-making. Anthropological expertise in mitigating the enormous amount of data provided by mass disasters and humanitarian investigations, and the ability to make the information accessible to families and the general public, have largely driven this trend. Working with families during and after disasters, or people whose loved ones have been 'disappeared' by their own government, requires not only knowledge of forensic techniques but also a solid social science background. Today, forensic anthropologists have leadership positions in governmental, inter-governmental and non-governmental organizations, whose roles include policy-making, first assessments and family liaisons.

The founding members of the Physical Anthropology Section of the American Academy of Forensic Sciences could not have anticipated how large and diverse this discipline could become in just four decades. They initiated a path forward that has bifurcated into multiple pathways to service science and the public, both in the USA and abroad. International collaboration and infrastructure building among forensic anthropologists is one key to maintaining a forward path in forensic science.

Acknowledgements

I wish to thank Jane Buikstra and Amy Mundorff for reading earlier versions of this chapter, and to Christina Fojas, who helped compile the citations and bibliography.

References

1. Dwight T. The sternum as an index of sex and age. *Journal of Anatomy* 1881; 15: 327–330.
2. Dwight T. The sternum as an index of sex, height, and age. *Journal of Anatomy* 1889/1890; 24:527–535.
3. Dwight T. The closure of sutures as a sign of age. *Boston Medical and Surgical Journal* 1890; 122:389–392.
4. Dwight T. The size of the auricular surfaces of the long bones as characteristic of sex: an anthropological study. *Journal of Anatomy* 1904/1905; 4:19–32.
5. Ousley SD, Jantz RL. *FORDISC 3.0: personal computer forensic discriminant functions.* University of Tennessee; 2005.
6. Spradley MK, Jantz RL. Sex estimation in forensic anthropology: skull versus postcranial elements. *Journal of Forensic Sciences* 2011; 56(2):289–296.
7. Todd TW. Age changes in the pubic bone. I. The male white pubis. *American Journal of Physical Anthropology* 1920; 3:285–339.
8. Todd TW. Age changes in the pubic bone. II. The pubis of the male Negro-White hybrid. *American Journal of Physical Anthropology* 1921; 4:1–26.
9. Brooks S, Suchey J. Skeletal age determination based on the *os pubis*: a comparison of the Acsadi-Nemeskeri and Suchey-Brooks methods. *Human Evolution* 1990; 5(3):227–238.
10. McKern TW, Stewart TD. *Skeletal age changes in young American males. Quartermaster research and development command.* 1957; Technical Report EP-45.
11. Lovejoy CO, Meindl RS, Pryzbeck TR, Mensforth RP. Chronological metamorphosis of the auricular surface of the ileum: a new method for the determination of adult skeletal age at death. *American Journal of Physical Anthropology* 1985; 68:15–28.
12. Meindl R, Lovejoy CO. Ectocranial suture closure: a revised method for the determination of skeletal age at death based on the lateral-anterior sutures. *American Journal of Physical Anthropology* 1985; 68:57–66.
13. İşcan MY, Loth SR, Wright RK. Metamorphosis at the sternal rib end: A new method to estimate age at death in White males. *American Journal of Physical Anthropology* 1984a; 65:147–156.
14. İşcan MY, Loth SR, Wright RK. Age estimation from the rib by phase analysis: White males. *Journal of Forensic Sciences* 1984b; 29(4):1094–1104.
15. İşcan MY, Loth SR, Wright RK. Age estimation from the rib by phase analysis: White females. *Journal of Forensic Sciences* 1985; 30(3):853–863.
16. Lamendin H, Baccino E, Humbert JF, Tavernier JC, Nossintchouk RM, Zerilli A. A simple technique for age estimation in adult corpses: the two criteria dental method. *Journal of Forensic Sciences* 1992; 37(5):1373–1379.

17. Prince DA, Ubelaker DH. Application of Lamendin's adult dental aging technique to a diverse skeletal sample. *Journal of Forensic Sciences* 2002; 47:107–116.
18. Stout SD. The use of bone histomorphometry in skeletal identification: the case of Francisco Pizarro. *Journal of Forensic Sciences* 1986; 39(3):778–784.
19. Stout SD, Paine RR. Histological age estimation using rib and clavicle. *American Journal of Physical Anthropology* 1992; 87:111–115.
20. Wittwer-Backofen U, Gampe J, Vaupel JW. Tooth cementum annulation for age estimation: results from a large known-age validation study. *American Journal of Physical Anthropology* 2004; 123(2):119–129.
21. Condon K, Charles DK, Cheverud JM, Buikstra JE. Cementum annulation and age determination in Homo sapiens. 2. Estimates and accuracy. *American Journal of Physical Anthropology* 1986; 71:321–330.
22. Boldsen JL, Milner GR, Konigsberg LW, Wood JW. Transition analysis: a new method for estimating age from skeletons. In: Hoppa RD,Vaupel JW, editors. *Paleodemography.* Cambridge: Cambridge University Press; 2002; 73–106.
23. Bethard JD. *A test of the transition analysis method for estimation of age-at-death in adult human skeletal remains* [thesis]. University of Tennessee; 2005.
24. Pearson K. Mathematical contribution to the theory of evolution: on the reconstruction of the stature of prehistoric races. *Philosophical Transactions of The Royal Society Of London. Series B: Biological Sciences* 1899; 192:169–244.
25. Fully G, Pineau H. Détermination de la stature au moyen du squelette. *Annales de Médecine Légale* 1960; 40:145–154.
26. Raxter MH, Auerbach BM, Ruff CB. Revision of the Fully technique for estimating stature. *American Journal of Physical Anthropology* 2006; 130:374–384.
27. Raxter MH, Ruff CB, Auerbach BM. Technical note: revised Fully stature estimation technique. *American Journal of Physical Anthropology* 2007; 133:817–818.
28. Konigsberg LW, Hens SM, Jantz LM, Jungers WL. Stature estimation as calibration: maximum likelihood and Bayesian approaches in physical anthropology. *Yearbook of Physical Anthropology* 1998; 107:65–92.
29. Willey P, Falsetti T. Inaccuracy of height information on driver's licenses. *Journal of Forensic Sciences* 1991; 36(3):813–819.
30. Maples WR. The identifying pathology. In: Rathbun TA,Buikstra JE, editors. Human identification: case studies in forensic anthropology. Springfield: Charles C. Thomas, 1984; 336–370.
31. Cunha E. Pathology as a factor of personal identity in forensic anthropology. In: Schmitt A, Cunha E, Pinheiro J, editors. *Forensic anthropology and medicine*. Towana, NJ: Humana Press; 2006; 333–358.
32. Wilson RJ, Bethard JD, DiGangi EA. The use of orthopedic surgical devices for forensic identification. *Journal of Forensic Sciences* 2011; 56(2):460–469.
33. Ubelaker DH, Jacobs CH. Identification of orthopedic device manufacturer. *Journal of Forensic Sciences* 1995; 40(2):168–170.
34. Jain AK, Chen H. Matching of dental X-ray images for human identification. *Pattern Recognition* 2004; 37:1519–1532.
35. Schuller-Götzburg P, Suchanek J. Forensic odontologists successfully identify tsunami victims in Phuket, Thailand. *Forensic Science International* 2007; 171: 204–207.

36. Dirkmaat DC, Cabo LL, Ousley SD, Symes SA. New perspectives in forensic anthropology. *Yearbook of Physical Anthropology* 2008; 51:33–52.
37. Dirkmaat DC, Adovasio JM. The role of archaeology in the recovery and interpretation of human remains from an outdoor forensic setting. In: Haglund W, Sorg M, editors. *Forensic taphonomy: the postmortem fate of human remains*. Boca Raton, FL: CRC Press; 1997; 39–64.
38. Hochrein MJ. An autopsy of the grave: recognizing, collecting, and preserving forensic geotaphonomic evidence. In: Haglund W, Sorg M, editors. *Advances in forensic taphonomy: method, theory, and archaeological perspectives*. Boca Raton, FL: CRC Press; 2002; 45–70.
39. Sauer NJ, Lovis WA, Blumer ME, Fillion J. The contributions of archaeology and physical anthropology to the John McRae case: a trial and a retrial. In: Steadman D, editor. *Hard evidence: case studies in forensic anthropology*. 2nd ed. Upper Saddle River: Prentice Hall; 2009; 122–132.
40. Symes SA, Berryman HE, Smith OC. Saw marks in bone: introduction and examination of residual kerf contour. In: Reichs K, editor. *Forensic osteology: advances in the identification of human remains* 2nd ed. Springfield, IL: Charles C. Thomas; 1998; 389–409.
41. Symes SA, Williams JA, Murrary EA, Hoffman JM, Holland TD, Saul JM, *et al.* Taphonomic context of sharp-force trauma in suspected cases of human mutilation and dismemberment. In: Haglund W, Sorg M, editors. *Advances in Forensic Taphonomy: Method, Theory, and Archaeological Perspectives*. Boca Raton, FL: CRC Press; 2002; 403–434.
42. Berryman HE, Symes SA. Recognizing gunshot and blunt cranial trauma through fracture interpretation. In: Reichs K, editor. *Forensic osteology: advances in the identification of human remains*. 2nd ed. Springfield, IL: Charles C. Thomas; 1998; 333–352.
43. Baumer TG, Passalacqua NV, Powell BJ, Newberry WN, Fenton TW, Haut RC. Age-dependent fracture characteristics of rigid and compliant surface impacts on the infant skull—a porcine model. *Journal of Forensic Sciences* 2010; 55(4): 993–997.
44. Daegling DD, Warren MW, Hotzman JL, Self CJ. Structural analysis of human rib fracture and implications for forensic interpretation. *Journal of Forensic Sciences* 2008; 53(6):1301–1307.
45. Wieberg DAM, Wescott DJ. Estimating the timing of long bone fractures: correlation between the postmortem interval, bone moisture content, and blunt force trauma fracture characteristics. *Journal of Forensic Sciences* 2008; 53(5):1028–1034.
46. Marks MK, Love JC, Dadour IR. Taphonomy and time: estimating the postmortem interval. In: Steadman D,editor. *Hard evidence: case studies in forensic anthropology*. 2nd ed. Upper Saddle River, NJ: Prentice Hall; 2009; 165–178.
47. Mann RW, Bass WM, Meadows L. Time since death and decomposition of the human body: variables and observations in case and experimental field studies. *Journal of Forensic Sciences* 1990; 35(1): 103–111.
48. Galloway A. The process of decomposition: a model from the Arizona-Sonoran desert. In: Haglung W, Sorg M, editors. *Forensic taphonomy: the postmortem fate of human remains*. Boca Raton, FL: CRC Press; 1997; 139–150.

49. Carter DO, Tibbett M. Does repeated burial of skeletal muscle tissue (*Ovis aries*) in soil affect subsequent decomposition? *Applied Soil Ecology* 2008; 40: 529–535.
50. Damann FE. *Human decomposition ecology at the University of Tennessee Anthropology Research Facility* [dissertation]. University of Tennessee, 2010.
51. Stokes KL, Forbes SL, Tibbett M. Freezing skeletal muscle tissue does not affect its decomposition in soil: evidence from temporal changes in tissue mass, microbial activity and soil chemistry based on excised samples. *Forensic Science International* 2009; 183(1–3):6–13.
52. Tibbett M, Carter DO, editors. *Soil analysis in forensic taphonomy: chemical and biological effects of buried human remains*. Boca Raton, FL: CRC Press; 2008.
53. Vass AA, Barshick SA, Sega G, Caton J, Skeen JT, Love JC. Decomposition chemistry of human remains: a new methodology for determining the postmortem interval. *Journal of Forensic Sciences* 2002; 47:542–553.
54. Vass AA, Bass WM, Wolt JD, Foss JE, Ammons JT. Time since death determination of human cadavers using soil solution. *Journal of Forensic Sciences* 1992; 37(5):1236–1253.
55. Maijanen H. Testing anatomical methods for stature estimation on individuals from the W. M. Bass Donated Skeletal Collection. *Journal of Forensic Sciences* 2009; 54(4):746–752.
56. Simpson EK, James RA, Eitzen DA, Byard RW. Role of orthopedic implants and bone morphology in the identification of human remains. *Journal of Forensic Sciences* 2007; 52(2):442–448.
57. Komar D, Lathrop S. Frequencies of morphological characteristics in two contemporary forensic collections: implications for identification. *Journal of Forensic Sciences* 2006; 51(5):974–978.
58. Konigsberg LW, Herrmann NP, Wescott DJ. Commentary on McBride *et al*. Bootstrap methods for sex determination from the *os coxae* using the ID3 algorithm. *Journal of Forensic Sciences* 2002; 47:424–426.
59. Steadman DW, Adams BJ, Konigsberg LW. Statistical basis for positive identification in forensic anthropology. *American Journal of Physical Anthropology* 2006; 131:15–26.
60. Adams BJ. Establishing personal identification based on specific patterns of missing, filled, and unrestored teeth. *Journal of Forensic Sciences* 2003a; 48(3):487–96.
61. Adams BJ. The diversity of adult dental patterns in the United States and the implications for personal identification. *Journal of Forensic Sciences* 2003b; 48(3):497–503.
62. Culbert WL, Law FL. Identification by comparison of roentgenograms of nasal accessory sinuses and mastoid processes. *Journal of the American Medical Association* 1927; 88:1634–1636.
63. Christensen AM. Testing the reliability of frontal sinuses in positive identification. *Journal of Forensic Sciences* 2005; 50(1):18–22.
64. Besana JL, Rogers TL. Personal identification using the frontal sinus. *Journal of Forensic Sciences* 2010; 55(3):584–589.
65. Wiersema JM. *The petrous portion of the human temporal bone: potential for forensic individuation* [dissertation]. Texas A&M University, 2006.

66. Wiersema JM, Love JC, Naul LG. The influence of the Daubert guidelines on anthropological methods of scientific investigation in the medical examiner setting. In: Steadman D, editor. *Hard evidence: case studies in forensic anthropology.* 2nd ed. Upper Saddle River, NJ: Prentice Hall; 2009; 80–90.
67. Armelagos GJ, Goodman AH. Race, racism, and anthropology. In: Goodman A, Leatherman T, editors. *Building a new biocultural synthesis: political-economic perspectives on human biology.* Ann Arbor, MI: University of Michigan Press; 1998; 359–377.
68. Brace CL. Region does not mean "race"—reality versus convention in forensic anthropology. *Journal of Forensic Sciences* 1995; 40(2):171–175.
69. Kennedy KAR. But Professor, why teach race identification if races don't exist? *Journal of Forensic Sciences* 1995; 40(5):797–800.
70. Ousley SD, Jantz RL, Fried DL. Understanding race and human variation: why forensic anthropologists are so good at identifying race. *American Journal of Physical Anthropology* 2009; 139(1):68–76.
71. Sauer NJ. Forensic anthropology and the concept of race: if races don't exist, why are forensic anthropologists so good at identifying them? *Social Science and Medicine* 1992; 34(2): 107–111.
72. Shriver MD, Smith MW, Jin L, Marcini A, Akey JM, Deka R, *et al.* Ethnic-affiliation estimation by use of population-specific DNA markers. *American Journal of Human Genetics* 1997; 60:957–964.
73. Kittles RA, Weiss KM. Race, ancestry, and genes: implications for defining disease risk. *Annual Review of Genomics and Human Genetics* 2003; 4:33–67.
74. Williams FLE, Belcher RL, Armelagos GJ. Forensic misclassification of ancient Nubian crania: implications for assumptions about human variation. *Current Anthropology* 2005; 46(2):340–346.
75. Elliott M, Collard M. Fordisc and the determination of ancestry from cranial measurements. *Biology Letters* 2009; 5(6):849–852.
76. Fried DL, Spradley MK, Jantz RL, Ousley SD. The truth is out there: how not to use Fordisc. *American Journal of Physical Anthropology* 2005; S40:103.
77. Birkby WH, Fenton TW, Anderson BE. Identifying southwest Hispanics using nonmetric traits and the cultural profile. *Journal of Forensic Sciences* 2008; 53 (1):29–33.
78. Rhine S. Nonmetric Skull Racing. In: Gill G,Rhine S,editors. *Skeletal Attribution of Race Maxwell Museum of Anthropological Papers No 4.* Albuquerque, NM: University of New Mexico; 1990; 9–20.
79. Hefner JT. Cranial nonmetric variation and estimating ancestry. *Journal of Forensic Sciences* 2009; 54(5):985–995.
80. Hughes CE, Juarez CA, Highes TL, Galloway A, Fowler G, Chacon S. A simulation for exploring the effects of the "trait list" method's subjectivity on consistency and accuracy of ancestry estimations. *Journal of Forensic Sciences* 2011; 56(5):1094–1106.
81. Walker PL. Sexing skulls using discriminant function analysis of visually assessed traits. *American Journal of Physical Anthropology* 2008; 136(1):39–50.
82. Schultz JJ, Collins ME, Falsetti AB. Sequential monitoring of burials containing large pig cadavers using ground-penetrating radar. *Journal of Forensic Sciences* 2006; 51(3):607–616.

83. Schultz JJ, Dupras TL. The contribution of forensic archaeology to homicide investigations. *Homicide Studies* 2008; 12(4):399–413.
84. Taylor RE, Suchey JM, Payen LA, Slota PJ. The use of radiocarbon (14C) to identify human skeletal materials of forensic science interest. *Journal of Forensic Sciences* 1989; 34(5):1196–205.
85. Ubelaker DH, Buchholz BA, Stewart JEB. Analysis of artificial radiocarbon in different skeletal and dental tissue types to evaluate date of death. *Journal of Forensic Sciences* 2006; 51(3):484–488.
86. Bytheway JA, Ross AH. A geometric morphometric approach to sex determination of the human adult *os coxa*. *Journal of Forensic Sciences* 2010; 55(4): 859–864.
87. Kimmerle EH, Ross AH, Slice DE. Sexual dimorphism in America: geometric morphometric analysis of the craniofacial region. *Journal of Forensic Sciences* 2008; 53(1):54–57.
88. Mundorff AZ. Anthropologist-directed triage: three distinct mass fatality events involving fragmentation of human remains. In: Adams B, Byrd J, editors. *Recovery, analysis, and identification of commingled human remains*. Towana, NJ: Humana Press; 2008; 123–44.
89. Mundorff AZ, Shaler RC, Bieschke E, Mar-Cash E. Marrying anthropology and DNA: essential for solving complex commingling problems in cases of extreme fragmentation. In: Adams B, Byrd J, editors. *Recovery, analysis, and identification of commingled human remains*. Towana, NJ: Humana Press; 2008; 285–300.
90. Egaña S, Turner S, Doretti M, Bernardi P, Ginarte A. Commingled remains and human rights investigations. In: Adams B, Byrd J, editors. *Recovery, analysis, and identification of commingled remains*. Towana, NJ: Humana Press; 2008; 57–79.
91. Olmo D. Crimes against humanity. In: Schmitt A, Cunha E, Pinheiro J, editors. *Forensic anthropology and medicine: complementary sciences from recovery to cause of death*. Towana, NJ: Humana Press; 2006; 409–430.
92. Bernardi P, Fondebrider L. Forensic archaeology and the scientific documentation of human rights violations: an Argentinean example from the early 1980s. In: Ferllini R, editor. *Forensic archaeology and human rights violations*. Springfield, IL: Charles C. Thomas; 2007; 205–232.
93. Steadman DW, Haglund WD. The scope of anthropological contributions to human rights investigations. *Journal of Forensic Sciences* 2005; 50(1): 23–30.
94. Ferllini R. The role of forensic anthropology in human rights issues. In: Fairgrieve S, editor. *Forensic osteologic analysis: a book of case studies*. Springfield, IL: Charles C. Thomas 1999; 287–301.
95. Komar D. Lessons from Srebrenica: the contributions and limitations of physical anthropology in identifying victims of war crimes. *Journal of Forensic Sciences* 2003; 48(4): 713–716.
96. Ross AH, Konigsberg LW. New formulae for estimating stature in the Balkans. *Journal of Forensic Sciences* 2002; 47(1): 165–7.
97. Kimmerle EH, Jantz RL, Konigsberg LW, Baraybar JP. Skeletal estimation and identification in American and East European populations. *Journal of Forensic Sciences* 2008; 53(3):524–532.
98. Schaefer MC, Black SM. Comparison of ages of epiphyseal union in North American and Bosnian skeletal material. *Journal of Forensic Sciences* 2005; 50 (4): 777–784.

99. Slaus M, Strinović D, Skavić J, Petrovecki V. Discriminant function sexing of fragmentary and complete femora: standards for contemporary Croatia. *Journal of Forensic Sciences* 2003; 48(3):509–512.
100. Cattaneo C. Forensic anthropology and archaeology: perspectives from Italy. In: Blau S, Ubelaker D, editors. *Handbook of forensic anthropology and archaeology*. Walnut Creek, CA: Left Coast Press; 2009; 42–48.
101. Doretti M, Snow CC. Forensic anthropology and human rights: the Argentine experience. In: Steadman D, editor. *Hard evidence: case studies in forensic anthropology*. 2nd ed. Upper Saddle River, NJ: Prentice Hall; 2009; 303–320.
102. Rissech C, Estabrook GF, Cunha E, Malgosa A. Using the acetabulum to estimate age at death of adult males. *Journal of Forensic Sciences* 2006; 51(2):213–229.
103. Rissech C, Estabrook GF, Cunha E, Malgosa A. Estimation of age-at-death for adult males using the acetabulum, applied to four western European populations. *Journal of Forensic Sciences* 2007; 52(4):774–778.
104. Rougé-Maillart C, Joussett N, Vielle B, Gaudin A, Telmon N. Contribution of the study of acetabulum for the estimation of adult subjects. *Forensic Science International* 2007; 171(2–3):103–110.
105. Rougé-Maillart C, Vielle B, Joussett N, Chappard D, Telmon N, Cunha E. Development of a method to estimate skeletal age at death in adults using the acetabulum and the auricular surface on a Portuguese population. *Forensic Science International* 2009; 188(1–3): 91–95.
106. Kemkes A, Göbel T. Metric assessment of the "mastoid triangle" for sex determination: a validation study. *Journal of Forensic Sciences* 2006; 51(5): 985–989.
107. Rösing FW, Graw M, Marré B, Ritz-Timme S, Rothschild MA, Rötzscher K, *et al.* Recommendations for the forensic diagnosis of sex and age from skeletons. HOMO 2007; 58(1):75–89.
108. Wahl J, Graw M. Metric sex differentiation of the *pars petrosa ossis temporalis*. *International Journal of Legal Medicine* 2001; 114(4–5): 215–223.
109. Durić M, Rakocević Z, Donić D. The reliability of sex determination of skeletons from forensic context in the Balkans. *Forensic Science International* 2005; 147: 159–164.
110. Oettlé AC, Steyn M. Age estimation from sternal ends of ribs by phase analysis in South African Blacks. *Journal of Forensic Sciences* 2000; 45(5):1071–1079.
111. Ferguson E, Kerr N, Rynn C. Race and ancestry. In: Black S, Ferguson E, editors. *Forensic anthropology: 2000 to 2010*. Boca Raton, FL: CRC Press; 2011.
112. Lieberman L, Kaszycka KA, Fuentes AJM, Yablonsky L, Kirk RC, Strkalj G, *et al.* The race concept in six regions: variation without consensus. *Collegium Antropologicum* 2004; 2:907–921.
113. Smedley A, Smedley BD. Race as biology is fiction, racism as a social problem is real: anthropological and historical perspectives on the social construction of race. *American Psychologist* 2005; 60(1):16–26.
114. Adams BJ, Herrmann NP. Estimation of living stature from selected anthropometric (soft tissue) measurements: applications for forensic anthropology. *Journal of Forensic Sciences* 2009; 54(4):753–760.
115. Telmon N, Gaston A, Chemla P, Blanc A, Joffre F, Rougé D. Application of the Suchey–Brooks method to three-dimensional imaging of the pubic symphysis. *Journal of Forensic Sciences* 2005; 50(3):507–712.

116. Baines KN, Edmond S, Eisma R. Stature. In: Black S, Ferguson E, editors. *Forensic anthropology: 2000 to 2010*. Boca Raton, FL: CRC Press; 201195–117.
117. DeAngelis D, Sala R, Cantatore A, Poppa P, Dufour M, Grandi M, *et al.* New method for height estimation of subjects represented in photograms taken from video surveillance systems. *International Journal of Legal Medicine* 2007; 121(6):489–492.
118. Baccino E. Forensic anthropology: perspectives from France. In: Blau S, Ubelaker D, editors. *Handbook of forensic anthropology and archaeology.* Walnut Creek, CA: Left Coast Press; 2009; 49–55.
119. Cox M. Forensic anthropology and archaeology: past and present—a United Kingdom perspective. In: Blau S, Ubelaker D, editors. *Handbook of forensic anthropology and archaeology.* Walnut Creek, CA: Left Coast Press; 2009; 29–41.
120. Prieto JL. A history of forensic anthropology in Spain. In: Blau S, Ubelaker D, editors. *Handbook of forensic anthropology and archaeology.* Walnut Creek, CA: Left Coast Press; 2009; 56–66.
121. Rissech C, Steadman DW. The demographic, socioeconomic and temporal contextualisation of the Universitat Autonoma de Barcelona Collection of Identified Human Skeletons (UAB Collection). *International Journal of Osteoarchaeology* 2011; 21(3):313–322.
122. Jantz LM, Jantz RL. Allometric secular change in the long bones from the 1800s to the present. *Journal of Forensic Sciences* 1995 September; 40(5):762–767.
123. Jantz LM, Jantz RL. Secular change in long bone length and proportion in the United States, 1800–1970. *American Journal of Physical Anthropology* 1999; 110(1): 57–67.
124. Wheatley BP. Perimortem or postmortem bone fractures? an experimental study of fracture patterns in deer femora. *Journal of Forensic Sciences* 2008; 53(1): 69–72.
125. Anderson BE. Ventral arc of the os pubis: anatomical and developmental considerations. *American Journal of Physical Anthropology* 1990; 83(4):449–458.
126. Schmidt CW, Symes SA, editors. *The analysis of burned human remains*. New York: Academic Press; 2008.

Further reading

Adams BJ, Byrd JE, editors. *Recovery, analysis, and identification of commingled human remains*. Towana, NJ: Humana Press; 2008.

Bass WM. Time since death: a difficult decision. In: Rathbun T,Buikstra J, editors. *Human identification: case Studies in forensic anthropology.* Springfield, IL: Charles C. Thomas; 1984.

Berryman HE, Smith OC, Symes SA. Diameter of cranial gunshot wounds as a function of bullet caliber. *Journal of Forensic Sciences* 1995; 40(5):751–754.

Berryman HE, Symes SA. Recognizing gunshot and blunt cranial trauma through fracture interpretation. In: Reichs K,editor. *Forensic osteology: advances in the identification of human remains*, 2nd ed. Springfield, IL: Charles C. Thomas; 1998; 333–352.

Blau S,Ubelaker DH, editors. *Handbook of forensic anthropology and archaeology*. Walnut Creek, CA: Left Coast Press; 2009.

Byrd JH, Caster JL, editors. *Forensic entomology: the utility of arthropods in legal investigations*. Boca Raton, FL: CRC Press; 2009.

Catts EP, Haskell NH, editors. *Entomology and death: a procedural guide*. Clemson, SC: Forensic Entomology Specialties; 1990.

Christensen AM. The impact of Daubert: implications for testimony and research in forensic anthropology (and the use of frontal sinuses in personal identification). *Journal of Forensic Sciences* 2004; 49(3):427–430.

Christensen AM, Crowder CM. Evidentiary standards for forensic anthropology. *Journal of Forensic Sciences* 2009;54(6):1211–1216.

Cox M, Malcolm M, Fairgrieve SI. A new digital method for the objective comparison of frontal sinuses for identification. *Journal of Forensic Sciences* 2009; 54(4): 761–772.

Crowder CM,Stout SD, editors. *Bone histology: an anthropological perspective*. Boca Raton, FL: CRC Press; 2011.

DiMaio VJM. *Gunshot wounds: practical aspects of firearms, ballistics, and forensic techniques*. Boca Raton, FL: CRC Press; 1999.

Fairgrieve S. *Forensic cremation: recovery and analysis*. Boca Raton, FL: CRC Press; 2008.

Galloway A, editor. *Broken bones: anthropological analysis of blunt force trauma*. Springfield, IL: Charles C. Thomas; 1999.

Haglund WD, Sorg MH, editors. *Forensic taphonomy: the postmortem fate of human remains*. Boca Raton, FL: CRC Press; 1997.

Haglund WD, Sorg MH, editors. *Advances in forensic taphonomy: method, theory, and archaeological perspectives*. Boca Raton, FL: CRC Press; 2002.

Hart GO. Fractural pattern interpretation in the skull: differentiating blunt force trauma from ballistics trauma using concentric fractures. *Journal of Forensic Sciences* 2005; 50(6):1276–1281.

Kimmerle EH, Baraybar JP, editors. *Skeletal trauma*. Boca Raton, FL: CRC Press; 2008.

Komar D, Buikstra JE, editors. *Forensic anthropology: contemporary theory and practice*. New York: Oxford University Press; 2008.

Lynnerup N, Kjeldsen H, Zweihoff R, Heegaard S, Jacobsen C, Heinemeier J. Ascertaining year of birth/age at death in forensic cases: a review of conventional methods and methods allowing for absolute chronology. *Forensic Science International* 2010; 201(1–3): 74–78.

Saville PA. Cutting crime: the analysis of "uniqueness" of saw marks on bone. *International Journal of Legal Medicine* 2007; 121:349–357.

Schmitt A, Cunha E, Pinheiro J, editors. *Forensic anthropology and medicine*. Towana, NJ: Humana Press; 2006.

Schwarcz HP, Agur K, Jantz LM. A new method for determination of postmortem interval: citrate content of bone. *Journal of Forensic Sciences* 2010; 55(6): 1516–1522.

Thompson TJU. Heat-induced dimensional changes in bone and their implications for forensic anthropology. *Journal of Forensic Sciences* 2005; 50(5):1008–1015.

6

Forensic toxicology: scope, challenges, future directions and needs

Barry K. Logan[1] **and Jeri D. Ropero-Miller**[2]

[1]*NMS Labs, Willow Grove, Pennsylvania*

[2]*RTI International, Research Triangle Park, North Carolina, USA*

6.1 Introduction

Toxicology is the study of the adverse effects of drugs and chemicals on biological systems. Forensic toxicology deals with the application of toxicology to cases and issues where those effects have administrative or medicolegal consequences, and where the results are likely to be used in court or in another quasi-judicial setting. This might include a professional licensing hearing or sports tribunal. Forensic toxicology is a thoroughly modern science, based on published and widely accepted scientific methods and practices for analysis of drugs, poisons and toxicants in biological materials, and for the interpretation of those results.

Forensic toxicology is built on a foundation of sub-disciplines that support the validity of the analytical practices, including instrumental analysis, analytical chemistry, biochemistry and immunology. Many of the methods employed in forensic toxicology have been derived from innovations in clinical medicine and academic laboratories, and these have been tested, peer-reviewed and refined as these sciences have matured. The interpretive aspects of the discipline, i.e. providing context for the analytical results, rely on discoveries in pharmacology, human and animal toxicology, drug discovery and development, medical and emergency room toxicology, epidemiology and behavioral science.

Thousands of articles related to forensic toxicology methods, instrumentation and interpretation are published in hundreds of peer-reviewed journals annually, increasing the understanding of the benefits, risks and dangers associated with the use, misuse and abuse of illicit and recreational drugs, medications, dietary supplements and

Forensic Science: Current Issues, Future Directions, First Edition. Edited by Douglas H. Ubelaker.

alcohol. This wide range of resources ensures multiple points of reference, comparison and validation, and it allows a high degree of scientific confidence in the measurement and interpretation of drug and toxicants in human beings.

Most applications of forensic toxicology can be classified under one of the following headings.

6.1.1 Death investigation toxicology

Forensic toxicologists work with pathologists, medical examiners and coroners in helping to establish the role of alcohol, drugs and poisons in the causation of death. The toxicologist evaluates the history and circumstances of a case and the available samples, and then detects, identifies and quantifies the presence of drugs and toxic chemicals in blood and tissue samples. This is done using state-of-the-art chemical and biomedical instrumentation that is capable of detecting small amounts of toxic materials, positively identifying them and accurately measuring how much is present.

Comprehensive scope, accuracy and reliability are essential, because this information is used in the legal determination of cause and manner of death. While that determination is ultimately the prerogative of the medical examiner, pathologist or coroner, the toxicologist is a key member of the team of experts that assist in reaching that conclusion, consulting on:

- pharmacology;
- drug kinetics and interactions;
- drug metabolism;
- adverse and idiosyncratic reactions;
- drug tolerance;
- post-mortem artifacts;
- drug stability;
- and various other factors.

The pathologist considers this information in the context of the investigative and medical history of the case and the findings of disease or other medical conditions found at autopsy. Accurately establishing the appropriate cause and manner of death has serious implications for public health and safety, and forensically reliable toxicology is a critical component of that process.

Death investigation toxicology is performed by both public and private laboratories. Some specialized private and academic laboratories supplement local government resources by providing particular expertise and esoteric testing that is not available in smaller laboratories. Two professional organizations that specifically promote standards ensuring the reliability of death investigation toxicology laboratories are the American Board of Forensic Toxicology (ABFT) and the College of Pathology (CAP). These organizations offer laboratory accreditation, professional certification, proficiency testing, on-site inspections and quality assurance programs.

6.1.2 Human performance toxicology

Human performance toxicology deals with the effects of alcohol and drug use on human performance and behavior, and the medicolegal consequences of drug and alcohol use. This could include investigations of impaired driving, vehicular assault and homicide, sexual assault and other drug-facilitated crimes, and aircraft, and motor vehicle and maritime collision investigations.

Forensic toxicologists perform analysis of drugs and alcohol in biological samples – typically blood and urine but also, increasingly, in other matrices such as hair and oral fluid – for the purposes of determining the timing, extent and impairment resulting from different patterns of drug and alcohol use. The toxicologist uses analytical methods, some of which are the same ones found in research and hospital laboratories, to isolate drugs from complex biological samples, prepare them for analysis through extraction and concentration and determine the identity and the amount of drug present.

Following the analytical phase, forensic toxicologists provide interpretation of the results with respect to whether the concentration represents therapeutic or recreational use, misuse or abuse, and can provide opinions about the likely effects of these patterns of use. This can include performance enhancement, such as may occur following the use of stimulants, and impairment from recreational or prescription medication use and misuse. Forensic toxicologists frequently testify in court, both to their analytical findings and to the interpretation of those findings.

Much blood alcohol and drug testing in police cases are performed in accredited private or academic forensic toxicology laboratories that voluntarily observe the same standards that are in place in the public sector. The International Council on Alcohol, Drugs and Traffic Safety (ICADTS), the National Highway Traffic Safety Administration (NHTSA), the US Food and Drug Administration (FDA), various arms of the US Department of Transportation (DOT) and similar governmental agencies worldwide promote the use of sound and proper forensic toxicology testing to support public health and safety.

6.1.3 Doping control

Governing bodies of most competitive and intramural sports have derived rules regarding performance-enhancing drug use to protect the health and welfare of the athletes, to maintain a fair and even competitive standard and to avoid wagering fraud. This applies to both human and animal sports. International groups such as the International Olympic Committee (IOC), the World Anti-Doping Agency (WADA) and the International Federation of Horseracing Authorities (IFHA) work to update and maintain these lists as patterns of drug use change.

Enforcing the rules requires periodic off-season, random and event-focused drug testing for performance-enhancing drugs and other medications that appear on the organizations' prohibited substances lists. Forensic toxicologists in this field use many of the same high-performance analytical methods to detect current and historical use of banned substances, including stimulants, anabolic steroids and diuretics.

6.1.4 Forensic workplace drug testing

Use of drugs by employees in the workplace has significant safety and economic consequences. Drug users have higher rates of absence, health benefit claims, workers' compensation claims and workplace accidents, as well as lower rates of productivity. Consequently, in most countries, workers in safety-sensitive positions are prohibited from using recreational drugs or taking certain medications without a prescription.

Enforcing these standards requires pre-employment, random and for-cause urine drug testing, such as following an accident or a transportation collision. In the United States, members of the armed forces and employees working for the federal government or their contractors are required to comply with the standards set forth in the Drug-Free Workplace Act, adopted by Congress in 1988. Forensic toxicologists perform testing of urine samples in laboratories, which are regulated and inspected on behalf of the federal government, in a program managed by the Substance Abuse and Mental Health Services Administration (SAMHSA). During testing for five major classes of abused drugs and their metabolites, scientists employ highly uniform and well-defined techniques and methods to minimize the risk of errors, and to ensure that employees are treated fairly and that testing is done to the highest forensic standards.

Other organizations that monitor workplace drug testing practices include the US Nuclear Regulatory Commission (USNRC), the European Workplace Drug Testing Society (EWDTS), the Joint Technical Committee of Australia and New Zealand Standards (AUG), the Swiss Working Group for Drugs of Abuse Testing Guidelines (AGSA) and the Health Science Authority of Singapore (HSA). This testing is comparable in quality and procedure to diagnostic testing for drugs of abuse performed in clinical and university laboratories.

As noted in these examples, forensic toxicology is practiced internationally in academic, hospital, government and private laboratories, each of which brings different strengths and resources to the profession, and this diversity has contributed to the maturity and sound scientific basis of the discipline. The body of scientific knowledge continues to grow, and the profession is active in looking to improve the practice of toxicology, fill knowledge gaps and meet new analytical and interpretive challenges as the range of available drugs and intoxicants expands. In the following sections, we consider where the needs and opportunities for development in forensic toxicology are, and what activities in the international community are ongoing to meet those needs.

6.2 Toxicology and the NAS Report

The history and provenance of all forensic science disciplines has become of increasing interest since the United States National Academy of Sciences (NAS) issued its critique of the profession in 2009 [1]. Academics, lawyers and non-practitioners of forensic science listened to input from forensic science practitioners and reported, with some dismay, the *ad hoc* manner in which forensic science had evolved.

The authors of the NAS report unfavorably contrasted forensic science to the evolutionary process of scientific discovery that takes place in an academic environment. In the latter, they describe the 'self-correcting nature of science' where replication of results, refinement of approach and deepening understanding move knowledge and science forward, while correcting false beliefs and assumptions from the past. They concluded that many of these checks and balances were missing from the forensic sciences, as it was more closed and inwardly focused, instead of being outward-looking and integrated with the larger scientific community.

In contrast to many of the long-established disciplines in forensic science, the NAS Committee held out the process of discovery, validation, standardization, regulation and ongoing refinement in DNA analysis techniques which have become routine forensic science tools over the last two decades. Overall, the Committee's assessment of the forensic sciences was not favorable, and they identified a series of opportunities to address what they perceived as deficiencies.

Forensic toxicology was not singled out for specific criticism in the NAS report. Rather, it was characterized as a laboratory-based technique and grouped with DNA and drug chemistry as disciplines that were largely validated and uniformly practiced (1, p. 7). The report identified that toxicology laboratories were short of equipment, had not adopted automation to improve efficiency (1, p. 59) and were understaffed (1, p. 219). It included forensic toxicology among a few disciplines believed to be built on solid bases of theory and research, in contrast to other techniques characterized as experience-based techniques, which rely more on heuristics, 'rule of thumb', educated guesswork, intuitive judgment and 'common sense'.

The NAS report acknowledged the leadership role taken by the ABFT in creating accreditation and certification programs for laboratories and individuals, respectively (1, p. 200). It recommended advanced post-graduate training for individuals entering toxicology, among other disciplines (1, p. 227), and for legal professionals to enable them to understand toxicological evidence (1, p. 234). It noted the need for greater availability of toxicology support for the death investigation system

(1, p. 250), and pointed out that only 37% of medical examiners have in-house toxicology laboratories.

Although forensic toxicology fared well in the NAS review, the report had several overarching concerns and some general recommendations from which forensic toxicology can learn and continue to improve. These included the following:

- *Terminology and laboratory report format should be standardized and should include a certain minimum amount of information* (1, p. 21).

The NAS report reminds us that the majority of users of the information generated in the laboratory are not toxicologists and need to be educated regarding the meaning of the report. The results must reference what has been tested for (i.e. what has been ruled in or out), and with what level of sensitivity. It should reflect the technique(s) that were used for the identification, confirmation and quantitation of the substances. Currently, no standards or guidelines exist for uniform report format or minimum report content in forensic toxicology reports.

- *'Research is needed to address issues of accuracy, reliability and validity in the forensic science disciplines'* (1, p. 22).

Accreditation programs currently in effect are placing increasing requirements on laboratories to document the validation of newly developed methods and to validate commonly used legacy methods that have been in place for a long time. Additionally, there is increasing pressure on laboratories reporting any quantitative result to state the precision and accuracy of the result and the amount of uncertainty inherent in any measurement. Few, if any, forensic toxicology laboratories have comprehensive estimates of the uncertainty in their quantitative analyses beyond referencing the performance of controls and proficiency samples. Alcohol may be the major exception to this. Consensus in this area needs to be developed, and more work is needed around the optimum approach.

- *'Laboratory accreditation and individual certification of forensic science professionals should be mandatory, and all forensic science professionals should have access to a certification process'* (1, p. 25).

There is a great diversity in the resources available in forensic toxicology laboratories and in the training and expertise of the people who work in them. Inevitably, this creates variability in the quality of the laboratory's analytical work and its ability to provide interpretation of its results. This is reflected in the high variability in performance for laboratories currently participating in voluntary proficiency testing.

Ensuring that laboratories follow at least minimum standards for laboratory quality – including written standard operating procedures (SOPs), proficiency testing, employee training, method validation, use of controls and corrective action programs – and ensuring they comply with their adopted standards, will improve the overall quality of analytical forensic toxicology and will increase the level of confidence in results. Implicit in this is another recommendation of the NAS report, that forensic laboratories should establish routine quality assurance and quality control procedures to ensure the accuracy of forensic analyses and the work of forensic practitioners (1, p. 26).

The NAS report also recommends certification of key individuals in the laboratory, based on their education, training, experience and competency, as assessed by examination and evaluated by an accredited board, as a necessary complement to laboratory accreditation. In the United States, the ABFT has provided opportunities for both accreditation and certification. Several US states have a mandatory requirement for laboratory accreditation, specifically New York, Oklahoma and Texas. Mandating accreditation and certification would undoubtedly raise the standards across the field, but many laboratories do not have the personnel or physical resources to meet accreditation standards. Not all key personnel currently employed in forensic toxicology laboratories would meet current standards for certification.

In 2009, the Society of Forensic Toxicologists (SOFT), the Toxicology Section of the American Academy of Forensic Sciences (AAFS) and the ABFT formed a Forensic Toxicology Council (FTC). The Council is a sponsoring organization of a Scientific Working Group for Forensic Toxicology (SWGTOX), which is developing recommendations to strengthen the current accreditation and certification standards. SWGTOX will be discussed later.

- *'. . . [A] national code of ethics for all forensic science disciplines . . . ' should be developed, and individual societies should be encouraged '. . . to incorporate this national code as part of their professional code of ethics'* (1, p. 26).

All of the sponsoring organizations of SWGTOX have codes of conduct and ethical requirements, and SWGTOX has published a Standard for Codes and Guides of Professional Conduct, which is currently under review. ABFT, AAFS and SOFT have policies on ethics, which members must affirmatively accept with their renewal of membership or recertification.

- *Graduate education programs should be developed ' . . . to cut across organizational, programmatic, and disciplinary boundaries . . . Emphasis should be placed on developing and improving research methods and methodologies applicable to forensic science practice and on funding research programs to attract research universities and students in fields relevant to forensic science . . . [L]aw school administrators and judicial education organizations . . . ' should establish ' . . . continuing legal education programs for law students, practitioners, and judges'* (1, p. 27).

This is a key recommendation that is critical to the development of forensic toxicology, both in the short and the long term. Forensic toxicology, like other forensic disciplines, is highly applied, but an individual's success in the field is tied to having a thorough understanding of core scientific principles in chemistry, biology, biochemistry, molecular biology, physics and statistics. While these can be supplemented at the undergraduate level with courses that introduce legal and medicolegal principles and forensic applications, it is at the level of graduate education where students truly develop the academic maturity to be able to apply scientific principles to forensic problem-solving, rather than be trained to perform a particular test.

Research needs in forensic toxicology are discussed later, but that research will inevitably take place largely in academic institutions. The development of relationships between practicing accredited forensic toxicology laboratories and accredited academic forensic science programs creates an ideal venue for the introduction and assessment of new technologies, validation of procedures, identification of new drugs and metabolites and discovery of new interpretive strategies. All forensic toxicology laboratories should look for this opportunity for collaboration. Such collaboration may also help to address an issue referenced later, regarding access for practitioners to the latest scientific peer-reviewed research.

A second part of this NAS recommendation, which is often overlooked, is the need to educate attorneys, judges, police officers and other criminal justice professionals about the strengths and limitations of test results, and to look beyond the four corners of the laboratory report and understand the limitations imposed by current knowledge and the nature of the sample. Most attorneys are not well equipped to assess adequately the strengths or weaknesses of a particular toxicology test. This can be addressed through toxicologists' participation in law school courses and in continuing legal education programs.

Finally, in the education arena, a key component to enabling professionals to be current, and to adapt to changes in technology, drug-use patterns and the integration of new scientific discovery into forensic casework, is an urgent and increasing need for continuing professional education. Opportunities for this continuing education exist through professional organizations such as AAFS, The International Association of Forensic Toxicologists (TIAFT) and SOFT, and funding that is primarily available through the National Institute of Justice (NIJ). Continuing professional education is a requirement in order to maintain professional certification, and the demand for this will increase once mandatory certification is implemented.

- *'To improve medicolegal death investigation'* (1, p. 29).

In this area, the NAS report makes several recommendations to address a lack of uniformity and quality in death investigation. Forensic toxicology is a key part of a comprehensive death investigation. In many jurisdictions, the pathologist and the toxicologist work closely together to ensure that appropriate testing is performed. In other offices, however, drug- or poison-related deaths may go unidentified, either through a lack of funds, expertise or training. Improving the process would require better communication between the toxicologist and pathologist, guidelines for what tests are appropriate in what kinds of cases, and increased awareness by pathologists, medical examiners and coroners of the limitations and scope of particular tests. A Scientific Working Group in Medicolegal Death Investigation (SWGMDI) has recently been established, and this creates an opportunity for closer collaboration with toxicology professionals.

Clearly, although forensic toxicology was not singled out for specific criticism in the NAS report, some of the general challenges faced by forensic science in terms of education and training of our professionals, individual certification and laboratory accreditation, method validation, measurement uncertainty and research into new technology are additional areas of opportunity for growth and quality improvement in toxicology.

6.3 Research priorities and resource needs for the future of forensic toxicology

The field of forensic toxicology benefits from discoveries and research in multiple fields. Forensic toxicology is well-integrated with other scientific disciplines, and many forensic toxicologists attend meetings of groups such as the American Association for Clinical Chemistry (AACC), Society of Toxicology (SoT), the International Association for Chemical Testing (IACT), the International Association for Therapeutic Drug Monitoring and Clinical Toxicology (IATDMCT), the American Society for Mass Spectrometry (ASMS) and the Association for Mass Spectrometry Applications in the Clinical Lab (MSACL).

On an international level, forensic toxicologists communicate through organizations including The International Association of Forensic Toxicologists (TIAFT), the Society for Hair Testing (SoHT), the German Society of Toxicological and Forensic Chemistry (GFTCh), the Canadian Society of Forensic Science (CSFS), the Asian Forensic Sciences Network (AFSN) and the Société Française de Toxicologie Analytique (SFTA). This diversity of professional organizations involving scientists from both within and outside the field, and outside the United States encourages innovation and promotes technological development.

Major recent milestones in forensic toxicology from an analytical perspective have included the development of affordable benchtop gas chromatography-mass spectrometry (GC-MS) and automation of homogeneous enzyme immunoassay in the 1980s, which together made possible the introduction of high-volume workplace urine drug testing.

The 1990s saw widespread adoption of MS, now within the economic reach of any serious forensic toxicology laboratory. In the area of interpretive toxicology, the 1990s saw the first publications regarding the post-mortem redistribution of drugs, which created greater uncertainty regarding the interpretation of post-mortem toxicology results – a challenge that continues today.

The early 2000s were characterized by a proliferation of hyphenated MS techniques, most notably liquid chromatography-mass spectrometry (LC-MS) and secondary fragmentation (liquid chromatography-tandem mass spectrometry, or LC-MS-MS) techniques. Refinements in high-vacuum pumping systems and small particle chromatographic columns have made LC-MS accessible to many more laboratories. Moving into the future, high-resolution quadrupole time-of-flight (TOF) mass spectrometry is moving out of the 'research' realm and into use in casework by forensic toxicology laboratories.

Current areas of opportunity for research and development and ongoing challenges for forensic toxicology laboratories are discussed below.

6.3.1 Address new interpretive challenges resulting from improvements in analytical detectability

With improved technology, more and more things can be measured at lower concentrations. LC-MS-MS, GC-MS-MS and LC-TOF techniques lower the background in analytical detection systems, and improve signal-to-noise to the point that detection limits in the low picograms per milliliters are now common. These limits of detection should be compared to the typical effective concentration ranges for drugs in biological systems, typically in the nanograms per milliliters to micrograms per milliliters range.

The meaning of these low levels with respect to effect or toxicity is not clear, and the presence of sub-effective concentrations of drugs raise additional questions about passive exposure, endogenous substances and absorption of drugs from food or the environment. Just because it is now possible to measure trace amounts of drugs or chemicals in casework samples, does not mean that these always have forensic significance, so caution needs to be taken with respect to interpretation regarding 'ingestion' or knowing use.

Enhanced detection limits may offer greater time windows to demonstrate drug use or exposure but, conversely, excretion profiles of some lipid soluble drugs may have extended terminal elimination phases, where typical pharmacokinetic approaches do not apply. Often, the data to distinguish where that happens do not exist. The increased sensitivity and lower signal-to-noise ratio of the latest generation of instruments does, however create the opportunity to produce better performance of assays with respect to analytical precision, accuracy and specificity, reducing the risk of inaccurate quantitations and false identifications. Forensic toxicologists need to work to document typical ranges of blood drug concentrations associated with abuse, therapeutic use and adverse effects, and to tailor technology and protocols accordingly.

6.3.2 Promote measurement of drugs and toxicants in multiple and alternative matrices

For some time, drugs and toxicants have been identified in alternate matrices beyond blood and urine. The most common matrices in post-mortem testing include liver, kidney, vitreous humor, brain, bile and gastric contents. Some investigators have reported drug concentrations in psoas muscle, skin and cerebrospinal fluid. However, the interpretive value of most drugs in these tissues has been limited.

Alcohol in vitreous humor, volatiles in brain tissues and antidepressants in liver tissues are a few examples of using drug concentrations in multiple matrices effectively. Additionally, there is general agreement that peripheral blood (iliac or femoral vein) concentrations are preferred over heart, subclavian or vena cava blood for interpretive purposes, as this minimizes the impact of post-mortem redistribution.

For the most part, however, when a drug is identified in an additional biological matrix beyond blood, the quantitative value does not offer much assistance to the toxicologist's interpretation, because relevant reference concentrations do not exist; models in which drug concentrations in different tissues can be interrelated to improve their interpretive value have not been developed. Additional research data on site-dependent differences and time-dependent changes in drug concentration would assist in the assessment of the significance of ante- and post-mortem changes in drug concentration.

The field of toxicology can be advanced by the collection and publication of drug concentrations in multiple tissues, with ancillary data on cause and manner of death, co-administered drugs, time since death and known or suspected dosage regimens. Centralized data collection systems, wherein these data can be shared and analyzed, would be extremely beneficial.

In forensic toxicology applied to testing samples from living subjects, there is established and increasing interest in testing in alternative matrices, beyond the traditional use of blood and urine. This includes compliance testing, impairment testing and employment testing.

Compliance testing for the court system, such as probation and parole, child protective services and drug courts, are now using oral fluid, sweat and hair to verify that an individual has abstained from drug-use in compliance with the terms of his or her sentencing or court mandated observation period. These alternate matrices have the benefits of being easier to collect, being less invasive to collect in comparison to blood or direct observation urine collection, not requiring same-gender collectors or specially prepared facilities, and providing a longer window of drug detection. Conversely, these sample types can be more difficult and expensive to analyze, and they still have some controversies such as the potential for environmental exposure with respect to sweat and inter-individual variation with respect to drug incorporation into hair based on pigmentation.

Oral fluid testing is seeing increasing use internationally as a specimen for testing in impaired driving cases. Belgium, France, Germany and Australia are using either oral fluid on-site testing, or oral fluid collection and analysis, as a confirmatory technique. Fourteen US States (Alabama, Arizona, Colorado, Indiana, Kansas, Louisiana, Missouri, New York, North Carolina, North Dakota, Ohio, Oregon, South Dakota and Utah) have approved the use of oral fluid for DUID investigations. Oral fluid has demonstrated potential for confirming both drug use and corroborating the cause of observed impairment. The ease of collection makes it less expensive, and it is possible to collect the sample proximate to the time of driving, thus making it highly relevant to the officer's observations.

Although not without its limitations, blood remains the best sample for relating observed impairment to different patterns of drug use, and for comprehensive detection of the broadest scope of agents. Currently, in the United States, workplace drug testing not regulated by the federal government (SAMHSA) can use alternative matrices for pre-employment and random testing programs. However, as of early 2012, this can not be done in the regulated arena, which only uses urine.

6.3.3 Keep up with changes in the emerging designer drug marketplace

Drug testing laboratories have played a long-standing cat-and-mouse game with laboratories that supply the illicit international drug market with new and novel chemical entities. From the advent of meperidine and fentanyl analogs in the 1980s, to the introduction of benzylpiperazines, synthetic cannabinoids and cathinone derivatives between 2009 and 2012, these drugs are frequently sought out by users looking for the latest variation on the euphoric, entactogenic, sedative, dissociative, hallucinogenic or stimulating high, and they are frequently available to the market before they appear in drug databases or libraries.

Examples of drug classes that are emerging in products targeted at recreational drug users in 2012 include naphthoylindoles, benzoylindoles, aminoindanes, aminotetralins, cathinones, benzylpiperazines, tryptamines, benzofurans, pyrovelarones and many others, as recently reported by the European Monitoring Center on Drugs and Drug Abuse (EMCDDA).

Careful monitoring of samples from medical examiner and impairment populations, and being attentive to the appearance of unexplained peaks in chromatograms and expending resources to identify them, helps to ensure that toxicology remains current. The proliferation of synthetic cannabinoid, cathinone and tryptamine designer drugs reflects the increasing sophistication of illicit drug chemists, who alter molecular structures one atom at a time in an attempt to maintain or fine tune drug potency and effect, while changing the structure sufficiently to keep one step ahead of the scheduled drug analog laws. This trend challenges drug testing laboratories to keep up with the market, to quickly develop and validate new assays for emerging drugs and to collect data to assist with analytical interpretation. Better communication between forensic toxicology and drug chemistry laboratories internationally will additionally assist in this effort.

6.3.4 Interpretive toxicology must be strengthened by further research

Currently the major research focus in forensic toxicology is in analytical method development, increased reliability, sensitivity and accuracy of testing in various sample types, using the latest technology. This work is important and must continue.

Significant advances in analytical capability, including use of high mass accuracy instruments and novel fragmentation techniques, are currently available mostly in the research environment, and they need to translate to production and practice of forensic toxicology.

Reliable information about the identity and amount of a drug or toxicant in blood or tissue is the basis for any meaningful interpretation. Interpretation of toxicological results is based on comparison to prior cases, to published data from relevant case series, with demographic, medical and dosing histories of the subjects, and to drug or toxicant concentrations in appropriate comparable matrices. Other comparable populations are subjects in clinical trials, subjects arrested for being under the influence of drugs and/or alcohol, adverse event reports in the literature from emergency rooms and treatment physicians, and compilations of post-mortem data.

There are weaknesses in any one of these resources. Examples include:

- published data may be in blood, but the source of the blood is not indicated;
- the case may involve whole blood, but a clinical study involving the analyte of interest reports serum and there is no recognized blood/serum ratio;
- the central-to-peripheral blood ratio may be reported for a drug in a few cases, but not at the concentration relevant to a particular case in question – and the analytical precision of the laboratory may be unknown, meaning the significance of a central and peripheral blood pair may be difficult to interpret;
- tolerance to the adverse effects of drugs develops, but there is currently no chemical marker for identifying drug tolerance in an individual.

Interpretively, awareness among toxicologists of pharmacogenetic factors such as low Km[1] forms of enzymes, that account for poor metabolism and elevated blood drug concentrations, and super-metabolizer individuals, whose genes express multiple copies of enzymes that accelerate metabolism and prevent patients from reaching expected drug concentrations for a given dose. All of these things make interpretive forensic toxicology a vocation honed by experience; it must be practiced with care and it takes many years of scrutinizing post-mortem cases, reading case histories and reviewing toxicology results to become truly expert. It also requires continued familiarity with the clinical and forensic literature on a wide range of drugs.

While this process is heavily science and evidence-based, it requires (to name a few):

- the application of heuristic skills and filters to weigh various pieces of information;
- collaboration with the pathologist, the investigator and/or other ancillary professionals;

[1] Km is the Michaelis constant for a particular enzyme and it represents the concentration of substrate that leads to half-maximal velocity for the enzyme. A small Km indicates that the enzyme requires only a small amount of substrate to become saturated. Hence, the maximum velocity is reached at relatively low substrate concentrations. Km is expressed in units of concentration, usually in molar units.

- consideration of the subject's drug history; and
- allowances for naïve and chronic drug use.

The process is analogous to that used in court, where the toxicologist is the evaluator of the available facts. The toxicologist is charged with determining whether the evidence supports a particular interpretation or, in the event that there is more than one explanation, whether one is more likely or if, in fact, neither meets a standard of reasonable scientific certainty.

Interpretive toxicology will never be a mathematical process. However, strengthening the basis and resources on which toxicologists make these decisions on a daily basis is key to the continued development and reliability of the field. Additional research and publication on post-mortem change, site-dependent differences in drug concentrations, adverse effects associated with particular drugs or drug combinations, the value of pharmacogenetic testing, and behavioral adverse effects associated with licit and illicit drug use, will all help improve the toxicologist's decision-making process and will make more results suitable for interpretation. These efforts can best be accomplished by collaborative research efforts, within toxicology, between clinical and forensic science and with international partners.

6.3.5 Standardize approaches and improve quality standards in forensic toxicology

Some fields in analytical chemistry, such as environmental testing and sports drug testing, have standardized on specific methods down to the technique, analytical conditions and sample preparation method. However, forensic toxicology laboratories have largely maintained their individual independence and have widely varied approaches to testing for the same drugs. This has the advantage of encouraging innovation and improvement, the rapid addition of newly emerging drugs, the inclusion of regionally important drugs and the flexibility to upgrade to new technology. Unfortunately, it has the disadvantage of creating inter-laboratory differences in both the scope and sensitivity of analysis, which can impact on the thoroughness of analysis and equal application of justice, and impact also on the consolidation of medical examiner or traffic safety toxicology data into regional or national databases.

Furthermore, as discussed in the NAS report, non-standardized forensic toxicology reports make it difficult to know what was analyzed for by the laboratory, and with what accuracy or reliability the laboratory could measure the drug or analyte. In the United States, the forensic toxicology community initiated efforts in the early 1980s to standardize the basis for sound and forensically supportable methods, promoted by the AAFS and SOFT and implemented through the voluntary laboratory accreditation program run by ABFT. Various groups published standards for testing in impaired driving and sexual assault casework.

In 2009, AAFS, SOFT and ABFT joined forces to form the Forensic Toxicology Council, which sponsored the creation of the Scientific Working Group in Forensic Toxicology (SWGTOX). The mission of SWGTOX is to develop and disseminate consensus standards for the practice of forensic toxicology [2]. As part of the establishment and by-laws of SWGTOX, overall objectives of the membership are to establish:

- laboratory standards, practices and protocols for quality assurance and quality control;
- laboratory accreditation and professional certification;
- education and training requirements;
- research, development, testing and evaluation requirements and recommendations.

In addition, SWGTOX provides guidance for the development of a code of professional conduct for forensic toxicologists and laboratories, as well as promoting public awareness of the field. All SWGTOX recommendations are developed on a consensus basis from working subgroups with US members and international advisors, and proceed through a tiered review process which includes:

- internal review and approval by SWGTOX members;
- a public comment period in which documents are posted on the SWGTOX website and distributed to all professional organizations that have a connection to forensic toxicologists;
- post-public comment edits and reviews by the appropriate SWGTOX sub-committee;
- final document approval by the entire SWGTOX membership.

SWGTOX is developing recommendations to strengthen current forensic toxicology practices, including publication of a Standard for Developing a Guide and Code of Professional Conduct in Forensic Toxicology, with expected publication in early 2012 and proposed standards for practice for method validation which began internal review in late 2011. Additional standards for certification boards, accreditation, research and education are also in development. SWGTOX will also work towards recommendations on expressing measurement uncertainty in a meaningful way.

The formation of SWGTOX represents a major accomplishment by the US forensic toxicology community, and its deliberations will go a long way to standardize approaches to laboratory analysis and improving confidence in the reliability, comparability and interpretability of results.

6.3.6 Improve access to scientific literature for practicing forensic toxicologists

In Europe, much forensic toxicology is practiced in academic centers or in government departments with strong ties to academia. As a result, this is where much of today's international research efforts in forensic toxicology are originating.

A key benefit of this relationship is ready access to the clinical and forensic literature through academic libraries.

Most government-run forensic toxicology laboratories in the US have limited journal access. Due to the extraordinary cost of maintaining and managing journal subscriptions for individual laboratories, many toxicologists rely instead on data compendia in books, or outdated in-house lists, for purposes of interpretation. While these are a good place to start, they are by necessity summaries and are frequently out of date, which jeopardizes the quality and reliability of the interpretation.

Although most libraries in US public universities are open to the public, service laboratories cannot readily access online journals and consequently choose not to, even when these journals may contain critically important information to a case. Electronic journal access significantly enhances their use and promotes faster adoption of, and proficiency with, new technology and the ability to find and to incorporate recent clinical and toxicological advances into interpretive opinions.

Government and academic support for practicing forensic laboratories (not just in toxicology), with collaborative teaching and research programs, adjunct faculty appointments and access to online electronic journals, would improve the basis on which forensic toxicology casework decisions are made every day and improve expertise within the field. Building a common resource of journals and research papers for the entire forensic science community is one critical component of improving the scientific basis of all forensic science, and it should be a priority of the government's response to the NAS Report. While this is important for all disciplines, it is especially so for forensic toxicology, which is highly reliant on advances in so many different medical and scientific fields. Some of the key forensic journals will be discussed later. Both SOFT and AAFS have recently made their annual meeting abstracts available on-line.

6.3.7 Promote more international collaboration and development

As alluded to several times in this review, the field of forensic toxicology continues to mature internationally, and there is much that the US learns from our colleagues in the rest of the world. TIAFT, in particular, promotes growth and development of forensic toxicology, and additionally supports developing countries. TIAFT established the Developing Countries Fund (DCF), which provides grants to professionals from those countries to attend the annual meeting and also provides a library of donated textbooks from members to interested DCF members. TIAFT DCF membership currently includes nine developing countries in South America, 12 in Africa and 14 in Asia. In addition, in 2011, forensic toxicologists from newly industrialized countries such as India, the Philippines, Thailand, Malaysia, Brazil and South Africa also became TIAFT members.

There are many other international organizations that promote forensic toxicology. The International Union of Toxicology (IUTOX) consists of over 20,000 toxicologists from more than 50 national and regional member organizations from countries worldwide. IUTOX has a Developing Countries Committee that coordinates and oversees activities of 'toxicological' interest to member societies from developing countries. It establishes relationships with toxicologists and groups of toxicologists in developing countries which do not have their own toxicology societies, but which potentially could have a national toxicological society with assistance. IUTOX also established the Congress of Toxicology in Developing Countries, which is an internationally recognized and prestigious forum for discussing the toxicological problems facing developing countries. The 8th Congress will be held in Bangkok, Thailand, in 2012.

The Korean Society of Toxicology (KST), The Toxicology Society of South Africa (TOXSA) and The Spanish Association of Toxicology (AETOX) are toxicological societies that have a section devoted to forensic toxicology, in addition to clinical environmental, veterinary, food safety and alternative methods that contribute to the development of these specialized fields.

Forensic toxicology journals are another measurement of international involvement in the field. Five journals are especially important to our scientific knowledge, three of which are published in countries other than the US:

- The *Journal of Forensic Sciences* is the official journal of AAFS, which represents membership in all 50 United States, Canada and 54 other countries worldwide [3]. *JFS* provides the latest information on more than ten disciplines represented in forensic sciences.
- *Forensic Science International* of Amsterdam (Elsevier) publishes contributions in the many forensic science disciplines, including forensic toxicology [4].
- The *Forensic Science Review* of Taiwan publishes review articles on technological applications based on basic research, techniques widely used in forensic science and innovations holding promise for the future [5].
- Published in Japan, *Forensic Toxicology* provides an international forum for publication of studies on toxic substances, drugs of abuse, doping agents, chemical warfare agents, and their metabolisms and analyses which are related to laws and ethics [6,7].
- Drug Testing and Analysis launched in 2009. This journal features analytically-focused papers, specifically for drug analysis, including many forensic applications in toxicology and drug chemistry.

All of these journals include original articles, reviews, mini-reviews, short communications and case reports.

Summary

Forensic toxicology has many applications with administrative or medicolegal consequences, for which its findings are used in court settings. Death investigation, human performance, doping control and forensic workplace drug testing are practiced in academic, hospital, government and in private laboratories worldwide, working in collaboration with other forensic sciences to support the criminal justice system.

The maturity and sound scientific basis of the discipline have evolved over many decades, but its future directions and needs have been assessed and refined here in light of the findings of the NAS report. The challenges that forensic toxicologists continue to face and address through research, collaboration and publication include:

- analytical detectability and its effect on interpretation;
- use of alternate biological matrices for testing;
- maintaining vigilance in pursuit of changes in the illicit drugs market;
- increased laboratory accreditation/professional certification;
- standardization of methods and quality standards in forensic toxicology;
- improvements in access to scientific literature for professionals; and
- improvements in international collaboration and development in forensic toxicology.

References

1. National Research Council. *Strengthening Forensic Science in the United States: A Path Forward.* Washington, DC: The National Academies Press; 2009.
2. Scientific Working Group on Forensic Toxicology. http://www.swgtox.org (accessed June 11, 2012).
3. Journal of Forensic Sciences. http://www.wiley.com/bw/journal.asp?ref=0022-1198 (accessed June 11, 2012).
4. Forensic Science International. http://www.elsevier.com/wps/find/journaldescription.cws_home/505512/description (accessed June 11, 2012).
5. Forensic Science Review. http://www.forensicsciencereview.com/ (accessed June 11, 2012).
6. Forensic Toxicology. http://www.springer.com/medicine/forensic/journal/11419 (accessed June 11, 2012).
7. Drug Testing and Analysis. http://onlinelibrary.wiley.com/journal/10.1002/(ISSN)1942-7611/homepage/Permissions.html (accessed June 5, 2012).

Further reading

European Monitoring Center for Drugs and Drug Abuse (EMCDDA), Briefing Paper: Online Sales of New Psychoactive Substances/Legal Highs: Summary of results from the 2011 Multilingual Snapshots. www.emcdda.europa.eu/attachements.cfm/att_143801_EN_SnapshotSummary.pdf (accessed January 6, 2012).

Ropero-Miller JD, Goldberger BA, editors. *Handbook of Workplace Drug Testing*. 2nd ed. AACC Press, Washington DC; 2009.

Farrell LJ, Kerrigan S, Logan BK. Recommendations for toxicological investigation of drug impaired driving. *Journal of Forensic Sciences* 2007; 52(5):1214–8.

LeBeau MA. Laboratory management of drug-facilitated sexual assault cases. *Forensic Science Review* 2010; 22:113.

Kelleher C, Christie R, Lalor K, Fox J, Bowden M, O'Donnell C (editors). *An Overview of New Psychoactive Substances and the Outlets Supplying Them*. National Advisory Committee on Drugs (NACD): Centre for Social and Educational Research, Dublin Institute of Technology 2012. http://www.nacd.ie/publications/Head_Report2011_overview.pdf (accessed January 6, 2012).

Peters FT. Recent advances of liquid chromatography-(tandem) mass spectrometry in clinical and forensic toxicology. *Clinical Biochemistry* 2011 Jan; 44(1):54–65.

US Department of Health and Human Services (HHS). Substance Abuse and Mental Health Administration. Mandatory guidelines and proposed revisions to mandatory guidelines for federal workplace drug testing programs. *Federal Register* 2010 April 30; 75(83):22809.

US Department of Health and Human Services (HHS). Substance Abuse and Mental Health Services Administration, Division of Workplace Programs. 1988. Mandatory guidelines for federal workplace drug-testing programs. *Federal Register* 1988 April 11; 53:11979.

US Department of Health and Human Services (HHS). Substance Abuse and Mental Health Administration. Mandatory guidelines and proposed revisions to mandatory guidelines for federal workplace drug testing programs. *Federal Register* 2004; 69 (71):19644.

Werner G. The measurement of uncertainty. *Clinical Pharmacology and Therapeutics* 1961 Mar-Apr; 2:143–6.

7

Odontology – dentistry's contribution to truth and justice

Iain A. Pretty[1], Robert Barsley[2], C. Michael Bowers[3], Mary Bush[4], Peter Bush[5], John Clement[6], Robert Dorion[7], Adam Freeman[8], Jim Lewis[9], David Senn[10] and Frank Wright[11]

[1]*School of Dentistry, University of Manchester, Manchester, United Kingdom*
[2]*School of Dentistry, Louisiana State University, New Orleans, Louisiana, USA*
[3]*The Ostrow School of Dentistry of the University of Southern California, Deputy Medical Examiner,Ventura, California USA*
[4]*Laboratory for Forensic Odontology Research, State University of New York, Buffalo, New York, USA*
[5]*South Campus Instrument Center, State University of New York, Buffalo, New York, USA*
[6]*Melbourne Dental School, University of Melbourne, Melbourne, Australia*
[7]*Forensic Dentistry Program, McGill University, Montreal, Quebec, Canada*
[8]*College of Dental Medicine, Columbia University, New York, New York, USA*
[9]*Alabama Department of Forensic Sciences, Auburn, Alabama, USA*
[10]*University of Texas Health Science Center, San Antonio Dental School, San Antonio, Texas, USA*
[11]*Hamilton County Coroner's Office, Cincinnati, Ohio, USA*

7.1 The discipline

Forensic odontology, the analogous term for forensic dentistry, is an art-science that utilizes dental and forensic knowledge to resolve civil and criminal legal issues [1]. As in many forensic disciplines, the principle of uniqueness or individualizing characteristics underlies much of the discipline [2]. The collection, assessment, and analysis of individualizing dental features are the basis of forensic dentistry. In the broadest terms, forensic dentists are responsible for two main types of activity – the identification of found human remains and the assessment of bitemark injuries [3,4]. However, these broad categories contain a multitude of complexities and sidelines that are ever-changing.

Forensic Science: Current Issues, Future Directions, First Edition. Edited by Douglas H. Ubelaker.

Forensic dentistry, like other forensic disciplines, changes to reflect the needs of those commissioning the activity. For example, in Europe, many forensic dentists will spend a significant amount of their time providing age estimates for living individuals seeking political asylum [5].

7.1.1 Scope of practice

Table 7.1 shows the activities with which forensic dentists will concern themselves. Each of these areas, with the exception of civil litigation – an area so large and varied that it would warrant a chapter of its own [6,7] – are covered in detail throughout this chapter, examining the state of the art and the challenges for the future, as well as considering the international aspect of the work.

7.1.2 Who employs forensic dentists?

Depending upon the jurisdiction and the pertinent local (municipal), regional (county, canton, state, province) and national (federal) laws, the forensic dentist may be called upon by a coroner, medical examiner, procurator fiscal, *juge d'instruction*, *médecin légiste*, the police or some other legal entity to perform specific services. Few forensic dentists are employed full time, although there are some dental academics whose work is based solely on odontology practice and research. The vast majority of the odontology workforce is based in primary care general dental practice – both a strength and a weakness within the discipline. The vast experience and capacity within primary care enables a large number of skilled and dedicated professionals to assist in, for example, mass fatality situations. However, the capacity for research development is restricted to a few locations worldwide, and the opportunities for regular practitioners to participate in such work is limited.

7.1.3 How are forensic dentists trained?

Forensic dentists will hold a dental degree. Some of these will be doctorate degrees (e.g. from North America), while others will be undergraduate (e.g. the United Kingdom and Australia). The use of skill mix within forensic dentistry has been slow to develop; dental hygienists [8] and others are involved with supporting dental identification teams, but the principal work still lies with dentally qualified individuals [9]. Until recent times, forensic dentistry was learned by reading articles/textbooks, by attending seminars/conventions and, in particular, by apprenticeship with sound mentorship; there was no formal training provided by a degree-granting educational institution [10]. Today, a handful of universities worldwide, who may or may not be affiliated with a forensic establishment, offer courses

Table 7.1 Forensic dentists will concern themselves with the following main activities:

Activity	Aspect	Notes
1. Human identification	**a.** Of deceased individuals	• Often in cases where bodies are badly damaged by fire, trauma, chemical assault or decomposition (see Figures 7.1 and 7.2).
	b. Of living individuals	• In cases of amnesia, senility or fraud.
	c. Of individuals within mass fatality incidents	• Often where body damage and commingling of remains has occurred.
	d. Age assessment of living and deceased individuals	• For living individuals, this is often undertaken to assess if they have reached the relevant jurisdiction's age of majority. • For deceased individuals as part of a dental profile.
2. Bitemark analysis	**a.** To establish if an injury has been caused by teeth	
	b. If those teeth were human or animal, adult or child	
	c. If those teeth can be compared to a suspect's dentition to establish an identification or exclusion	
3. Assessment of peri-oral injuries and abuse in child, elder and spousal cases	**1.** Injuries related to abuse seen in the head/neck region	
	2. Neglect of the teeth and oral care	
	3. Intra-oral signs of sexual and physical abuse	
4. Civil litigation matters, often involving standards of care issues	**a.** Failure of dental professionals to provide adequate treatment	
	b. Poor or neglectful treatments	
	c. Estimate of ongoing treatment costs following accidents	

Table 7.1 Forensic dentists will concern themselves with the following main activities:

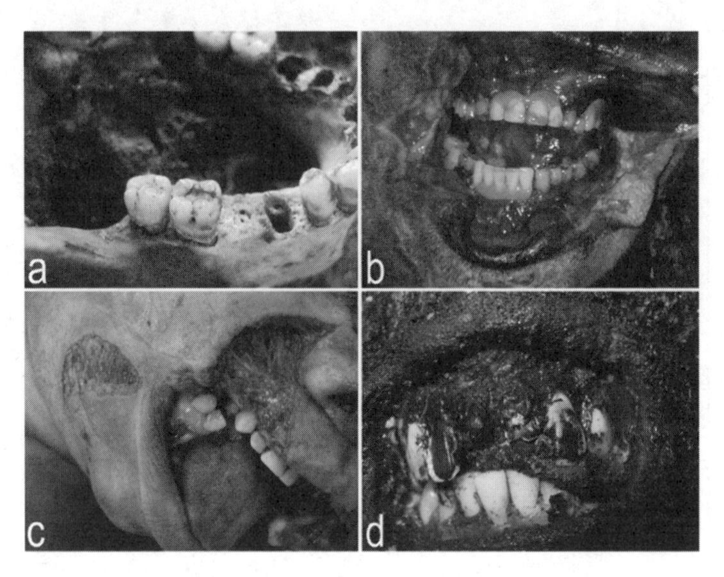

Figure 7.1 Typical presentations of body conditions in forensic dental identifications.

a. Skeletal remains – note that this individual has teeth missing post- and ante-mortem. Note the presence of an amalgam restoration – each 'filling' is custom-made and hence an individualizing feature.
b. Gross decomposition and mummification – this individual has both natural teeth and prostheses. The odontological community has been campaigning, with some success, to ensure that all such devices are labeled with patient-identifying features.
c. Trauma – typical of that seen in road traffic accidents. The teeth are useful for identification, not only because they are repositories of individual characteristics (whether restored or not), but because enamel is the hardest structure in the human body and, therefore, will resist most post-mortem and perimortem assaults.
d. Fire – while the maxillary anterior teeth appear to be badly damaged by exposure to heat, radiographs of the area will reveal a wealth of information within the root structures. Mandibular teeth, and those located posteriorly, are undamaged by the fire.

leading to certificates in specific fields of forensic dental practice, often leading to either master's or PhD qualifications.

7.1.4 How are forensic dentists recognized?

Board certification (or specialist status) has become an issue in both the forensic and judicial communities in North America. The organization that offers certification in forensic dentistry in North America is the American Board of Forensic Odontology, Inc. (ABFO), founded in 1976 under the auspices of the National Institute of Justice (NIJ). The ABFO is accredited by the Forensic Specialties Accreditation Board (FSAB). The accredited specialist uses the designation DABFO. Board certification is not a guarantee of the truth in results, but it is a means of distinguishing minimum

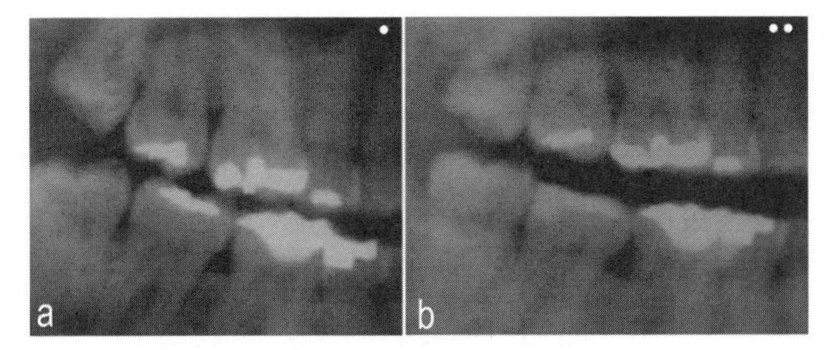

Figure 7.2 A typical comparison identification.
The radiograph in (a) is ante-mortem and will have been collected from the dentist of record; that in (b) is the post-mortem image, taken by the forensic dentist in the mortuary. The process of comparative identifications begins by assessing the similar features in tooth structure, form, morphology and clinical treatment. The next stage is to identify any discrepancies between the ante- and post-mortem views. On many occasions, discrepancies are explainable, frequently due to temporal effects. However, a single unexplained discrepancy would lead to a negative, or exclusion, identification. In this case, a small amalgam restoration on the mandibular second molar has been replaced by a larger resin (white) composite restoration. This is biologically plausible and hence explainable. This is a positive identification.

competence. Indeed, there have been cases of poor bitemark analysis practice (leading to wrongful convictions) that have featured board and non-board certified forensic dentists.

Internationally, there are no equivalent certification bodies. For example, in the United Kingdom, the General Dental Council does not recognize forensic dentistry as a specialty. National forensic odontology groups are, however, lobbying various accreditation organizations to develop competency frameworks for odontology, and there are obvious benefits to this approach.

Given that any dentist can claim to be a forensic expert, board certification in a forensic discipline assures the trier of fact of basic knowledge, education, experience and training of the expert in that forensic discipline. Moreover, certain ethical standards are maintained by that individual, the person can be peer-reviewed, he or she must keep current in knowledge, standards and procedures of the discipline, and he/she must be recertified on a timely basis.

7.2 Historical background

Historically, teeth have long been utilized for identification purposes [11]. In the first century C.E., Julia Agrippina, the mother of Nero and sister of Caligula, relied upon the arrangement of teeth in a skull presented to her to be satisfied that her rival and divorced wife of Nero, Lollia Paulina, was dead. History records numerous cases of heroes who fell in battle being identified for proper burial by their teeth and oral structures.

The first recorded instance utilizing dental expert testimony occurred in Scotland in 1814, in an attempt to prove the identity of a body used at the College Street Medical School. The first recorded American case arose some 35 years later, also using a denture to prove the identity of the body of Dr. George Parkman, murdered by his colleague Dr. John Webster at Harvard University. Also, during the mid-to-late nineteenth century, two attempts to identify the remains of King Louis XVII, who perished at age 10 in 1795 were made. In each case, a coffin was disinterred from the French church where he was supposedly buried and, in each case, the skeletal remains were determined to be at least sixteen years of age – ruling out the Dauphin in an early case of dental age estimation.

In May of 1897 in Paris, France, several dentists banded together to assist in the identification of the more than 125 individuals (mostly women) who perished in a fire that erupted when an early movie film projector exploded at a society charity event. In the first recognized text about Forensic Odontology – *L'Art Dentaire en Medecine Legale* – Dr. Oscar Amoedo revealed that the idea for using dentists came neither from the dentists nor from the physicians (who were unable to make any identifications due to the extensive thermal damage present on most victims), but rather from a member of the diplomatic corps. More than 100 victims were identified [12,13]. This was odontology's first recorded foray into disaster victim identification (DVI), occurring nearly contemporaneously with C.E. Kells, Jr.'s development of the dental radiograph, which would play a large role in future dental identifications.

The first documented American case involving a bitemark was heard in court in 1870. In 1905 (Germany) and in 1906 (Britain), cases involving bites into cheeses left at the scenes of crimes were heard in court. A similar case in America occurred in Texas in 1954, resulting in the first American conviction involving a bitemark [14].

As interest in the field burgeoned, dentists working with law enforcement attended meetings of other forensic scientists and specialists. The first course in forensic odontology was given by the US Armed Forces Institute of Pathology (AFIP) in 1962 and, at the AFIP's second offering in 1970, the six founding members of the American Society of Forensic Odontology were present and formed the society.

The year before, dental members of the American Academy of Forensic Sciences (AAFS) had, at the urging of then AAFS President-elect Cyril Wecht and Secretary/Treasurer Arthur Schatz, recruited a sufficient number of members in the AAFS to form the Odontology Section, the Academy's eighth section. Six years later, the American Board of Forensic Odontology was formed by one Canadian and eleven American pioneers [15].

Equivalent organizations exist throughout the world. The International Organization for Forensic Odonto-Stomatology provides leadership and has national affiliates from Australia and New Zealand, through Africa and into Europe. The international interest in odontology demonstrates the importance of the discipline, but also the global nature of forensic science today – not least when considering mass fatality incidents.

7.3 Key issues in odontology

Every forensic discipline needs to assess and appraise itself constantly in light of developments, both within science and also the judicial systems within which it operates. The practice of identification of humans from their dental records is well accepted, and the key issues surrounding this are based upon capacity development, training, information technology and logistics.

The theories and principles underlying the techniques are not threatened. Bitemark analysis, however, finds itself under a critical spotlight. As described earlier, there is a paucity of academic capacity within forensic dentistry and perhaps this, combined with the complexity and expense of forensic research, has led, in part, to a highly critical review of bitemark analysis within the National Academy of Sciences (NAS) Report into Forensic Science [16]. Much needs to be done within this area of practice, and it is encouraging to note that there has been a resurgence in research activity aiming to provide an evidence base for the practice and, perhaps, to restrict or redefine how such marks should be used in forensic applications.

7.4 Bitemarks

Human bitemarks in skin are often associated with serious crimes such as assaults, homicides, abuse and sex crimes [17]. Evidentiary documentation of the bitemark is achieved via photography, DNA harvesting, impressions of the bitemark and, occasionally, re-section of the tissue from deceased victims [18]. The bitemark evidence is then used to investigate any potential links between a suspected biter and the bitemark.

The value of bitemark evidence involves issues of physical contact between the biter and the person bitten, the pain associated with the biting and, finally, a linking of the biter to the bitemark [19]. The means of establishing a link can be salivary DNA matching and/or the resultant bitemark analysis between the biter's dentition and the bitemark [20,21]. It is worth noting that not every bitemark will have sufficient forensic significance to enable physical comparisons, although there may still be value in the injury [22]. For example, in child abuse cases, the fact that an adult has bitten a child may be sufficient evidence for a family court to consider (see Figure 7.3).

It is certainly true that salivary DNA evidence collected from a bitemark is the optimal means of assessing a suspect's involvement in the injury [20,21]. However, bitemarks also demonstrate that physical contact between a biter and the victim occurred – as such intent can be shown [23]. Some odontologists argue that, when a well-defined bitemark and a small closed population of suspected biters (each of whom present with distinctive dentitions) are seen, such injuries can be used to discriminate between individual suspected biters and the bitemark. This argument for limiting the scope of bitemark analysis has been expanded by some who believe that bitemark comparisons should be limited to simply including or excluding individuals [24,25].

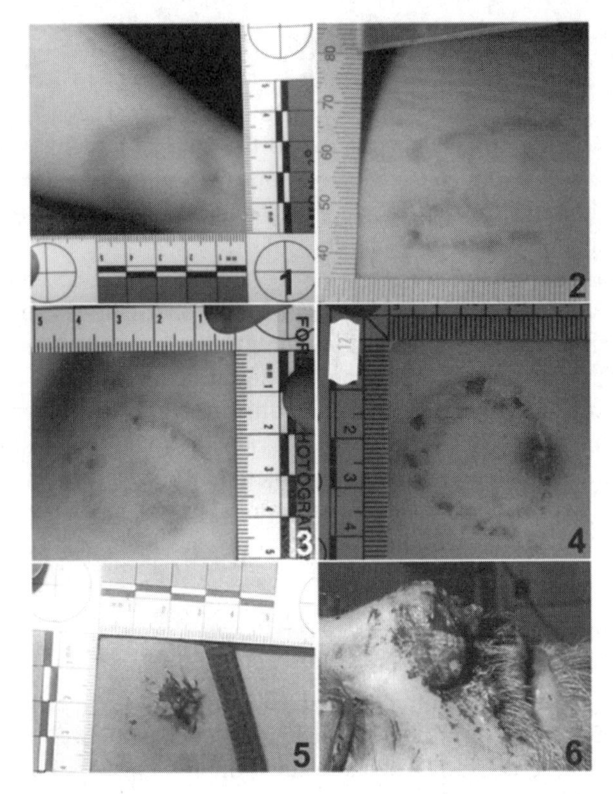

Figure 7.3 The bitemark severity and significance scale.
The use of scales has been advocated to enable the forensic significance of an injury to be assessed and an appropriate level of analysis to be undertaken:

1. Very mild bruising, no individual tooth marks present, diffuse arches visible, may be caused by something other than teeth – *low forensic significance*.
2. Obvious bruising, with individual, discrete areas associated with teeth, skin remains intact – *moderate forensic significance*.
3. Very obvious bruising, with small lacerations associated with teeth on the most severe aspects of the injury, likely to be assessed as definite bitemark – *high significance*.
4. Numerous areas of laceration, with some bruising; some areas of the wound may be incised. Unlikely to be confused with any other injury mechanism – *high forensic significance*.
5. Partial avulsion of tissue, some lacerations present, indicating teeth as the probable cause of the injury – *moderate forensic significance*.
6. Complete avulsion of tissue, possibly some scalloping of the injury margins, suggesting that teeth may have been responsible for the injury. May not be an obvious bite injury – *low forensic significance*.

An increasing number of issues have been raised about the use of bitemarks to identify individuals, extending even to the use of salivary DNA. For example, there have been known cases of non-biter DNA harvested from bitemarks in child abuse cases, rendering caution in absolute biter identity solely via DNA matching. Further concerns surround the increasing evidence base that suggests that bitemark patterns in skin are highly susceptible to postural, anatomical and photographic distortion [26–28]. This distortion can impact on the ability of forensic dentists to discriminate adequately between biters with similar dentitions. Evidence is also being developed that brings into question the often agreed [29], but never proved, contention that the human dentition (in relation to those aspects seen within bitemarks) is unique [30–33].

Many odontologists who support bitemark analysis state that it is not the science, but rather the practitioners, who are at fault. Rogue odontologists who undertake analysis of injuries without sufficient forensic detail have sullied the discipline, and risk it being rejected by judicial systems [34]. There are certainly cases that support this. However, the NAS report and the increasing voice within the forensic and legal communities requires that a robust assessment and clear guidance on the use of this evidence is required [24].

It can be argued that there is a paradigm shift occurring within bitemark assessments, driven in part by laboratory-based research but also by research assessing the construct validity of the science [24–25]. In tandem, these pieces have suggested that a highly cautious approach to bitemarks is required. The vast majority of bitemarks in human skin are not suitable for bitemark analysis because of distortion of the skin or a lack of detailed individual tooth markings in the pattern. Yet bitemarks lacking evidentiary quality continue to be used in bitemark analysis; the development of scales and interpretative guidelines is required [35,36].

Many odontologists have argued for more restrictive bitemark analysis guidelines and standards that clearly delineate the criteria that must be met for a bitemark to be subjected to bitemark analysis. A change in the methodology used in teaching bitemark analysis is also required, so that analysts understand when bitemark analysis can be utilized and when it should be avoided. The development of a bitemark 'analysis pathway', based on forensic significance, is required [17].

7.5 Disaster victim identification

Disaster victim identification (DVI) or multiple fatality incidents (MFI) are most commonly termed 'mass disasters' [37]. A mass disaster is defined as any multiple fatality incident that overwhelms the local or convening authority [37]. This can be different from one locality to another; some large cities have greater capability to deal with MFIs than smaller towns, states, or even countries.

The goal of any DVI team is to recover, autopsy (when needed), scientifically identify and then to release the victims of mass fatality incidents to their families, or repatriate them to their country of origin [38]. The protocols developed to accomplish these goals may differ between DVI teams. However, a constant is that it is a team effort, comprised of many individuals with differing specialties. Recently, the Federal Bureau of Investigation has started a DVI Scientific Working Group (DVISWG), which, in cooperation with Interpol, is developing best practices for use by DVI teams worldwide.

The Odontology Section of the DVI team is broken up into three sub teams: an ante-mortem team, a post-mortem team and a comparison team. Dr. Keiser-Nielsen developed a statistical model using the presence or absence of a tooth, as well as whether any of the remaining teeth are restored [39,40]. For example, the possible combinations of a person missing six teeth and having five restored teeth are 59,609,309,760, or the probability of another individual having the same combination is 1 in over 59 billion.

This work has been further developed by the Joint POW/MIA Accounting Command (JPAC) with software called *OdontoSearch* [41]. This program features not a collection of unknown missing or unidentified individuals, but rather a database of 40,108 individuals who have voluntarily participated in dental health studies. The goal of the system is to provide a representative sample of the dental treatment of the adult US population [41]. It allows the user to run searches of specific dental chartings to determine their likely incidence in the US population.

For example, if a user searched for a specific dental charting pattern and this was observed in only 24 of the individuals within the database, then this would indicate that the pattern would be seen 1 in 1666 individuals, and it can therefore give the user an empirical probability that can be quantified for use in a report or in a court of law [41].

In the United States, pursuant to the section under Emergency Support Function (ESF #8), the Disaster Mortuary Operational Response Team (DMORT) was founded and is now under the mandate of the Department of Health of Human Services (HHS), which oversees the National Disaster Medical System (NDMS). DMORT is a direct asset of NDMS. DMORT's role is to aid the local authority in the collection of ante-mortem information through a Family Assistance Center and post-mortem data via forensic examination, identification and release of the victims of mass fatality incidents to loved ones [42–44].

DMORT, as an asset, can be deployed within 24 hours and can be activated under several legal authorities, including the National Response Plan, the Public Health Services Act, the Aviation Disaster Family Assistance Act, Presidential Mandate and federal and state existing agreements. DMORT has three Disaster Portable Morgue Units (DPMUs), which can be transported by truck, rail or plane. Each contains within it all the equipment necessary to operate a morgue operation.

DMORT is broken down into ten regional teams. Each team consists of a commander, deputy commanders, administrative officers and a training officer, and includes individuals with expertise in all areas of morgue operations and family assistance. The dental team is overseen by regional Dental Section chiefs, who provide leadership at the time of a disaster. They will appoint dentists from the deployed team, who will take on the role of post-mortem, ante-mortem and comparison section chiefs. If a regional team is deployed, the initial roster is approximately 35 individuals, with approximately 12 of those being part of the dental team. The roster can be scaled up or down, depending of the size and scale of the MFI, either by increasing the size of the deployed regional team or by bringing in other regional teams. Typically, deployments are for two-week intervals, and therefore teams need to have redundancy in each position for extended deployments of DMORT.

The ante-mortem team is responsible for obtaining the dental records for all the missing persons, to decipher and transcribe them using a predetermined coding method, and to enter the information either directly into a computer system or onto forms [38]. In the United States, there is a Health Insurance Portability and Accountability Act (HIPAA) exemption 45 CFR 164.512(g) for Medical Examiners and Coroners, allowing physicians and dentists to release medical and dental records when requested. The ante-mortem teams must be familiar with the significant variation in dental charting practices employed internationally [40]. The increasing internationalization of disaster work is driving the development of globally agreed protocols and systems.

The post-mortem team will examine, chart, photograph and radiograph the dentition of all the recovered victims of the MFI, taking special care to memorialize the evidence available to them [39,45,46]. It is important to remember that post-mortem team members are handling evidence of potential crimes. They may get only one chance to memorialize the post-mortem information accurately, so attention to detail is therefore paramount. Post-mortem teams record their findings using a predetermined set of codes onto forms or, in some cases, directly into a computer system [39,45,46]. The use of hand-held, battery-operated x-ray generators and digital radiographic sensors have eliminated the need for darkrooms, chemicals and processors. It has also increased the level of efficiency at which DVI teams work [47–49].

The digital environment has also allowed for the enhancement of images and for instantaneously viewable images, giving immediate feedback to those taking the radiographs and providing the ability to improve the angulation of the radiograph. The use of IT has changed the archival process from rooms filled with boxes of charts, as in New York City in the wake of the attacks on the World Trade Center, to what now can be archived on a flash memory data storage device. The need for a well-networked system with proper security and a robust real-time backup system cannot be understated.

The comparison team evaluates the information garnered by the ante-mortem team and the post-mortem team in order to establish the identity of the victims [37,50]. There are several computer programs designed to help cut down the numbers of potential matches so that the odontologist does not have to search manually through hundreds, or even thousands, of records. In the United States, the DMORT uses a program called WinID3. In 1997, The Victorian Institute of Forensic Medicine in Australia developed a program called **D**isaster **A**nd **V**ictim **ID**entification, or DAVID [51–54]. The city of New York has recently developed a more robust program designed to aid in the management of the entire mass fatality incident, called Unified Victim Identification System, or UVIS. Within UVIS are several modules, including a dental module called UVIS Dental Module, or UDIM.

All of these programs allow for the storage of ante-mortem and post-mortem information, digital radiographs and digital images. Using sophisticated algorithms, the programs develop a potential match list. However, it should be noted that it is still the forensic odontologist who makes the identifications through a comparison of the ante-mortem and post-mortem information.

Dental identification is a relatively fast, inexpensive modality to scientifically identify human remains [37,38]. It has been used thousands of times in individual identifications, as well as MFIs. From the year 2000 through 2010, there have been over 500,000 victims of mass disasters.

The United States has an ongoing mass disaster in missing and unidentified persons. The FBI houses and runs the National Crime Information Center (NCIC), which has approximately 90,000 individual entries. This problem is not unique to the United States; such databases of missing individuals exist worldwide. In an attempt to address this issue, the Department of Justice recently developed a new web-based system called NamUS. NamUS is accessible by all in an attempt to rectify this ongoing mass disaster by providing a repository for both missing persons information and unidentified post-mortem details.

There is a growing need for well-trained dedicated individuals with the ability both to deal with the recognized stress of the mission and to withstand the possibility of austere environments. Developing best practices for disaster victim identification is a step in the right direction. Understanding the cultural and religious needs of the families affected is paramount to the success of any operation. Often, families describe the anguish of the unknown as the worst part; it is therefore incumbent on those involved in identifying victims of mass fatality incidents to get it right each and every time, as one can only imagine the distress to a family with even one misidentification.

7.6 Aging

The legal interest for scientific dental age assessment arose from the need to estimate the chronological age of children to enforce the British penal code and child labor laws of the early 1800s. Dr. A. T. Thomson, a medicolegal expert, began suggesting that dentition would be useful in child age estimation in 1836 [55]. One year later, Dr. Edwin Saunders published a pamphlet with tables that could be utilized by 'relatively untrained people' to estimate the age of children for the purpose of enforcing the Factory Act of 1833 [56].

The purpose and value of forensic dental age assessment has expanded in today's society to include:

- the narrowing of search possibilities of unidentified individuals;
- differentiation of cluster victims in mass disasters;
- age estimation at the time of death;
- and, for the living without birth documentation, aid in determining social benefits and legal age of majority for immigration and criminal issues.

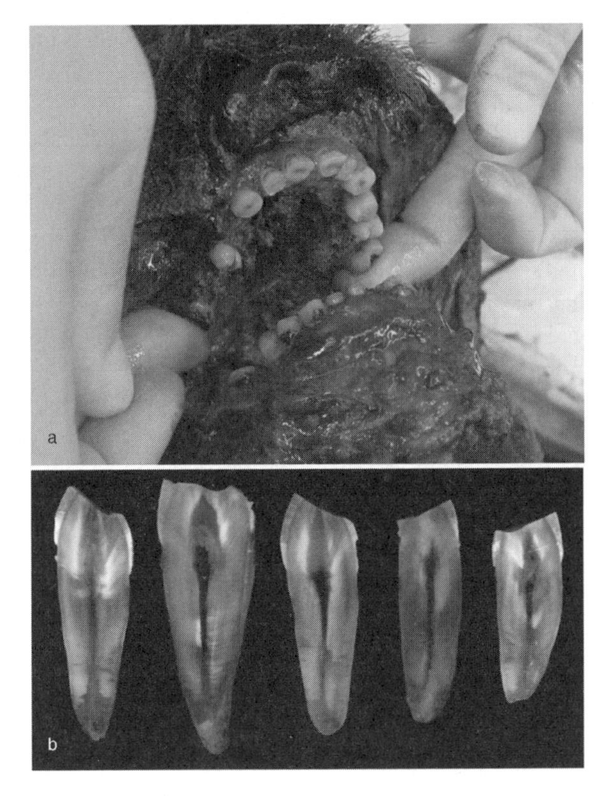

Figure 7.4 Example of age assessment using the human dentition.
In this example – translucent dentine is being employed to determine age at death (a). Dentine translucency begins at the apex of the teeth and then progresses, with age, towards the coronal portion. The assessment of translucency has been shown to be a very effective means of ageing deceased individuals (b).

Tooth formation and growth, post-formation changes and biochemical changes form the scientific rationale in performing dental age estimations. Child and adolescent dental age assessment methods utilize tooth formation and growth changes described by a tooth's progressive morphological development of the crown, root and apex. Once skeletal growth has ceased, dental age assessment has traditionally relied on methods that involve the evaluation of gross anatomical and histological changes of teeth (see Figure 7.4). Biochemical techniques, aspartic acid racemization and radioactive carbon-14 dating are new methods which have the advantage of being useable on individuals of any age [57].

7.6.1 Key issues

Aging is an individual process; therefore, age assessment is problematic because it is an attempt to estimate chronological age on the basis of physiological age. Regardless, numerous studies have demonstrated that dental age assessment is more accurate than other methodologies, particularly in sub-adults [58–62]. Unfortunately, our ability to accurately assess human age diminishes as true chronological age increases.

Ancestral and sexual dimorphisms are factors long established as important criteria in dental age assessment accuracy [63–65]. Although tooth development and mineralization is primarily under genetic control, current dental age assessment guidelines and standards recommend the use of population-specific studies to minimize the effects of ancestral admix and local environmental factors. Environmental factors that may potentially affect the accuracy of age assessment include climate, nutritional health, disease, medical and dental treatments, habits, addictions, occupation and dental and skeletal abnormalities [65,66].

In order to ensure the best estimate of chronological age, the most current, valid, relevant and reliable studies available must be applied, strict adherence to technique processes must be followed, and all available information needs to be considered. This includes the utilization of a multi-disciplinary approach. The particular age assessment technique(s) selected as most suitable for use depends on the specific circumstances of the case [67–69].

Age estimation of the living necessitates the use of non-invasive methods for moral, ethical, cultural and religious reasons, limiting the investigator to atlas-style data sets or radiographic evaluation. Some have questioned the use of radiographic evaluation of the living due, to potential effects of ionizing radiation [70]. However, when accuracy is paramount, the utilization of radiographic methodology is necessary, because the most reliable sub-adult methods utilize the evaluation of morphologic developmental changes of the teeth. Sub-adult techniques that rely upon emergence pattern as the age assessment criteria are renowned for inaccuracy, and atlas style methods have the inherent problems of non-sex differentiation and a relatively high degree of inter-observer disagreement.

In adult dental age assessment, root translucency and secondary dentine formation have proven to be the most reliable post-formation dental variables [71]. Both are minimally affected by ancestry and sex; but, root translucency is the only factor not affected by external factors including dental restorative therapy, orthodontic treatment, trauma, hyper and hypo-occlusion [72,73]. Therefore, the best adult age estimation methods rely on the evaluation of non-restored teeth in normal occlusion and, they utilize root transparency and secondary dentine as primary variables. Most dental post-formation techniques require tooth extraction and, sometimes, sectioning. When ethical, moral or religious issues prevent tissue destruction, secondary dentine can be evaluated radiographically and age assessed using the Kvaal *et al.* [74] technique.

7.6.2 Dental age estimation research

The majority of current dental age estimation research involves population-specific studies of well-established techniques. Specifically, studies utilizing the Demirjian dental morphologic staging system [75] for children and studies similar to Mincer *et al.* [76] for adolescent age assessment are being published. Recent adult age estimation research includes volumetric secondary dentine studies using x-ray computed tomography (CT) and cone beam computed tomography (CBCT), as well as population-specific studies of the Lamendin [77] method.

Initial results from aspartic acid racemization (AAR) and radioactive carbon-14 (^{14}C) studies are most promising. Racemization is a naturally occurring *in vivo* process in dentinal collagen where the L-isomer of the amino acid, aspartic acid, steadily converts to the D-isomer over time until equilibrium has been reached. This process slows considerably after death, allowing for assessment of age at death. AAR accuracy has been achieved to within $\pm$ 3 years when proper sampling and analytic techniques are followed [78].

^{14}C dental techniques can be used to estimate the date of birth. Atmospheric ^{14}C increased from 1955 through 1963 as a result of above-ground nuclear testing. The carbon dioxide, in turn, became incorporated into plants through photosynthesis and was eventually consumed by animals. Therefore, human tissue mimics the atmospheric ^{14}C levels at any given time. Dental enamel contains carbon, it forms during a narrow time frame for any particular tooth during childhood development and there is no metabolic remodeling of tooth enamel, so estimation of the date of birth is possible and has been shown to be accurate to $\pm$ 1.6 years [79]. The ^{14}C method is only valid for individuals born after 1943, since third molar enamel completes formation by age 12.

When AAR and ^{14}C data are mutually considered, an estimated date of death can calculated. Although accurate, both AAR and ^{14}C methods require sacrifice of tooth structure and can be time-consuming and more expensive than traditional dental age estimation techniques.

7.6.3 Education and certification

The American Board of Forensic Odontology (ABFO) will institute dental age estimation requirements for board certification beginning in 2014. Currently, there are three independent workshops available for dental age assessment education. In the United States, the ABFO offers a workshop at the annual American Academy of Forensic Sciences Meeting in odd-numbered years and the Southwest Symposium on Forensic Odontology, held at the University of Texas health Science Center in San Antonio, Texas, in even-numbered years. In Europe, a workshop is periodically offered by the forensic odontology group from Katholieke Universiteit, Leuven.

7.7 The international perspective

7.7.1 Global identification issues

It is worth stating that the odontology community is strongly committed to the principle that, as a fundamental human right, everyone is entitled to an identity in life and after death [80]. History has demonstrated that once this has been lost sight of, atrocious acts can occur. Despite the wide and increasing range of the work of the forensic odontologist, the identification of the dead remains our most important task. The 20th and 21st centuries have been marred by massacres and genocide, resulting in millions of deaths worldwide. As previously described, the rationale and evidence base for the comparative identification process is robust and has not been challenged [81].

As means of communication have become faster, cheaper and more ubiquitous, the impromptu photographic documentation and dissemination of cataclysmic events by people caught up in them has become commonplace. It is now the exception if acts of civil unrest, interethnic violence, state-organized oppression or terrorist atrocities go unrecognized worldwide within hours. However, the subsequent investigation of such events can be delayed for years, during which time those responsible for massacres often try to conceal their crimes with exhumations and the reburial of victims at different locations. This leads to commingling of remains or the separation of body parts across different sites, as occurred in the former Yugoslavia. The scale of the task confronting investigators vastly exceeds the capacity of any agency formed to cope only with the day-to-day tasks of a well-regulated society, so other agencies need to be called upon to provide sufficient expertise for very long-term deployments.

One such agency is the International Committee of the Red Cross (ICRC), which has a long-term commitment to The Missing project, which seeks to unite families scattered in times of war or civil unrest. It also strives to identify the deceased for the benefit of the surviving relatives, and as means of reconciling former enemies and recent conflicts. Another more recent organization with similar aims is the International Commission for Missing Persons (ICMP), which was established by President Clinton in response to The Balkans conflict. In both cases, the imperative is the rights of the dead and their families, and obtaining evidence to be used as the basis of prosecutions at the International Criminal Court in The Hague or elsewhere remains of secondary importance.

The issue confronting agencies like the ICRC and ICMP is a logistical one. How can hundreds of thousands of deceased be identified in a timely manner by the small numbers of experts able to do so? This is where forensic odontology has a vital role to play. Looking across a wide range of mass disasters over the last few decades, forensic odontological expertise has been pre-eminent in the identification of the deceased. If ante-mortem records exist for comparison with post-mortem remains, then there is no comparable technique in terms of speed, accuracy or cost-effectiveness.

Unfortunately in lesser-developed societies such as Rwanda, and even the former Yugoslavia, dental records either do not exist or are not of sufficient standard to be useful. In such circumstances, the odontologist can only rely upon their anthropological knowledge to look for ancestral traits in the dentition or establish age at death among the young from knowledge of the chronology of tooth development. They can also draw upon their clinical experience to make inferences about habits, customs and lifestyles from environmental alterations to the adult dentition and unrecorded dental treatments. Such work is often called dental post-mortem profiling, and it is of great help in reducing the number of records that need to be examined from missing persons lists or, indeed, the production of an artist's impression or similar.

Recent work on the 3-D mapping of faces is on the verge of creating statistically-based morphometric systems that will permit the rapid computer generation of the most plausible facial features for any individual, following a topographic scan of the underlying skull retrieved from a mass grave; such images can be rapidly acquired by CT [82–84]. For many unsophisticated people, there is a bewilderment and distrust of DNA. Many derive more comfort from seeing clothing or personal possessions of their next of kin retrieved from a gravesite. To be able to generate unique, high-quality images of the appropriate ancestry with a good likeness is likely to be bring comfort to many, particularly as it was previously only necessary to identify scientifically a representative sample of the dead to mount a prosecution. This left the families of the remaining unidentified victims with nowhere to go.

The challenge for the odontologist then remains how to maximize the effectiveness of their skills in different circumstances. This can only be achieved by delegating many tasks, such as ante-mortem record collection, to others. Record collectors who may already be dental surgery assistants need to be trained in self-preservation to ensure their safety, and in their techniques of enquiry, so as not re-ignite deeply held passions or past pain in relatives.

Dental records need to be sought using every imaginable avenue and can present in many forms. For example, smiling photographs of the missing can be used for comparisons with skulls, either by 2-D/2-D superimposition, or 2-D against a particular projection and silhouette extracted from a 3-D CT image of the skull.

Of course, not all major losses of life are caused intentionally by humans. Natural disasters, epitomized by the recent 2004 tsunami in the Indian Ocean and the 2011 tsunami in Japan, killed many people. Similarly, the ever-increasing mass movements of people in aircraft, carrying more passengers with each new generation for economic efficiency, are certain to lead to the occasional disastrous loss of life on a large scale.

7.7.2 Education, training and certification

It is the preparedness to be able to scale the response and capability of the worldwide forensic odontological community to any set of prevailing demands that presents our biggest challenge. The solution to the paradox of small numbers of routine forensic cases, interposed for unpredictable periods between major disasters, can only come from structured education and training. Experience has shown that to insert the forensically untrained and psychologically unprepared dentist into a DVI situation is potentially disastrous, both for the dentist and for the investigation [85].

The key to the future rests with education and training, with those countries where forensic odontology is practiced at a high standard taking some responsibility to pass their expertise to neighboring countries still developing such expertise. Such education and training obligations can be supported financially, either by direct governmental support or through international NGOs. Such investments can strengthen investigatory expertise in neighboring countries, and this effectively extends border security for the donor nations. It also establishes strong professional bonds between colleagues who would normally rarely meet, but who one day may have to work side by side on a mass disaster investigation.

National societies all have an important role to play. Most began at a similar time in the last century, as loose associations of practitioners all having an interest in forensic odontology. In the ensuing decades, they have now all become sufficiently developed to be challenged by the need to establish standards, accredit or certify practitioners and seek specialist status for recognized experts in forensic odontology. However, this important transition to fully professional bodies also enables some standardization to be agreed and implemented. Where this has occurred, and practitioners who are normally widely geographically separated are drawn together in a shared task, that task becomes much easier, because everyone knows not only what to do themselves, but what their counterparts are doing to support them.

7.7.3 Attitudes to bitemarks

While there is a wide consensus on almost every aspect of forensic odontological practice, the issue of bitemark analysis remains problematic and controversial [24,25]. The issues have been described above, and some of the research solutions follow, but it is difficult to agree on how to deal with such evidence from an international perspective. The problem is that opinions seem to be polarized between those practitioners who have given opinions in many bitemark cases, and others who are so skeptical of the value of bitemark evidence that they consider it to be practically worthless. This is an unfortunate situation, and further research is certainly needed to improve current morphological comparisons between the dentition of biters and the bitten.

The situation is not helped by the fact that there have been important cases of people who have been convicted for long periods, and who were even facing execution, for murders they were later shown not to have committed when DNA analysis has excluded them from involvement with the victim [86]. There is also the uneasy feeling that, in many cases where opinions on bitemark evidence have been given, there has been little or no expert challenge by the legal defense team to refute the assertions of the prosecution. This may add to a list of successful prosecutions, but it does not validate the methodology employed.

The understandable backlash in some countries almost 'threw the baby out with the bathwater' and, in some jurisdictions, bitemark evidence has almost become inadmissible. This overlooks much that can be gleaned from a thorough investigation of a suspected bitemark injury. Most importantly, the ability to use a comparison of bitemark and the dentition of the accused to give unequivocal evidence to show the accused could never have made the injury cannot be overlooked. The exculpatory value of bitemark evidence must ensure that such evidence is documented and collected on every occasion.

The continuing use of bitemark evidence to positively identify the biter will require considerable research and judicial and peer assessment before universal acceptance – or indeed rejection – can take place. However, the current position of polarized views cannot continue.

7.8 Research objectives and barriers

7.8.1 History

Historically, hypothesis-driven scientific studies in forensic odontology have been limited. There are logical reasons for this. As described previously, many forensic odontologists are practicing dentists and do not engage in research activities. With some of the more important questions to be answered requiring access to large population data, research facilities and equipment, experiments are nearly impossible to perform outside of an academic institution. In addition, studies require ethical board review, which will severely limit the types of studies that can be performed. This, again, is difficult to accomplish outside of an academic environment. Input and involvement from academia, with adequate models, must therefore form the basis of future research. Traditionally, however, this input has been limited.

The main reason that has hindered academic involvement has been lack of funding opportunities in this area. The National Institute of Justice (NIJ) has been the main granting source in the forensic sciences for the United States. Other agencies that fund dental research, such as the National Institute of Health, National Science Foundation and the National Institute of Dental and Craniofacial Research, do not fund dental *forensic* projects. Given the absence of adequate funding, academic-based researchers cannot muster resources, and may not be attracted to perform the necessary studies to advance the field.

Some analysis of NIJ funding is supplied in the 2009 National Academy of Sciences (NAS) report [16]. It was noted that 21 projects were funded in 2007, but that none of the open questions about common forensic science methods were addressed in those projects. A conclusion of the report is that the level of support is well short of what is necessary.

In forensic odontology, a large portion of the literature has been based on case reports and method development for the practitioner. While these can be informative, they do not usually add to the research base and legal relevance in a way that investigatory empirical studies would. In spite of the difficulties, this trend is changing. In the last few years, there has been an increase in the number of empirical scientific studies from academic sources, creating an encouraging outlook as to the scientific basis of this discipline.

7.9 Current state of research

There are three main areas of forensic odontology that represent possible research avenues, and each of these areas has been described previously:

1. *Victim identification* (including the important areas of disaster victim identification and missing/unidentified persons).
2. *Age estimation.*
3. *Bitemark analysis.*

7.9.1 Victim identification

Victim identification is essentially based on comparison of ante-mortem and post-mortem information, including radiographs, written information in dental charts, or even facial photographs (the latter has become more important in recent world turmoil in countries where victims may not have dental records – see above in section 7.7: The international perspective) [87].

The area of victim identification as a whole was not criticized in the 2009 NAS report for having a lack of scientific foundation, and it is considered to be a well-established means of identification [16]. However, as the identification process still involves elements of pattern recognition, it may be subject to issues of bias and selection [35]. This may potentially be more significant in DVIs in which multiple subjects may have similarities in the dentition, or which lack high-quality and complete ante-mortem dental records. In the USA, there is no established minimum number of concordance points necessary to determine identity, with only a single study examining the performance of odontologists in identification cases [81]. These, then, represent the frontiers of research yet to be established in the arena of dental victim identification.

Recent research has focused on the presence, analysis and recognition of the types and brands of dental materials present in the dentition. Development of dental products has resulted in the potential presence of a wide array of materials, having varied chemical and physical properties. These properties can be used to determine brand or source of material, potentially adding another level of certainty or point of concordance to the process [88].

Other advances in the field have included digital radiography and the development of computerized matching systems that have revolutionized the processing of victims in mass disasters [51,52]. These advances, tied with the evolution of missing/unidentified person systems, point to the continuing fundamental value of dental contribution to forensic identification and avenues of future research.

7.9.2 Age estimation

Dental age estimation concerns the issues of whether an individual might be considered an adult in a given jurisdiction, and also on estimation of the age of a decedent. The NAS report made no reference to this aspect of forensic dentistry in its review, and no area in need of research was identified in the report.

Research in this area has been confounded by the variety of proposed methods of assessment, making meta-analysis of the existing data difficult at best. It has been suggested that age estimation data have population dependence but, due to the small sample size of most studies, it may be difficult to separate the effect of outliers from population effects. At best, current methods may return a probability that an individual is within a certain age range, but they can have low precision as the estimated range might be large, depending on the methods used.

A recent search on the Web of Science database using the terms 'dental age estimation' yielded 183 citations, 150 of which were published in the last decade, illustrating the level of recent scholarly activity in this field. The most promising research direction in this area would be to combine dental methods with other anthropological measures [57], such as craniometrics and wrist development. Further research in this area would require large population studies, using multiple age estimation methods in collaboration with forensic anthropologists [89].

7.9.3 Bitemark analysis

While the areas of victim identification and age estimation did not receive criticism in the NAS report, the discipline of bitemark analysis was scrutinized. The evaluation and summary assessment of bitemark analysis outlined the areas lacking in foundational study, stressing that this area was in great need of exploration.

With regard to bitemark analysis, there are two basic postulates: that human dentition is unique and that those unique features reliably transfer to the skin. It can be said that these are not two separate issues [19]. For example, if the human dentition were found to be unique, and it was determined that that unique detail does not transfer to the skin, then uniqueness of the dentition with regard to a bitemark is immaterial. Conversely, if the patterns of the teeth transfer accurately, but the human dentition is not unique to each individual, then the fidelity of transference is of no consequence. Recent research results have suggested that both of these statements require investigation [26,30,32].

Study into the second postulate of accurate detail transfer is hampered by strong limitations posed by ethical review boards. Performing research on live human volunteers in an academic setting is highly regulated; thus, the variables associated with the wound response, bruising and changes/distortion associated with these variables cannot be investigated. Some of the variables in the living or perimortem interval include age, health factors, nutrition, hydration, time and force, coagulation status, lividity, decomposition, temperature, time delays, bitemark aging, and healing or decomposition changes in living vs. dead. It is clear that there are significant barriers to research in these areas.

Human Subject Institutional Review Boards (HSIRB), which review proposals of scientific study prior to their start, determine whether they meet ethical requirements for human involvement. Volunteer participation is carefully reviewed and monitored, limiting the types of studies that can be performed, and particularly those that may cause possible harm to an individual. Bitemarks are usually inflicted during violent altercations and with enough force to cause a substantial wound. Replicating this violent act in a study on a volunteer is simply not possible.

In order to acquire HSIRB approval, informed consent must be obtained from the volunteer. Informed consent must describe to the volunteer what will happen to them during the study and any foreseeable adverse events that may occur. This allows the volunteer to decide whether or not they are willing to take the risks posed by engaging in the study. In bitemark analysis, there is no research that allows for proper description of what these risks may be, or when they may occur during the study. The foundation is not in place even to form the basis for a proper HSIRB proposal, regardless of replication of the violent act.

For example, the questions of how much force needs to be applied to bring the volunteer to the pain threshold, how much to cause a bruise to form and how much to avulse the skin are unanswered. Severe harm could be inflicted on a volunteer without this prior knowledge; scarring, infection or worse could result. This is why HSIRB review and approval are necessary.

The NAS report stated that investigation into these areas would not be infeasible. Clearly, however, there are exceptions to this statement. In the discipline of bitemark analysis, this *is* infeasible. There are other areas of forensic science that are also hampered by the inability to perform research on living subjects. The questions surrounding shaken baby syndrome are a prime example, as experimenting with this act on a child is not possible. These limitations have stressed the need for models and, while not exact or sometimes even suitable substitutes, they would allow investigation into areas that otherwise would not be open to exploration.

In bitemark research, various models have been employed to study the transfer of teeth to a substrate. Media such as wax and Styrofoam have been commonly used, but these lack the properties found in skin, undergoing permanent plastic deformation rather than visco-elastic change as would be seen with human tissue. Skin is also anisotropic (having properties dependent on direction) and rebounds, making wax and Styrofoam poor analogs for comparison. These materials have, however, provided for some simple baseline studies in pattern analysis.

Animal model studies, in particular using live pigs or pigskin, have also been used as a substitute for live human beings. These studies have claimed the advantage of biting into skin. Studies performed on live animals potentially allow investigation into the vital response, permitting study of bruising, wounding and changes associated with these variables [90]. Though performed on an animal model, HSIRB review is still required. Proper and ethical use of animals is highly regulated.

The main disadvantage of a pigskin model is that it lacks the proper anatomic configuration of a human. As stated earlier, skin is anisotropic – that is, it has different properties in different directions on the anatomy. In the direction parallel to tension lines, the skin is inherently tighter and, while in the perpendicular direction it is looser. This affects how much the skin can stretch, and thus distort, in any given direction when bitten. These patterns are not the same in an animal as they would be in a human.

Cadaver models have also been utilized. These offer the advantage of proper anatomical configuration but lack the vital response. Bruising and wounding can obviously not be studied. However, indentation of skin and resulting distortion can be investigated, and lack of wound response may be seen as an advantage as this allows for study of clear indentations. With the variable of wound response eliminated, distortion of the tissue itself can be observed, as many of the biomechanical factors of skin are retained in properly stored cadavers. Thus, the viscoelastic, anisotropic nature of skin that undergoes a nonlinear response to stress can be researched [33].

Cadaver models have been used to study the biomechanical properties of skin for well over 150 years, and still represent perhaps one of the best possible models for continued research into some of the variables associated with bitemarks. Some of these variables include body location, elasticity and contour, intermediate material (i.e. clothing), post-mortem positioning, and also the effect of orthodontic and other dental treatment on bite appearance and distortion.

The other premise of bitemark analysis focuses on the uniqueness of the human dentition. Studies that have explored this issue have used sample sizes that have been too small, have lacked a formal statistical approach or have used a flawed statistical method. To date, statistical studies both in 2-D and 3-D have been performed [30–32]. Demonstrating uniqueness may not be attainable [91], since it is impossible to measure all of the dentitions that exist and have existed [92]. Rather than consider uniqueness, the parameters that describe and define closeness of fit might better be studied. An emerging direction is to combine a quantitative measure of dental similarity with a quantitative measure of variation in bitemarks, to provide a realistic assessment of evidentiary threshold.

A 2010 study has already shown the dramatic effect of orthodontic treatment on dental similarity. This result may have been anticipated, given that the goal of orthodontic treatment is to attain a homologous dental alignment. The results of other studies have also shown that dental match rates are population-dependent [30,93].

Concomitant with the development of research models, there is a need for experimental model validation, determination of quantitative measurement error in teeth and bitemarks and, also, development of frequency statistics of dental characteristics. Similarly, establishment of numeric thresholds in evidence quality and interpretation guidelines and thresholds could be considered under-represented in the scientific literature [22].

Further issues that warrant study include the effects of perceptual and cognitive bias, expert vs. non-expert performance, validation trials, operator concordance rates, empirical error rates and operator reproducibility. Such studies have the benefit of utilizing authentic forensic materials and using construct validity to assess efficacy. For example, if forensic dentists cannot agree that an injury is a bitemark, then the actual causation is irrelevant – the study has demonstrated a lack of reliability. Such construct studies are being undertaken currently and are providing worrying evidence about the robustness of the processes undertaken by forensic dentists in their approaches to bitemark evidence.

7.9.4 Future direction

In order to attract academic-based research to the field of forensic odontology, funding must be made available. The future of forensic odontology research will depend on this. Hypothesis-driven scientific study will help to provide and fill many of the voids in forensic odontology, but this can only be accomplished if resources are made available. Future research direction should continue to focus on development of appropriate models, in which variables can be identified and controlled. Once established, such models can be used for educational and testing purposes, to provide validation for the field.

7.10 The future of forensic odontology

By the end of the 20th century, forensic odontology was an accepted discipline in much of the world. The discipline had developed not as an outgrowth solely of academic dentistry, but by means that relied on the experience of relatively few individuals. The methods were largely adapted from those used in the practice of dentistry. Forensic odontologists operated more or less independently from others in their specialized field, and they often had associated strong influences from their coroners, medical examiners, police and judiciary. In the latter half of the 20th century, the number of cases increased and forensic dentists worldwide began to form professional associations and to develop guidelines and standards. Forensic odontology presentations at professional meetings, and the publication of books and articles, increased during those years. Those cases that dealt with dental identification and dental age estimation were largely based on traditional dental knowledge bases. The same cannot be said of those cases dealing with bitemarks.

This was new territory for most dentists – the identification of individuals from marks made by the teeth. Publications during these formative years were dominated by case presentations and descriptions of existing, new or proposed techniques. The published science dealing with bitemark analysis was limited. Although there were individuals within forensic odontology who urged caution and implored forensic dentists to apply scientific principles to their work, their advice was unheeded by some in the area of bitemark analysis, sometimes with disastrous results. The preceding content of this chapter describes the issues and the research efforts surrounding bitemarks.

It is the opinion of the authors that, if bitemark analysis is to survive and, by association, forensic odontology itself, forensic dentists must appreciate that the framework within which they operated during the last century is gone and that the discipline must change – and change rapidly. Change will require new skills and learning to work within a new framework of scientific and legal expectations. Many of these changes will apply to all aspects of forensic odontology.

With the adoption of new technologies in forensic pathology environments, such as routine full-body CT scanning, the numbers of traditional invasive autopsies may reduce or become much more targeted. This will enable (or oblige) virtual autopsies to be undertaken, and odontologists will have to become familiar with non-invasive approaches for their own observations.

The increasing maturity of three-dimensional morphometric analyses, coupled with the increasing use and affordability of 3-D scanners, will generate large amounts of normative population-specific data. These data are already being used to create facial archetypes. The practical utilization of sophisticated statistical algorithms for assessing similarities and differences in facial form will become commonplace to corroborate or discount identity. Forensic odontology credentialing bodies are already gearing their training and testing methods toward the use of these new technologies. These have obvious implications for expert witness testimony and how the courts weigh it, and indeed for the whole field of reconstructive anatomy in the quest for identification.

Currently, bitemark analysis is characterized by strongly divided opinions among the forensic odontology community. Opinions range widely from those who are very convinced of their ability to associate a particular biter to a specific injury, to those with deeply held skepticism about such claims. Such a wide divergence of opinions is not helpful to the collective reputation (or morale) of forensic odontologists. This problem must be addressed urgently in a very objective way, so that a consensus based upon a strong evidential base can emerge. This will require a deeper analysis than counting the number of 'successful' opinions given.

The adversarial systems of justice in various parts of the world have failed to challenge bitemark analysis evidence adequately to assure that those charged with crimes receive equal treatment under the law. Bitemark evidence is powerful and influential in courtrooms and must be able to withstand scrutiny from all viewpoints. This can only happen if the evidence is based on solid and sound principles, properly applied and equally available to all. The delayed exonerations of persons who were wrongly convicted, based on unequal availability of expertise or the use of questionable odontological evidence, is ample proof of the need for change.

The key to the continuing development of forensic odontology from a small base of keen (and frequently unremunerated) enthusiasts to a fully recognized specialty with a career path within dentistry, lies in formal education and training. This must then be coupled to certification of individual practitioners and accreditation of the certifying organizations by highly respected professional bodies that are recognized by scientific and governmental organizations.

Each country will have to develop its own path. In North America, the American Board of Forensic Odontology is accredited by the Forensic Specialties

Accreditation Board and offers certification to those practitioners who qualify, apply and pass the board examinations. In Australia, practitioners of forensic odontology are soon to have their expertise examined under the auspices of the Royal Australasia College of Pathologists before registration authorities will grant specialist status on the dental register. In the UK, the Council for the Registration of Forensic Practitioners (CRFP) made headway into the process before it ultimately collapsed. Other countries have varying mechanisms, but the trend is, appropriately, moving toward more rigorous oversight.

This movement is a great opportunity for international cooperation toward the harmonization of guidelines, standards and best practices worldwide. The movement of people around the world and across borders, both legally and illegally, has never been greater; it can be anticipated to continue to increase, as the world's population continues to increase while resources are finite or diminishing and some nations remain more prosperous than others.

The estimation of age from the developmental status of the dentition will require an almost continuous revision of standards and reference values for people of different ancestry and customs. Fortunately, with the advent of cone beam CT imaging and the low dose of radiation needed, the availability of data to build and refine reference ranges for the chronology of dental development should become a manageable task.

One of the great benefits of modern communications is the ability to use telemedicine technology as a teaching tool. A significant problem confronting educators in recent times has been the antagonistic relationship between their institutions competing for enrollments for financial reasons. The collegiality that would enable colleagues in different institutions to combine their respective expertise would provide a teaching resource second to none. Some universities around the world have become aware of such problems.

One large consortium of universities from the USA, Canada, the UK, Asia and Australia – Universitas 21 – is striving to break down the barriers to learning and is attempting to draw upon the forensic odontological expertise distributed unevenly across the consortium to develop e-learning packages so that forensic odontology can be taught at least at a basic level to all dental students in every dental school in the group. There are other universities worldwide with existing forensic odontology programs, but they are conspicuously missing from this consortium. Their addition, or the creation of a comprehensive consortium, could benefit all.

At a higher level, it is entirely conceivable that courses at the graduate level could be constructed which would bring together international experts to present cases or demonstrate techniques to colleagues around the world, both in real time and as recorded modules in an e-learning library for self-directed study. To assist with recognition and accreditation for participants, the contributing organization most suitable for the candidate's registration could award the degree/diploma/certificate of proficiency as appropriate. The e-learning courses could be independent of, part of, or in addition to existing programs.

These proposals may make some current practitioners uneasy, but they are inevitable and are being driven by the ever more rigorous requirements of science and the courts. It can be envisaged that a structure similar to other medical or dental specialties will come into existence. In this structure, general practitioners may still do the bulk of the identification tasks, but within a hierarchical structure.

A comprehensively qualified and suitably certified forensic odontologist employed by a government agency or a university with the agency's imprimatur could oversee the activities of the group of odontologists comprised of people with different levels of experience and expertise. Those odontologists in the leadership positions would supervise and mentor the developing odontologists and newcomers. This would ensure the continuance of a culture of lifelong learning and maintain a very functional group in perpetuity. Implicit in such a structure is both a career path and the obligation of governments to fund such structures. (Both are currently lacking.) This structure should be seen as a national and international resource to be relied upon in times of emergency. When a tsunami smashes or a plane crashes, it is too late to Google 'forensic odontology'!

Education and training are the keys to the future of forensic odontology. Universities must develop formal post-doctoral programs in forensic odontology. Such programs must be comprehensive, with strong odontology, head and neck anatomy, oral pathology, oral medicine, human development, diagnostic imaging, research and legal components. They must necessarily encompass the entirety of forensic odontology theory and techniques, from those well established to the cutting edge and state of the art.

Forensic odontologists worldwide must continue to encourage the development of rigorous certification programs that are vetted and accredited by appropriate bodies. New and existing certification bodies must develop, and continuously update, guidelines, standards and best practices based on sound principles. They must be willing to enforce the adherence to those standards by their certified members, including standards for ethics. Certified forensic odontologists must commit to continuous learning and must be able to demonstrate proficiency in all phases of the discipline. These are rational, prudent and manageable goals. With determined effort, the future of forensic odontology will be very bright.

References

1. Pretty IA. Forensic dentistry: 1. Identification of human remains. *Dental Update* 2007; 34(10):621–622, 624–626, 629–630.
2. Sweet D. Forensic dental identification. *Forensic Science International* 2010; 201(1–3):3–4.
3. Rothwell BR. Principles of dental identification. *Dental Clinics of North America* 2001; 45(2):253–270.
4. Rothwell BR. Bite marks in forensic dentistry: a review of legal, scientific issues. *Journal of The American Dental Association* 1995; 126(2):223–232.

5. Thevissen PW, Fieuws S, Willems G. Human dental age estimation using third molar developmental stages: does a Bayesian approach outperform regression models to discriminate between juveniles and adults? *International Journal of Legal Medicine* 2010; 124(1):35–42.
6. Wood RE. Forensic aspects of maxillofacial radiology. *Forensic Science International* 2006; 159 Suppl 1: S47–S55.
7. Chrz B. Forensic dentistry. Civil litigation and the narrative report. *Oklahoma Dental Association (ODA) Journal* 1997; 88(2):31, 42, 54.
8. Rawson RD, Nelson BA, Koot AC. Mass disaster and the dental hygienist: the MGM fire. *Dent Hygiene* 1983; 57(4):12, 17–18.
9. Sweet D. Why a dentist for identification? *Dental Clinics of North America* 2001; 45(2):237–251.
10. Lincoln HS, Lincoln MJ. Role of the odontologist in the investigation of domestic violence, neglect of the vulnerable, and institutional violence and torture. *Forensic Science International* 2010; 201(1–3):68–73.
11. Rollo F, Mascetti M, Cameriere R. Titian's secret: comparison of Eleonora Gonzaga della Rovere's skull with the Uffizi portrait. *Journal of Forensic Sciences* 2005; 50(3):602–607.
12. Zarranz MC. Dr. Oscar Amoedo's book and its influence on the creation of the Forensic Dental Institute in Japan. *Revista del Museo de la Facultad de Odontología de Buenos Aires* 1999; 14(28):19–20.
13. Amoedo O. Fire on the Charity Bazaar. *Revista del Museo de la Facultad de Odontología de Buenos Aires* 1998; 13(6):16–20.
14. Brummit P, Stimson PG. History of Forensic Dentistry. In: Senn DR, Stimson PG, editors. *Forensic Dentistry.* 2nd ed. Boca Raton, FL: CRC Press; 2010; 11–25.
15. ASFO. History of the ASFO. In: Herschaft EE, editor. *ASFO Manual of Forensic Odontology.* Albany, NY: Impress Printing & Graphics; 2006; 112–134.
16. National Academy of, Sciences. *Strengthening Forensic Science in the United States: A Path Forward.* Washington, DC: National Academies Press; 2009.
17. Hinchliffe J. Forensic odontology, part 4. Human bite marks. *British Dental Journal* 2011; 210(8):363–368.
18. Barry LA. Bite mark evidence collection in the United States. *Bulletin of The History Of Dentistry* 1994; 42(1):21–27.
19. Pretty IA. Forensic dentistry: 2. Bitemarks and bite injuries. *Dental Update* 2008; 35(1):48–50, 53–54, 57–58.
20. Sweet D, Shutler GG. Analysis of salivary DNA evidence from a bite mark on a body submerged in water. *Journal of Forensic Sciences* 1999; 44(5):1069–1072.
21. Sweet D, Hildebrand, D. Saliva from cheese bite yields DNA profile of burglar: a case report. *International Journal of Legal Medicine* 1999; 112(3):201–203.
22. Pretty IA. Development and validation of a human bitemark severity and significance scale. *Journal of Forensic Sciences* 2007; 52(3):687–691.
23. Pretty IA, Sweet D. The scientific basis for human bitemark analyses – a critical review. *Science and Justice* 2001; 41(2):85–92.
24. Pretty IA, Sweet D. A paradigm shift in the analysis of bitemarks. *Forensic Science International* 2010; 201(1–3):38–44.

25. Clement JG, Blackwell SA. Is current bite mark analysis a misnomer? *Forensic Science International* 2010; 201(1–3):33–37.
26. Bush MA, Thorsrud K, Miller RG, Dorion RB, Bush PJ. The response of skin to applied stress: investigation of bitemark distortion in a cadaver model. *Journal of Forensic Sciences* 2010; 55(1):71–76.
27. Bush MA, Cooper HI, Dorion RB. Inquiry into the scientific basis for bitemark profiling and arbitrary distortion compensation. *Journal of Forensic Sciences* 2010; 55(4):976–983.
28. Bush MA, Miller RG, Bush PJ, Dorion RB. Biomechanical factors in human dermal bitemarks in a cadaver model. *Journal of Forensic Sciences* 2009; 54(1):167–176.
29. Rawson RD, Ommen RK, Kinard G, Johnson J, Yfantis A. Statistical evidence for the individuality of the human dentition. *Journal of Forensic Sciences* 1984; 29(1):245–253.
30. Sheets HD, Bush PJ, Brzozowski C, Nawrocki LA, Ho P, Bush MA. Dental shape match rates in selected and orthodontically treated populations in New York State: a two-dimensional study. *Journal of Forensic Sciences* 2011; 56(3):621–626.
31. Sheets HD, Bush MA. Mathematical matching of a dentition to bitemarks: use and evaluation of affine methods. *Forensic Science International* 2011; 207(1–3):111–118.
32. Bush MA, Bush PJ, Sheets HD. Statistical evidence for the similarity of the human dentition. *Journal of Forensic Sciences* 2011; 56(1):118–123.
33. Miller RG, Bush PJ, Dorion RB, Bush MA. Uniqueness of the dentition as impressed in human skin: a cadaver model. *Journal of Forensic Sciences* 2009; 54(4):909–914.
34. Bowers CM. Problem-based analysis of bitemark misidentifications: the role of DNA. *Forensic Science International* 2006; 159 Suppl 1: S104–S109.
35. Page M, Taylor J, Blenkin M. Context Effects and Observer Bias-Implications for Forensic Odontology. *Journal of Forensic Sciences* 2011; 57(1):108–112.
36. Page M, Taylor J, Blenkin M. Reality bites-A ten-year retrospective analysis of bitemark casework in Australia. *Forensic Science International*. In press.
37. Hinchliffe J. Forensic odontology, part 2. Major disasters. *British Dental Journal* 2011; 210(6):269–274.
38. Hill AJ, Hewson I, Lain R. The role of the forensic odontologist in disaster victim identification: lessons for management. *Forensic Science International* 2011; 205(1–3):44–47.
39. Kvaal SI. Collection of post mortem data: DVI protocols and quality assurance. *Forensic Science International* 2006; 159 Suppl 1: S12–S14.
40. DeValck E. Major incident response: collecting ante-mortem data. *Forensic Science International* 2006; 159 Suppl 1: S15–S19.
41. Adams BJ. Establishing personal identification based on specific patterns of missing, filled, and unrestored teeth. *Journal of Forensic Sciences* 2003; 48(3):487–496.
42. Scanlon R. DMORT and the future of PADIT. *Pennsylvania Dental Journal* 2002; 69(2):39–40.
43. Rose DE, Williams JM. Walking humbly. Minnesota's DMORT team members remember 9–11. *Northwest Dentistry* 2002; 81(5):29–32.

44. Hampl P. Forensic dentistry. Beyond recognition. *Oklahoma Dental Association (ODA) Journal* 1997; 88(2):18–20.
45. Wood RE, Kogon SL. Dental radiology considerations in DVI incidents: A review. *Forensic Science International* 2010; 201(1–3):27–32.
46. Schuller-Gotzburg P, Suchanek J. Forensic odontologists successfully identify tsunami victims in Phuket, Thailand. *Forensic Science International* 2007; 171(2–3):204–207.
47. Pittayapat P, Thevissen P, Fieuws S, Jacobs R, Willems G. Forensic oral imaging quality of hand-held dental X-ray devices: comparison of two image receptors and two devices. *Forensic Science International* 2010; 194(1–3):20–27.
48. Pittayapat P, Oliveira-Santos C, Thevissen P, Michielsen K, Bergans N, Willems G, *et al.* Image quality assessment and medical physics evaluation of different portable dental X–ray units. *Forensic Science International* 2010; 201(1–3):112–117.
49. Hermsen KP, Jaeger SS, Jaeger MA. Radiation safety for the NOMAD portable X-ray system in a temporary morgue setting. *Journal of Forensic Sciences* 2008; 53(4):917–921.
50. Lain R, Taylor J, Croker S, Craig P, Graham J. Comparative dental anatomy in Disaster Victim Identification: Lessons from the 2009 Victorian Bushfires. *Forensic Science International* 2011; 205(1–3):36–39.
51. Lewis C, Leventhal L. Combining the Locator System with WinID3—identifying victims from dental remains in a large disaster. *Refuat Hapeh Vehashinayim* 2007; 24(3):6–11, 53.
52. Al-Amad SH, Clement JG, McCullough MJ, Morales A, Hill AJ. Evaluation of two dental identification computer systems: DAVID and WinID3. *The Journal of Forensic Odonto-Stomatology* 2007; 25(1):23–29.
53. McGivney J. Commentary on: Lewis C. WinID2 versus CAPMI4: two computer-assisted dental identification systems. *Journal of Forensic Sciences* 2002; 47 (3):536–538. *Journal of Forensic Sciences* 2003; 48(2):472.
54. Lewis C. WinID2 versus CAPMI4: two computer-assisted dental identification systems. *Journal of Forensic Sciences* 2002; 47(3):536–538.
55. Thompson AT. Lectures on medical jurisprudence now in course of delivery at London University. *The Lancet* 1836; 1:281–286.
56. Saunders E. *The teeth a test of age, considered with reference to the factory children: addressed to the Members of both Houses of Parliament*. London: Hansard; 1837.
57. Cunha E, Baccino E, Martrille L, Ramsthaler F, Prieto J, Schuliar Y, *et al.* The problem of aging human remains and living individuals: a review. *Forensic Science International* 2009; 193(1–3):1–13.
58. Schmeling A, Geserick G, Reisinger W, Olze A. Age estimation. *Forensic Science International* 2007; 165(2–3):178–181.
59. Gonzalez-Colmenares G, Botella-Lopez MC, Moreno-Rueda G, Fernandez–Cardenete JR. Age estimation by a dental method: a comparison of Lamendin's and Prince & Ubelaker's technique. *Journal of Forensic Sciences* 2007; 52(5):1156–1160.
60. Cardoso HF. Accuracy of developing tooth length as an estimate of age in human skeletal remains: the deciduous dentition. *Forensic Science International* 2007; 172(1):17–22.

61. Schmeling A, Reisinger W, Geserick G, Olze A. Age estimation of unaccompanied minors. Part I. General considerations. *Forensic Science International* 2006; 159 Suppl 1: S61–S64.
62. Olze A, Reisinger W, Geserick G, Schmeling A. Age estimation of unaccompanied minors. Part II. Dental aspects. *Forensic Science International* 2006; 159 Suppl 1: S65–S67.
63. Anderson DL, Thompson GW, Popovich F. Age of attainment of mineralization stages of the permanent dentition. *Journal of Forensic Sciences* 1976; 21(1):191–200.
64. Demirjian A, Levesque GY. Sexual differences in dental development and prediction of emergence. *Journal of Dental Research* 1980; 59(7):1110–1122.
65. Lewis JM, Senn DR. Dental age estimation utilizing third molar development: A review of principles, methods, and population studies used in the United States. *Forensic Science International* 2010; 201(1–3):79–83.
66. Pretty IA, Addy LD. Associated postmortem dental findings as an aid to personal identification. *Science and Justice* 2002; 42(2):65–74.
67. Willems G, Moulin-Romsee C, Solheim T. Non-destructive dental-age calculation methods in adults: intra- and inter-observer effects. *Forensic Science International* 2002; 126(3):221–226.
68. Soomer H, Ranta H, Lincoln MJ, Penttila A, Leibur E. Reliability and validity of eight dental age estimation methods for adults. *Journal of Forensic Sciences* 2003; 48(1):149–152.
69. Ritz-Timme S, Cattaneo C, Collins MJ, Waite ER, Schutz HW, Kaatsch HJ, *et al.* Age estimation: the state of the art in relation to the specific demands of forensic practice. *International Journal of Legal Medicine*, 2000. 113(3): p. 129–136.
70. Aynsley-Green A. Unethical age assessment. *British Dental Journal* 2009; 206(7):337.
71. Maples WR. An improved technique using dental histology for estimation of adult age. *Journal of Forensic Sciences* 1978; 23(4):764–770.
72. Bang G, Ramm E. Determination of age in humans from root dentin transparency. *Acta Odontologica Scandinavica* 1970; 28(1):3–35.
73. Johanson G. Age determinations from human teeth: a critical evaluation with special consideration of changes after fourteen years of age. *Odontologisk Revy* 1971; 22:122–126.
74. Kvaal SI, Kolltveit KM, Thomsen IO, Solheim T. Age estimation of adults from dental radiographs. *Forensic Science International* 1995; 74(3):175–185.
75. Demirjian A, Goldstein H, Tanner JM. A new system of dental age assessment. *Human Biology* 1973; 45(2):211–227.
76. Mincer HH, Harris EF, Berryman HE. The A.B.F.O. study of third molar development and its use as an estimator of chronological age. *Journal of Forensic Sciences* 1993; 38(2):379–390.
77. Lamendin H, Baccino E, Humbert JF, Tavernier JC, Nossintchouk RM, Zerilli A. A simple technique for age estimation in adult corpses: the two criteria dental method. *Journal of Forensic Sciences* 1992; 37(5):1373–1379.
78. Waite ER, Collins MJ, Ritz-Timme S, Schutz HW, Cattaneo C, Borrman HI. A review of the methodological aspects of aspartic acid racemization analysis for use in forensic science. *Forensic Science International* 1999; 103(2):113–124.

79. Spalding KL, Buchholz BA, Bergman LE, Druid H, Frisen J. Forensics: age written in teeth by nuclear tests. *Nature* 2005; 437(7057):333–334.
80. Hinchliffe J. Forensic odontology, Part 1. Dental identification. *British Dental Journal* 2011; 210(5):219–224.
81. Pretty IA, Pretty RJ, Rothwell BR, Sweet D. The reliability of digitized radiographs for dental identification: a Web-based study. *Journal of Forensic Sciences* 2003; 48(6):1325–1330.
82. Graham JP, O'Donnell CJ, Craig PJ, Walker GL, Hill AJ, Cirillo GN, *et al.*The application of computerized tomography (CT) to the dental ageing of children and adolescents. *Forensic Science International* 2010; 195(1–3):58–62.
83. Evans S, Jones C, Plassmann P. 3-D imaging in forensic odontology. *Journal of Visual Communication in Medicine* 2010; 33(2):63–68.
84. Jackowski C, Lussi A, Classens M, Kilchoer T, Bolliger S, Aghayev E, *et al.* Extended CT scale overcomes restoration caused streak artifacts for dental identification in CT – 3-D color encoded automatic discrimination of dental restorations. *Journal of Computer Assisted Tomography* 2006; 30(3):510–513.
85. Webb DA, Sweet D, Pretty A. The emotional and psychological impact of mass casualty incidents on forensic odontologists. *Journal of Forensic Sciences* 2002; 47(3):539–541.
86. Bowers CM, Pretty IA. Expert disagreement in bitemark casework. *Journal of Forensic Sciences* 2009; 54(4):915–918.
87. Bernstein ML. The application of photography in forensic dentistry. *Dental Clinics of North America* 1983; 27(1):151–170.
88. Bush MA, Miller RG, Norrlander AL, Bush PJ. Analytical survey of restorative resins by SEM/EDS and XRF: databases for forensic purposes. *Journal of Forensic Sciences* 2008; 53(2):419–425.
89. Martrille L, Ubelaker DH, Cattaneo C, Seguret F, Tremblay M, Baccino E. Comparison of four skeletal methods for the estimation of age at death on white and black adults. *Journal of Forensic Sciences* 2007; 52(2):302–307.
90. Avon SL, Victor C, Mayhall JT, Wood RE. Error rates in bite mark analysis in an *in vivo* animal model. *Forensic Science International* 2010; 201(1–3):45–55.
91. Cole SA. Forensics without uniqueness, conclusions without individualization: the new epistemology of forensic identification. *Law, Probability and Risk* 2009; 8(3):1–23.
92. Saks MJ, Koehler JJ. The individualization fallacy in forensic science evidence. *Vanderbilt Law Review* 2008; 199:199–219.
93. Bush MA, Bush PJ, Sheets HD. Similarity and match rates of the human dentition in three dimensions: relevance to bitemark analysis. *International Journal of Legal Medicine* 2011; 125(6):779–784.

8

Forensic psychiatry and forensic psychology

Stephen B. Billick[1] and Daniel A. Martell[2]

[1]*New York Medical College, Valhalla, New York, USA*
[2]*Park Dietz and Associates, Newport Beach, California, USA*

Forensic psychiatry and forensic psychology are the explication of psychiatric and psychological issues as they pertain to an issue in the legal arena. Mental health experts have been asked by the courts and attorneys to help them understand the implication of mental health status to individuals involved in the legal process for a very long time. In the United States, the expert may be retained either by opposing counsel or by the court itself.

Forensic psychiatry has been practiced as long as humans have been confounded by erratic human behavior. Some believe that the first expert evidence in a murder trial was in ancient Babylonia in 1850 BC, given by a midwife. The code of Hammurabi (1800 BC) established 'intent' as an important factor in criminal law. Regarding the issue of 'intent', Deuteronomy 19:1–13 (1200 BC) in Hebrew scripture describes the logic for establishing 'refuge cities', where a person who had accidentally killed someone would be safe from avenging relatives [1]. Understanding human behavior and mental illness has long been of interest. Hippocrates described the 'wandering uterus' to explain female psychosomatic complaints.

8.1 History of psychiatry in the united states

Psychiatry has been an important part of American medicine since before the American Revolution. Benjamin Rush, MD was one of the founders of the first hospital in what is now the United States – the Pennsylvania Hospital. Founded in 1752, the hospital's mission included treating mental illnesses. Dr. Rush wrote the first textbook on psychiatry in the US, *Medical Inquiries and Observations upon the Diseases of the Mind*, published in 1812 [2]. Isaac Ray, MD wrote his *A Treatise on the Medical Jurisprudence of Insanity* in 1838 [3].

Forensic Science: Current Issues, Future Directions, First Edition. Edited by Douglas H. Ubelaker.

The American Psychiatric Association (APA) was founded in 1844 (originally the Association of Medical Superintendents of American Institutions for the Insane), and began publishing the *American Journal of Psychiatry* (AJP) (originally called the *American Journal of Insanity*). Both the APA and AJP were established well before the American Medical Association and its scientific journal. Indeed, it was the creation of the first juvenile justice system in the USA, in Illinois in 1899, that necessitated the creation of the first child and adolescent psychiatric clinic in Cook County, Illinois in 1909, with the expressed intent to aid the new juvenile court.

Moving closer to the modern scientific era of psychiatry, regarding training and certifying competence, in 1925, Karl Menninger, MD, a psychiatrist, submitted the first report on legal aspects of psychiatry to the APA. In 1933, American medicine established what is now called the American Board of Medical Specialties (ABMS), which began certifying physicians as competent in their specialties. Two years later, in 1935, the American Board of Psychiatry and Neurology (ABPN) was founded to certify competence in psychiatrists and neurologists. 1948 was the year that the present American Academy of Forensic Sciences (AAFS) was founded, which has a Section on Psychiatry and Behavioral Sciences.

In 1969, The American Academy of Psychiatry and the Law (AAPL) was founded to provide education on forensic psychiatry to improve standardized assessments and testimony. This led to the recognition of forensic psychiatry as a formal subspecialty of psychiatry, with the development of a certifying examination for competence by ABPN, recognized by the ABMS. Concurrently, the Accreditation Council on Graduate Medical Education (ACGME) created standards for formal residency training in forensic psychiatry.

In 2003 and afterwards, certification by the ABPN in forensic psychiatry could only be attained by first completing an ACGME approved forensic psychiatry residency, then passing the ABPN certifying examination. Certification for forensic psychiatry is for a ten-year period, and then recertification for another ten-year period is granted after passing a recertification examination for continued competence in forensic psychiatry. This ensures that the psychiatrist practicing forensic psychiatry is up to date with scientific and legal advances.

Scientific advances regarding diagnosis and treatment are critical for any medical specialty with implications for legal processes. The first successful psychopharmacologic treatment of any psychiatric disorder was in 1937, when Dr. Bradley used benzedrine in treating a child with what is now called attention deficit hyperactivity disorder [4].

The first successful treatment for severe mental disorders (schizophrenia, mania, psychotic depression) was electroconvulsive therapy, first performed by Drs. Bini and Cerletti in Italy in 1938 [5]. Lothar Kalinowski, MD, a German psychiatrist educated in Germany and Austria, was present at this procedure, and he brought it to New York City as he was fleeing the Nazis (Dr. Kalinowski had a Jewish parent and had gone initially to Italy for safety).

Chlorpromazine was introduced into the US in 1955 to treat schizophrenia. The MAO inhibitors (MAOi) were used for depression in the 1950s and the tricyclic antidepressants (TCA) were used beginning in the 1960s. Lithium was approved for bipolar disorder in the early 1970s. With the advancement of scientifically based treatments, diagnostic precision became more important as different treatments were targeted for different psychiatric entities.

American psychiatry developed and published, in 1952, the *Diagnostic and Statistical Manual of Mental Disorders* (DSM-I) [6] to standardize the nomenclature of the various psychiatric disorders that were recognized at the time. DSM-II, in 1968 [7], revised and updated the recognized mental disorders. However, the differential treatment options (antipsychotic medications, MAOis, TCAs and lithium) required greater specificity and sensitivity in diagnostic acumen.

In the early 1970s, the National Institute of Mental Health developed the Research Diagnostic Criteria (RDC) precisely to establish sensitivity, specificity, validity and inter-rater reliability in psychiatric diagnosis of mental disorders. Using the findings from the RDC research, the APA developed DSM-III [8], which included for the first time inclusionary criteria and exclusionary criteria for individual diagnoses. With continued research, the APA updated the DSM to the DSM-IV [9] and now has the current DSM-IV-TR [10], where there was a revision of the text within the book.

Since 2000, the DSM-IV-TR has been the scientific standard for psychiatric diagnosis in the United States. In the courts and legal arena, this standard can only be modified by demonstrating intervening updated research accepted by the scientific community as valid and reliable. Currently, the APA is updating to a future DSM-V through scientific work groups and committees.

Similarly, there are scientific evidence-based treatments that are accepted within the scientific and academic psychiatric community. The Food and Drug Administration certified medications and treatments for specific conditions based on the reliable and valid research. Within the realm of psychotherapies, there has been extensive research establishing behavioral therapy, cognitive therapy, motivational interviewing, rational emotive behavior therapy, dialectical behavioral therapy, relapse prevention therapy (for addictive disorders) and some others as being indicated for specific disorders and efficacious. Unlike the DSM, which is the 'gold standard' for diagnosis, there is no single text for treatments which is of the same stature. There are many excellent texts on psychiatric treatment, and these are used, along with updated current scientific research, in peer-reviewed journals as the basis for many expert psychiatric opinions in the legal processes.

Similar to general psychiatry, each of the subspecialties in psychiatry have several leading but not definitive texts. Perhaps forensic psychiatry has Richard Rosner, MD's *Principles and Practice of Forensic Psychiatry* second edition [11] as the single foremost text in the US, with several other quite useful texts in addition. Within the field of child and adolescent forensic psychiatry, Elissa Benedek, MD has published with co-authors a series of books over her career focusing on the issues of juveniles and their psychiatric needs in the legal arena. Her latest book, *Principles and Practice of Child and Adolescent Forensic Mental Health*[12], is the leading text currently available.

8.2 History of psychology in the United States

The history of psychology in America can be traced to William James, who established the first psychological laboratory at Harvard University in 1875, where he studied the functioning of the brain, localization of function to the cerebral hemispheres and the connection between the brain and human consciousness, perception, cognition and emotion [13]. Seventeen years later, the American Psychological Association was founded by G. Stanley Hall, who gathered 30 psychologists together at Clark University in 1892.

Clinical psychology, with its focus on the assessment, diagnosis and treatment of mental disorders, emerged from the work of Lightner Witmer, who established the first psychological clinic at the University of Pennsylvania shortly thereafter in 1896. However, it was not until the 1940s, when World War II brought an increasing demand for the psychological treatment of returning war veterans, that clinical psychologists turned their attention from psychological assessment to the psychological treatment of mental disorders.

Forensic psychology's role in the courtroom dates back to 1908, when Hugo Munsterberg published *On the Witness Stand*, the first book to outline the application of psychology to legal issues. However, the modern era of forensic psychology is anchored to the founding of the American Psychology-Law Society in 1969, followed by the establishment of the American Psychological Association's Division 41 (Psychology and Law) in 1981, and the American Psychological Association's recognition of forensic psychology as a specialty by the Commission for the Recognition of Specialties in Professional Psychology (CRSPP) in 2001. The American Board of Forensic Psychology (ABFP), a specialty board of the American Board of Professional Psychology (ABPP), was established in 1978 to offer board certification to licensed psychologists who meet the qualifications required by ABPP and ABFP, who pass a written examination, who submit two practice samples that have been approved by our practice sample review committee, and who pass a three hour oral examination.

Forensic psychologists have developed and validated a variety of tests for use in the assessment of specific forensic issues, including competency to stand trial, mental state at the time of the offense, malingering and deception, interrogative suggestibility, competency to waive Miranda rights, trauma exposure and violence risk assessment, to name a few.

8.3 History of the Section on Psychiatry and Behavioral Sciences in AAFS

Originally known simply as the Psychiatry Section, Psychiatry and Behavioral Science was one of the original sections of the Academy upon its founding in 1948. Hence, the AAFS Psychiatry and Behavioral Sciences Section has the distinction of being the first professional organization in the United States committed to forensic behavioral science.

The founder of the section was Val Satterfield, MD, an expert in sex offenders and a professor of clinical psychiatry at Washington University in St. Louis. He became President of the AAFS in 1957–58 and succeeded AAFS organizer Rutherford B.H. Gradwohl, MD as the Director of the St. Louis Police Department's Crime Research Laboratory [14]. Subsequent Psychiatry and Behavioral Science fellows who have become President of the AAFS include Maier Tuchler, MD in 1968–69; and Richard Rosner, MD in 1996–97.

The AAFS Psychiatry and Behavioral Science Section also has a long history of cooperation with, and involvement, in the development of the American Academy of Psychiatry and the Law (AAPL), which was founded in 1969. AAFS and the Forensic Sciences Foundation were co-sponsors with AAPL of the American Board of Forensic Psychiatry, Inc (ABFP). Subsequently, AAFS and AAPL jointly sponsored the Accreditation Council on Fellowships in Forensic Psychiatry (ACFFP) [15]. When the ABPN and ABMS recognized forensic psychiatry as a subspecialty and created a credentialing examination, the ABFP sunsetted and ceased to exist. Similarly, when the ACGME created residency standards for forensic psychiatry residency training, the ACFFP also sunsetted and cease to exist. Technically speaking, the terms 'fellow' and 'fellowship' have no legal medical standing in forensic psychiatry; the correct terminology would be 'forensic psychiatry resident' and 'forensic psychiatry residency'.

In 1982, the members of the section voted to include forensic psychologists among its members, and the Psychiatry Section became the Psychiatry and Behavioral Sciences Section. Currently, the section boasts an international membership, including members from the United States, Canada, Spain, Italy and Germany.

8.4 Areas of legal importance for psychiatric and psychological testimony

Psychiatrists are asked to do mental health assessments in criminal proceedings to help the court to establish criminal responsibility. Defendants may be not guilty by reason of insanity, or may have diminished capacity. The legal standards vary from state jurisdiction to state jurisdiction and also to the federal courts. Some states have abolished the 'insanity' defense and have instituted 'guilty but mentally ill' as a possible finding.

Forensic psychological and psychiatric assessment is also used to aid in civil suits, where the plaintiff may be suffering psychiatric distress, wholly or in part due to the injury or negligence. In civil suits, the defendant may seek psychiatric evaluation to diminish their responsibility for the injury or negligence.

Many different areas of psychological research are related to the legal and criminal justice systems. The most common are clinical forensic psychologists, who study issues such as the ways that psychopathology relates to crime, violence risk, amenability to treatment and civil psychological damages. Many other academic disciplines in psychology are also active in the forensic arena:

- Developmental psychologists study the decision-making ability of children and teenagers with regard to critical psycho-legal issues such as abortion and birth control.
- Cognitive psychologists study the underlying memory and judgment processes related to issues such as child sexual abuse allegations, 'recovered' memories, employment discrimination and medical malpractice.
- Experimental psychologists studying sensation and perception look at the impact of environmental conditions on the reliability of eyewitness identifications.
- Social psychologists study issues such as jury decision-making and the social psychological phenomena contributing to false confessions.

Hence, there is a large and diverse universe of psychologists applying research methodologies to study relevant forensic problems.

Acknowledging the diversity of forensic psychiatric and psychological research endeavors, there are nonetheless several specific topics that currently seem to dominate both forensic practice and the international research agenda. These major international research tracks include:

- forensic neuroscience;
- the search for predictors of violence and criminality – closely linked with the study of psychopathy
- methods for detecting malingering and deception;
- the study of ways to improve police practices;
- research related to jury selection and decision-making.

8.4.1 Forensic neuroscience

The application of brain science in general – and brain imaging in particular – to forensic issues is an area that some have dubbed 'neurolaw' [16]. Forensic neuroscience is already bringing evidence from the study of brain structure and function into the courtroom. Its application in civil litigation is perhaps the least controversial, in that brain scans provide useful direct evidence of brain lesions and neuronal disease that may be relevant to establishing causation and the assessment of damages. Here, neuroimaging is being used as demonstrative evidence of brain damage in support of tort claims (e.g. head injury, exposure to neurotoxins, medical malpractice), to sustain findings of dementia among elders involved in testamentary capacity (e.g. competency to execute or change a will), and in undue influence cases. It is also being presented regularly as mitigation evidence in capital sentencing [17,18], to help explain violent criminal behavior in an effort to establish diminished moral culpability.

Forensic neuroscience is also relevant to a broader range of criminal law issues, for example as evidence of diminished competency to proceed or participate at various points in the criminal process (i.e. to waive Miranda rights, to confess, to stand trial, to be executed); establishing permanent incompetence to stand trial pursuant to the US Supreme Court's landmark decision in Jackson v. Indiana (1972); or providing evidence of vulnerability to interrogative suggestibility and the production of false confessions.

Proponents of brain imaging maintain that brain scans provide a window into the workings of the mind, permitting jurors to observe for themselves abnormalities in the architecture or functioning of the brain that are posited to underlie the criminal behavior or tort damages at issue in a given case. Critics respond that there is a dearth of sensitive, specific and reproducible findings associating brain scan patterns with specific psychiatric disorders, much less the complex cognitive functions and behaviors at issue in court [19]. Significant limitations in the current state of neuroimaging research constrain its ability to inform legal decision-making [20–22].

Another scientifically controversial and largely untested use involves the use of functional magnetic resonance imaging (fMRI) as a neuroscience 'lie detector', the logistics and practicality of which cast doubt on its viability, at least in the near term [23–25]. There is also a neuroscience interest in studying psychopathy and the ways in which the brains of psychopaths may differ from 'normal' brains, both structurally and functionally [26].

8.4.2 Psychopathy and risk assessment for dangerousness

Forensic behavioral science has long been on a quest for reliable and valid methods to predict – and thus prevent or control – violent and dangerous behavior. Various actuarial methods and risk assessment instruments have been developed, and their application to real-world prediction has been one of the most hotly studied areas in forensic behavioral science over the past 20 years [27,28].

One of the most promising predictor variables to emerge from this line of research has been the personality construct known as psychopathy. Closely associated with antisocial personality disorder, but more malignant in form, psychopathy has been identified as a powerful violence risk factor, and its proper assessment and true predictive utility in children, adolescents and adults are areas of enormous international research activity [29–31].

8.4.3 Detection of malingering and deception

The concern that defendants in criminal cases, or plaintiffs in civil litigation, may exaggerate their problems for secondary gain, is a major concern in forensic psychiatry and psychology. Psychiatric methods for the detection of malingering derive from the identification of aberrant, tell-tale findings during the forensic psychiatric examination [32].

Psychological assessment of malingering relies on a wide variety of specialized tests and measures that have been validated for this purpose. Some of these are free-standing, such as the Structured Interview of Reported Symptoms (SIRS); the Validity Indicator Profile (VIP); or the Test of Memory Malingering (TOMM). Others are embedded within larger tests, such as the validity scales of the Minnesota Multiphasic Personality Inventory – II (MMPI-2) [33,34]. Such tests have been particularly well developed in the field of forensic neuropsychology for the purpose of detected sub-optimal effort and exaggeration of cognitive and neurobehavioral impairments [35,36]. Additional research efforts in this area, beyond the development of refinement of the tests themselves, include studies of the impact of coaching on an individual's ability to 'fool' malingering tests [37–39].

8.4.4 Forensic behavioral science and police practices

Another area of keen interest in forensic psychiatry and psychology has to do with the critical analysis of police practices and methods for improving the reliability and accuracy of information gathered by law enforcement agencies. With the significant interest generated by the Innocence Project in the United States, the United Kingdom and elsewhere, increased research attention has been drawn to the issues of false confessions and the effects of police interrogation techniques on vulnerable populations, including juveniles and persons with mental illness or intellectual disabilities [40,41].

Substantial and programmatic research has been developed, looking at the extraordinary fallibility of eyewitness memory and its implications for the reliability of eyewitness testimony [42,43]. Other research in this area is helping to transform the manner in which police lineups and photo arrays are used in the process of eyewitness identifications [44,45].

8.4.5 Jury research

Emerging primarily from social psychology is an interest in juries and how they work. Major lines of behavioral science research have been developed in this area over the past 30 years. One such area has explored the science of jury selection, using population statistics and community survey research to aid in the identification of potentially sympathetic jurors and to help in crafting the presentation of a case to a particular group of jurors. Another research area has focused on the ways individual jurors perceive, interpret and remember evidence; the group processes involved in jury deliberation; and prediction of jury decision outcomes [46–48]. Other research is currently exploring the so-called 'CSI effect', studying the ways in which television programs have educated jurors about forensic science and changed their expectations about the evidence produced at trial [49].

8.5 International perspectives

A review of the recent topics being presented at the scientific sessions of the world's major forensic behavioral science organizations (including the AAFS at its annual meeting, the 4th International Congress on Psychology and Law, the American Psychology-Law Society, the Academy of Psychiatry and the Law, and the International Academy of Law and Mental Health, European Association of Psychology and Law, the Australian and New Zealand Association of Psychiatry, Psychology and Law and the Forensic Psychiatry Section of the World Psychiatric Association) reflects the rich and diverse array of subjects currently being addressed internationally in forensic behavioral science. Much of international forensic psychological and psychiatric research and practice is being guided by the major substantive issues described above. In turn, those topics are shaping the future directions for the field.

There is also, however, an international concern and focus on frank discussions of ethical and human rights issues. These arise, for example, in the controversies surrounding the role of psychiatry and psychology in interrogations and torture [50–52], research conducted on prisoners [53], and concerns about actual innocence, false confessions, and wrongful conviction [54,55].

8.6 Future directions and research

The scope of the research in forensic psychiatry and psychology reflects the breadth of the field and the richness of the subject matter that we deal with on a daily basis. There is no doubt that the field will continue to develop along the currently extant lines of research that have been described here. Findings from the forensic behavioral research laboratory are currently making their way into practice, as can be seen in reforms in police practices and new court precedents.

Research into the neurobiological substrates of violence, aggression and criminal behavior can be expected to develop dramatically in the next decade. However, brain science is not immune from the problems that plague most research on violent behavior. Low base rates of violent behavior and the obvious constraints inherent in designing ethical research protocols have led to small studies of samples of convenience, and these have largely precluded replication. The lack of large, well-controlled studies, with clear measurement strategies, hobbles what is currently knowable and presents the greatest challenge for this avenue of inquiry going forward.

Unambiguous results from empirical experimental studies designed specifically to examine the causative relationships between regional brain dysfunction, using any imaging modality and these types of complex behaviors, are needed before any introduction of functional or structural scans into the courts, as either exculpatory or mitigating evidence can be considered scientifically justified [56].

Continued refinements can also be expected in the study of violence risk assessment and psychopathy, much of which may look to neurobiological and behavior-genetic factors as underlying risk and protective variables that interact with environmental factors to affect behavioral outcomes. Young scholars interested in careers in forensic behavioral science would be well advised to obtain a strong foundation in neuroscience and brain-behavior relationships, as well as behavior genetics, in addition to the traditional forensic psychiatric or psychological curriculum.

References

1. Prosono M. History of forensic psychiatry. In: Rosner R,editor. *Principles and Practice of Forensic Psychiatry.* 2nd ed. London: Arnold; 1994:14–30.
2. Rush B. *Medical Inquiries and Observations upon the Diseases of the Mind.* Philadelphia: Kimber & Richardson, 1812.
3. Ray I. *A Treatise on the Medical Jurisprudence of Insanity.* Boston: Charles C. Little & James Brown, 1838.
4. Bradley C. The behavior of children receiving Benzedrine. *American Journal of Psychiatry* 1937; 94:577–585.
5. Prudic J. Electroconvulsive Therapy. In Sadock BJ,Sadock VA,Ruiz P,editors. *The Comprehensive Textbook of Psychiatry*, 9th ed. Philadelphia: Wolters Kluwer/ Lippincott Williams & Wilkins; 2009; 3285–3302.
6. American Psychiatric Association. *Diagnostic and Statistical Manual of Mental Disorders.* 1st ed. Washington, DC: American Psychiatric Association; 1952.
7. American Psychiatric Association. *Diagnostic and Statistical Manual of Mental Disorders.* 2nd ed. Washington, DC: American Psychiatric Association; 1968.
8. American Psychiatric Association. *Diagnostic and Statistical Manual of Mental Disorders.* 3rd ed. Washington, DC: American Psychiatric Association; 1980.
9. American Psychiatric Association. *Diagnostic and Statistical Manual of Mental Disorders.* 4th ed. Washington, DC: American Psychiatric Association; 1994.
10. American Psychiatric Association. *Diagnostic and Statistical Manual of Mental Disorders.* 4th ed., text rev. Washington, DC: American Psychiatric Association; 2000.
11. Rosner R, editor. *Principles and Practice of Forensic Psychiatry*, 2nd ed. London: Arnold; 2003.
12. Benedek EP, Ash P, Scott CL. *Principles and Practice of Child and Adolescent Forensic Mental Health.* Arlington, VA: American Psychiatric Publishing, Inc.; 2009.

13. James W. *The Principles of psychology, Vols 1 and 2*. New York: Henry and Holt; 1890.
14. Field KS. *History of the American Academy of Forensic Sciences*, 1948–1998. West Conshohocken, PA: ASTM; 1998.
15. Rosner R. Psychiatry and Behavioral Science Section Award-Winning Papers of 1998. *Journal of Forensic Sciences* 1999; 44(3):564.
16. Silva JA. The relevance of neuroscience to forensic psychiatry. *Journal of The American Academy Of Psychiatry and The Law* 2007; 35:6–9.
17. Seiden JA. The criminal brain: Frontal lobe dysfunction evidence in capital proceedings. *Capital Defense Journal* 2004; 16:395.
18. Snead OC. *Neuroimaging and the "complexity" of capital punishment*. Notre Dame Law School Legal Studies Research Paper No. 07–03. 2007 Feb. http://ssrn.com/abstract=965837.
19. Martell DA. Neuroscience and the law: Philosophical differences and practical constraints. *Behavioral Sciences and The Law* 2009; 27:123–136.
20. Mayberg HS. Functional brain scans as evidence in criminal court: An argument for caution. *Journal of Nuclear Medicine* 1992; 33(6):18N–25N.
21. Patel P, Meltzer CC, Mayberg HS, Levine K. The role of imaging in United States courtrooms. *Neuroimaging Clinics of North America* 2007; 17(4):557–567.
22. Reeves D, Mills MJ, Billick SB, Brodie JD. Limitations of brain imaging in forensic psychiatry. *Journal of The American Academy Of Psychiatry and The Law* 2003; 31(1):89–96.
23. Lee TM, Liu HL, Tan LH, Chan CC, Mahankali S, Feng CM, *et al*. Lie Detection by functional magnetic resonance imaging. *Human Brain Mapping* 2002; 15:157–164.
24. Mohamed FB, Faro SH, Gordon NJ, Platek SM, Ahmad H, Williams JM. Brain mapping of deception and truth telling about an ecologically valid situation: functional MR imaging and polygraph investigation—initial experience. *Radiology* 2006; 238:679–688.
25. Simpson J. Functional MRI Lie Detection: Too Good to be True? *Journal of The American Academy Of Psychiatry and The Law* 2008; 36(4):491–498.
26. Wahlund K, Kristiansson M. Aggression, psychopathy and brain imaging – Review and future recommendations. *International Journal of Law and Psychiatry* 2009; 32 (4):266–271.
27. Skeem JL, Monahan J. Current Directions in Violence Risk Assessment. *Current Directions in Psychological Science* 2011; 20(1):38–42.
28. Heilbrun K. *Evaluation for Risk of Violence in Adults*. New York: Oxford, 2009.
29. Reidya DE, Shelley-Tremblay JF, Lilienfeld SO. Psychopathy, reactive aggression, and precarious proclamations: A review of behavioral, cognitive, and biological research. *Aggression and Violent Behavior* 2011; 16(6):512–524.
30. Jürgen L, Müller MD. Psychopathy—an approach to neuroscientific research in forensic psychiatry. *Behavioral Sciences and The Law* 2010; 28(2):129–147.
31. Salekin RT, Lynam DR. Child and adolescent psychopathy: An introduction. In: Salekin RT, Lynam DR, editors. *Handbook of Child and Adolescent Psychopathy*. New York, NY: Guilford Press; 2010; 1–11.
32. Resnick PJ. The Detection of Malingered Psychosis. *Psychiatric Clinics of North America* 1999; 22(1):159–172.

33. Rogers R, editor. *Clinical Assessment of Malingering and Deception*. New York: Guilford Press, 2008.
34. Yoxall J, Bahr M, Barling N. Australian psychologists' beliefs and practice in the detection of malingering. In: Hicks RE, editor. *Personality and individual differences: Current directions*. Bowen Hills: Australian Academic Press; 2010; 315–326.
35. Heubrock D, Petermann F. Neuropsychological assessment of suspected malingering: Research results, evaluation techniques, and further directions of research and application. *European Journal of Psychological Assessment* 1998; 14(3):211–225.
36. Larrabee GJ. Introduction: Malingering, research designs, and base rates. In: Larrabee GJ, editor. *Assessment of malingered neuropsychological deficits*. New York, NY: Oxford University Press, 2007; 3–13.
37. Dunn TM, Shear PK, Howe S, Ris MD. Detecting neuropsychological malingering: effects of coaching and information. *Archives of Clinical Neuropsychology* 2003; 18:121–134.
38. Powell MR, Gfeller JD, Hendricks BL, Sharlandin M. Detecting symptom- and test-coached simulators with the test of memory malingering. *Archives of Clinical Neuropsychology* 2004; 19(5):693–702.
39. Ben-Porath YS. The ethical dilemma of coached malingering research. *Psychological Assessment* 1994; 6(1):14–15.
40. Fisher RP, Geiselman RE. The Cognitive Interview method of conducting police interviews: Eliciting extensive information and promoting Therapeutic Jurisprudence. *International Journal of Law and Psychiatry* 2010; 33(5–6):321–28.
41. Gudjonsson GH. *The Psychology of Interrogations and Confessions: A Handbook*. Chichester, UK: John Wiley & Sons, Ltd.; 2008.
42. Brewer N, Wells GL. Eyewitness Identification. *Current Directions in Psychological Science* 2011; 20(1):24–27.
43. Loftus E. *Eyewitness Testimony*. Cambridge: Harvard University Press; 1996.
44. Wells G, Penrod SD. Eyewitness Research: Strengths and Weaknesses of Alternative Methods. In: Rosenfeld B,Penrod SD, editors. *Research Methods in Forensic Psychology*. New York: John Wiley & Sons; 2011; 237–256.
45. Levi AM. Improving the Police Lineup. In: Evans S,editor. *Public policy issues research trends*. Hauppauge, New York: Nova Science Publishers; 2008; 167–210.
46. Bornstein BH, Greene E. Jury Decision Making: Implications For and From Psychology. *Current Directions in Psychological Science* 2011; 20(1):63–67.
47. Warren J, Kuhn D, Weinstock M. How do jurors argue with one another? *Judgment and Decision Making* 2010; 5(1):64–71.
48. Winter RJ, Robicheaux T. *Questions About the Jury: What Trial Consultants Should Know About Jury Decision Making*. Handbook of Trial Consulting 2011; (1):63–91.
49. Holmgren JA, Fordham J. The CSI Effect and the Canadian and the Australian Jury. *Journal of Forensic Sciences* 2011; 56(Suppl):S63–S71.
50. Matthews D. Psychiatry and torture. *World Psychiatry* 2006; (5):94–95.
51. Pope KS, Gutheil TG. The American Psychological Association & Detainee Interrogations: Unanswered Questions. *Psychiatric Times* 2008; (25):8.
52. Janofsky JS. Lies and coercion: why psychiatrists should not participate in police and intelligence interrogations. *Journal of The American Academy of Psychiatry and The Law* 2006; 34(4):472–478.

53. Taborda JGV, Arboleda-Flórez J. Forensic psychiatry ethics: expert and clinical practices and research on prisoners. *Revista Brasileira de Psiquiatria* 2006;28(Supl II):S86–S92.
54. Cutler BL. *Conviction of the innocent: Lessons from psychological research.* Washington, DC: American Psychological Association; 2012.
55. Kassin SM. False Confessions Causes, Consequences, and Implications for Reform. *Current Directions in Psychological Science* 2008; 17(4):249.
56. Mayberg H. *The Brain on Trial.* Invited address, Annual Meeting of the American Academy of Psychiatry and the Law, Tucson, AZ; 2010.

Further reading

Benedek EP, Ash P, Scott CL, editors. *Principles and Practice of Child and Adolescent Forensic Mental Health.* Washington, DC: American Psychiatric Publishing, Inc.; 2010.

Felthous AR, Saß, editors. *International Handbook of Psychopathic Disorders and the Law.* Vol. 2. Chichester, England: John Wiley & Sons, Ltd.; 2007.

Melton G, Petrila J, Poythress N, Slobogin C. *Psychological Evaluations for the Courts: A Handbook for Mental Health Professionals and Lawyers.* 3rd ed. New York: Guilford; 2007.

Rosner R, editor. *Principles and Practice of Forensic Psychiatry.* 2nd ed. London: Arnold; 2003.

Sadoff RL, editor. *Ethical Issues in Forensic Psychiatry: Minimizing Harm.* Oxford, UK: John Wiley & Sons, Ltd; 2011.

9

Forensic document examination

William M. Riordan[1], Judith A. Gustafson[2], Mary P. Fitzgerald[3] and Jane A. Lewis[4]

[1]*Questioned Document Unit, Internal Revenue Service, National Forensic Laboratory, Chicago, Illinois, USA*
[2]*United States Treasury Department, National Forensic Laboratory, Questioned Documents Unit, Chicago, Illinois, USA*
[3]*Internal Revenue Service National Forensic Laboratory, Chicago, Illinois, USA*
[4]*Alabama Department of Forensic Sciences, Auburn, Alabama, USA*

9.1 The field of forensic document examination

Forensic document examination involves the examination of documentary evidence in order to determine authenticity or authorship. It is a field that had its beginnings over a hundred years ago, starting with handwriting examination, and it has broadened over the past century to encompass the many expanding technologies related to the production of documents.

According to ASTM Standard E444-09, *Standard Guide for Scope of the Work of Forensic Document Examiners*, the forensic document examiner makes scientific examinations, comparisons, and analyses of documents in order to:

1. establish genuineness or non-genuineness, or to expose forgery, or to reveal alterations, additions or deletions;
2. identify or eliminate persons as the source of handwriting;
3. identify or eliminate the source of typewriting or other impressions, marks or relative evidence;
4. write reports or give testimony, when needed, to aid the users of the examiner's services in understanding the examiner's findings [1].

Typical problems for the forensic document examiner (FDE) are the examination and comparison of handwriting (cursive, hand printing and numerals), mechanical

Forensic Science: Current Issues, Future Directions, First Edition. Edited by Douglas H. Ubelaker.

impressions (such as typewriting and rubber stamps) and the output of other business machines such as printers, copy machines and facsimiles. Other problems include:

- the identification of ink, paper and writing instruments;
- the dating or sequence of preparation of documents;
- the examination and decipherment of alterations, obliterations and indented material on questioned documents, as well as the decipherment and restoration of torn, burned or water-soaked papers.

While handwriting examinations comprise the majority of examinations requested, in general the scope of the field may encompass any problem concerning a document.

Document evidence may prove to be important in both criminal and civil cases. Evidence types commonly submitted to document examiners include checks, credit card receipts, bomb threats, anonymous letters, bank robbery notes, business records, log books, medical records, prescriptions, suicide notes, hotel registrations, pawn tickets and graffiti images. In fact, any written or printed text on a substrate may be included in casework examined by a document examiner. Questioned documents are not only evidence in financial crimes; they may be present in death investigations, robberies, burglaries, arsons, bombings, auto thefts and other crimes. In civil cases, questioned documents may be associated with contested wills, malpractice suits, contract disputes, deceptions and various types of financial disputes.

9.1.1 Education and training

Presently, there is no college degree program in this field. Basic science education is an ideal foundation for entry into a career as an FDE. Most qualified FDEs have an undergraduate or master's degree. The FDEs receive apprenticeship-type training in a recognized questioned document laboratory under the direct supervision of a qualified and experienced FDE for a period of two to three years. The training covers the full range of document examinations.

ASTM *Standard Guide for Minimum Training Requirements for Forensic Document Examiners E2388-11* describes initial professional training for forensic document examiners.

9.1.2 Instrumentation

Specialized instruments used by FDEs include stereo microscopes and light sources (natural, fiber optic and incandescent). Computer scanners and digital cameras are used to record accurately detailed images of questioned documents. Video spectral comparison allows FDEs to separate visually similar inks with a camera sensitive to the ultraviolet, infrared and visible portions of the electromagnetic spectrum. Electrostatic detection devices develop latent indented writing on questioned documents. FDEs use specialized grids to assess spacing in typewritten documents and computer-generated texts and to detect inserted material. Ribbon analysis instruments are used to record text material present on carbon typewriter ribbons, and comparison microscopes enable side-by-side microscopic comparisons.

9.2 Principles of identification

9.2.1 Handwriting identification theory

Handwriting identification is based on the principle that no two people write exactly alike. Within each person's handwriting there is a distinct set of repeated characteristics and habits; these include not only the specific details of the way a person constructs a letter, but also the shape and manner of connections between letters, the size and height relations between letters, slant, spacing, and other factors such as pressure variations and line quality. It is the combination of these characteristics in a person's handwriting that make it unique and different from everyone else's, and therefore identifiable.

No person writes exactly the same way twice; each person has what is known as a natural range of variation. This is a natural range, through which the characters of a person's handwriting vary. A person constructs each letter following a basic structure or theme, but there is some variation because we are not machines and we do not produce things with machine-like precision.

Development of handwriting in an individual

When learning to write in school, there is a conscious effort to copy forms as demonstrated by a teacher or observed in a copybook. Once these forms are learned, the student no longer has to think consciously about how to make each form; it becomes a more subconscious or unconscious act, and handwriting becomes individual. For handwriting to be identifiable, it must contain sufficient individual identifying characteristics (as opposed to being generic copybook form). Generally, by the end of high school, a person's writing has stabilized and has developed sufficient individuality to become identifiable.

Source of uniqueness

Studies of various populations have been done in recent years, establishing the principle that people do not write exactly alike, and these are considered in a later section of this chapter. The working aspect of this principle arises from an understanding that each person is a unique individual and his or her handwriting reflects this. In the development of a mature writer, the variables that affect and shape their handwriting are numerous and complex and are not duplicated exactly in any other person. These may include a variety of physical, mental and environmental factors, such as size and strength of the hand, fingers and arm, writing posture, manner of holding a writing instrument, muscular and fine motor control, dexterity, eye-hand coordination, concentration ability, sense of proportion and neurological conditions. Other variables include the handwriting system they were taught and degree of that training, and personal modifications, whether through occupational influences or other life experiences.

9.2.2 Office machine/device identification theory

This category includes mechanical impressions (such as typewriting and rubber stamps) and the output of other business machines such as printers, copy machines and facsimiles. The identification of the product or output of an office machine or device back to an individual source is based on the presence of damage and wear characteristics that develop as the machine is used, and which appear on the product. Over time, as damage and wear characteristics continue to occur and accumulate, the combination of these characteristics becomes unique and different from other office machines/devices of the same type, and therefore become identifiable.

For the product or output to be identifiable, the machine or device must have developed sufficient damage and wear characteristics, as opposed to being in a new, unused or little-used condition. It should be noted that non-impact machines/devices have less potential for producing unique damage and wear characteristics. If insufficient individual characteristics are present, only a match back to a class of machines or devices (e.g. make, model) may be possible.

9.3 Forensic examinations and comparisons

An FDE applies scientific methods to questioned document problems. The examiner makes an objective comparison of the evidence submitted, beginning at a point of complete neutrality. The final results in an examination are directly determined by the quality and quantity of the material available to the FDE for the comparison.

Known standards, whether it be handwriting, typewriting, samples from a copier, etc. must be sufficient and comparable. One or two standards are generally not enough to form the basis for an identification. A sufficient quantity of known standards should be collected from the known writer or known device/machine to allow the FDE to establish its characteristics and study the range of natural variation where applicable. Standards should attempt to duplicate the conditions of the questioned document. For example, for a handwriting examination, known standards should contain similar letters and letter combinations to those seen in the questioned material and should be in a similar format (e.g. lined paper if the questioned document is lined, etc.). For a copier examination, a range of standards taken at various settings (dark-light) should be submitted to ensure that comparable samples are taken.

An additional requirement for known handwriting standards is that they be normal. The known writing should be freely and naturally written, as opposed to disguised, or altered by other influences which could hinder writing ability (e.g. drugs, alcohol, injury or an awkward writing position).

The terms 'requested' and 'collected' known writing generally refer to handwriting standards. Requested standards are handwriting specimens written by request, in which the text of the questioned document is duplicated on the requested standards. Collected standards are genuine writings of the subject, written during the course of daily activities (e.g. personal correspondence, canceled checks, employment records, legal, government or official documents, etc.). Collected standards aid in establishing the full, natural range of characteristics of the subject's writing, can aid in indicating whether the request writing is natural or disguised, and should be contemporaneous.

9.3.1 Handwriting comparison

The FDE conducts a handwriting examination by doing a side-by-side comparison of the questioned and known writings, directly comparing each letter and characteristic. The writer's habits, characteristics and range of variation are established by looking through the body of submitted known writing. The writing characteristics present in the questioned writing are compared and evaluated with those established in the known writing, determining if they fit within that person's range and can be fully explained by the known writing. An identification is not based on the similarity of one or two letter forms – it takes into consideration all of the features of the writing.

9.3.2 Typewriting comparison

It has been reported that the last typewriter was produced in 2011. Typewriters played a major role in questioned document examination for nearly a century. Today, however, the use of typewriters has greatly diminished. In earlier days, typebar machines contained 44 independently operating bars and, through use and abuse, they often developed their own set of alignment and typeface defects. With the onset of single element machines, there was less occurrence of developed defects. The comparison of a questioned typewritten document with sufficient standards taken from a known typewriter may result in an identification or elimination. In some cases, however, the questioned and known material is similar in design but lacking the individual characteristics (defects) needed for an identification.

The questioned document laboratory will contain extensive typewriter reference materials. With these references, the FDE can identify the make and model of the machine used to produce the questioned typewritten document in some cases. References also aid the examiner in the determination of the year (or time frame) when a type style became available, when a particular letter design changed within a type style, or when a particular typewriter feature became available, making it possible for the FDE to determine if a typewritten document has been fraudulently back-dated.

9.3.3 Computer-generated documents

Computers and various printing technologies generate most of the documents in everyday modern business. An FDE can assist in identifying the imaging process used to create a questioned document. Examinations of computer-generated documents include determinations regarding:

- if the questioned document is an original or a copy;
- the printing technology used and the possibility of it being dated;
- if the document was created with more than one printing technology;
- if there is evidence of page substitution in a multi-page document.

9.3.4 Examination of alterations

Although documents are altered for a wide variety of reasons, some are altered for the purpose of fraud or deception. Usual types of fraudulent alterations include:

- changing the date on memos, letters, wills or contracts;
- changing monetary amounts on checks or other financial documents;
- changing information on forms of identification and ownership;
- altering medical records, ledger books and other business records.

Common methods of alterations include additions, deletions, obliterations and substitutions. Altered documents may be handwritten, typewritten or various types of printed documents. Additions may be added strokes, characters, words, sentences, paragraphs, pages or a combination of these.

Alteration by addition in a handwritten document can be potentially detected by differences in handwriting, differences in writing instrument and crowding. In the case of a handwritten document written in ink, an examination of the ink may differentiate the added ink from the original handwriting. Ink may be examined visually, microscopically and instrumentally (with the aid of various filters and light sources). Inks that appear to be similar in the visible spectrum may appear to be very different in the invisible spectrum. A chemical analysis of inks is another option for the FDE.

Additions to typewritten documents may be potentially detected by differences in ribbon characteristics, differences in design or size of type, differences in alignment or differences in arrangement or format. An FDE has knowledge of typewriter examinations and would conduct a visual and microscopic comparison with the aid of specialized measuring plates, as well as typewriter reference standards.

Deletions are commonly made by erasing original material with the use of a rubber eraser. Rubbing with an abrasive material, scraping with a razor blade or application of a chemical are also means of erasure. Deletions are generally detected by the presence of paper disturbance, which may or may not have traces of graphite (pencil) or ink. Pencil erasures can be made to a document with little disturbance to the paper in some cases but, more often, paper disturbance is evident. In the case of ballpoint ink or typewriting, there are usually remaining ink traces that can possibly be deciphered. Attempts at chemical erasure often affect the appearance of the paper.

Deletions are examined and processed for evidence of and decipherment of latent writing or typewriting indentations. They are examined visually, microscopically and instrumentally (with the aid of various filters and light sources), in an effort to decipher erased material and/or to detect evidence of chemical processing.

Alteration by obliteration is achieved by covering with a material, which may be similar or different to what is being covered. Opaque correction fluid is a medium that is often used for obliteration, but pencil or various types of ink may also be commonly used. When the covering material is a different medium or a different ink from what it is covering, the possibility of differentiation and decipherment of the altered material exists. An extensive covering with the same ink used to write the document usually precludes the possibility of decipherment.

A substitution involves a two-step process. It is actually a deletion, or obliteration of material followed by an addition of material.

9.3.5 Ink analysis

The purpose of ink analysis is two-fold: to compare two or more inks to determine if they are of the same or different formulation; and for dating purposes, to determine whether the type of ink was in existence at the time the document was purportedly prepared.

For comparison work, non-destructive techniques are initially used to determine whether two or more inks are the same or different. Next, a chemical separation technique known as thin layer chromatography (TLC) is used. This technique requires the removal of small amounts of ink and paper, each about the size of a typewritten period.

In reporting conclusions, the FDE will state whether or not different ink formulations have been used. If no differences can be seen, the examiner may state that the formulas are similar throughout the tests conducted. In this instance, there still may be a chance that different formulas or pens were used. It is virtually impossible to say whether the same pen was used to write two or more entries. However, it can be concluded, if formula or batch differences are found, that different pens were used to write two or more entries.

Ink dating involves trying to determine the manufacturer of an ink and its introduction date. The same physical and chemical techniques described above are typically first used to determine how many different ink formulas are present on the documents. Next, each questioned ink separation (on the TLC plate) must be compared with a library of reference standards, called the International Ink Standards Library.

The ink library is the most comprehensive collection of writing inks in existence. It was started in 1968 and now contains over 10,000 samples obtained from all over the world, both from manufacturers and from the open market. During the process of comparing a questioned ink with others of the same type (ballpen or non-ballpen) and color, the examiner makes a list of possible matches from the library. The questioned inks and the matching known standards are then compared side by side on another TLC plate. The process continues until the questioned ink is found to match one or more of the standards.

Note: at this time it is not possible to determine how long an ink has been on a document, or how long it has been on a document relative to another ink entry.

9.3.6 Indentation examinations

Indented writing is an imprint or physical impression of writing that is left on underlying pages when the top sheet of paper is written upon. The writing imprint or physical impressions result from the pressure of the writing instrument. Indented writing is useful in linking or connecting evidence. It may add identifying information to an anonymous letter, connect a robbery note to a writing pad recovered from a suspect, or provide information from a document that was once associated with the questioned document in your possession. Indentations may be examined, deciphered and recorded with the aid of oblique light and photography, or by processing on an electrostatic apparatus designed to detect and produce a visual image of the indented writing on a transparent film. Indentations from typewriters and other types of mechanical impressions may also be detected, examined and deciphered.

9.3.7 Comparison of copies

In some investigations, it may be helpful to link different questioned documents to the same source. In other cases, one questioned document may prove to be an altered copy of another document. Documents may be copies of the same source document to which original handwriting or typewriting has been added. FDEs can conduct an examination and comparison to determine whether copies have a common source, if they can be identified as having been produced on a specific machine, or if they can be linked in any way.

9.3.8 Charred document examinations

Charred or burned documents may be stabilized, examined and photographed with the aid of various filters and light sources in an effort to decipher and record information that may not otherwise be visible on the charred document.

9.3.9 Torn document examinations

The FDE will examine the characteristics of the torn edges before restoring a document to its original arrangement. In some cases, torn portions of the same document are found in different places, establishing a link in an investigation. The examiner may prove that two or more pieces of paper were at one time part of the same document, by conducting an examination and comparison of the characteristics along the torn edges of the pieces of paper.

9.4 Forensic document examination past to present

9.4.1 Origins of questioned document examination

Disputes concerning the authorship and authenticity of documents have been occurring since ancient times. Laws in ancient Rome provided for the acceptance of expert testimony concerning documents [2]. In 1562, the English Parliament declared forgery to be a statutory offense and, in 1684, 'comparison of hands' was ruled to be 'good evidence in cases of treason'. A significant legal step occurred in 1762, when a British court ruled that handwriting is identifiable [3].

Early courts sometimes relied on 'recognition witnesses' – people who were familiar with a given writer's handwriting and would testify to it in court [2]. The use of someone with specialized skill in comparing known standards with questioned handwriting gradually became the more accepted practice. The first case in English-speaking courts in which a specially-qualified witness testified was

Goodtitle d. Revett v. Braham in 1792, in which a direct comparison between questioned and known writings was conducted rather than reliance on recognition. In this case, two experts were admitted, their special qualifications being that they had experience as inspectors of franks [2].

While there is little written history of document examination in America until the late 19th century, there are records of court decisions, particularly after 1800, which show that there were individuals testifying about handwriting identification. Names of the experts, however, do not seem to have been recorded until after 1870 [3,4]. An 1808 Louisiana code (code of 1808, p. 306, art. 226) stated that, in resolving a question of authorship, a questioned signature 'must be ascertained by two persons having skill to judge of handwriting', who would then determine authorship based on comparing it to known writings. Just a few years later, in the Louisiana case of *Sauve v. Dawson, 2 M.R. 203 (1812)*, a signature on a promissory note was compared to an appeal bond executed by the defendant and was determined to be genuine [5].

These early cases conformed to rules derived from English common law, which decreed that questioned writings could only be compared to known writings that were already in evidence in a case. Decisions in Massachusetts in 1814, and later in Connecticut and Vermont, removed this restriction and allowed the admittance of other writing samples [4]. This greatly expanded the potential to pursue handwriting comparisons as a form of evidence in cases.

The profession of document examination gained significant recognition in the mid-19th century with a case involving the Junius Letters. These were a series of letters written and published around 1770, which discussed many controversial issues of the time dealing with liberty, politics and government in England. The author used the pen name Junius, and much speculation occurred for decades regarding the true identity of Junius. Charles Chabot, one of the earliest practitioners in the field of questioned document examination, concluded from a handwriting examination that the author was Sir Philip Francis [6]. His clear and thorough presentation of his evidence and conclusions helped document examination to gain acceptance in the English legal system [3].

Legal decisions of the 1880s and 90s affirmed the importance of handwriting evidence and the use of experts, saw ink examination accepted as testimony and also saw the first testimony on typewriter identification in 1893 [3]. Several early examiners wrote books on questioned documents that were published in the 1890s: *A Treatise on Disputed Handwriting* (1894) by William Elijah Hagan, *Manual for the Study of Documents* (1894) by Persifor Frazer, and *Ames on Forgery* (1899, 1900) by Daniel T. Ames.

9.4.2 Albert S. Osborn and the formation of the American Society of Questioned Document Examiners (ASQDE)

Albert S. Osborn (1858–1946) is widely considered to be the founder of modern document examination. He began a career as a penmanship instructor at the

Rochester Business Institute in New York. In those days it was common practice for an attorney to consult a penmanship instructor to examine and draw an opinion about disputed handwriting cases. Osborn is recorded to have been working as a document examiner in 1887, and he began publishing articles on the topic [3,4,7].

In 1910, he wrote *Questioned Documents,* a classic work which brought together the fundamental principles of the field formulated over the previous century, and he broadened the scope of questioned document examination to include typewriting, ink and paper examinations. Through his writings, lectures and testimony, handwriting identification began to gain wider acceptance in the courts and in society [2,4].

He revised *Questioned Documents* in 1929 and, throughout his career, he wrote several other books and many articles on document examination and expert testimony. He also wrote extensively about courtroom procedures, expert testimony, court presentation and legal decisions, publishing *The Problem of Proof* in 1922 and *The Mind of the Juror* in 1937 [4].

Also, during this time period, document examination gained additional attention and acceptance through the writings of Professor John H. Wigmore of Northwestern University School of Law, who also wrote the introduction for several of Osborn's books. Wigmore, a noted authority on American evidence law, contributed greatly to the acceptance of expert testimony and document examination as a forensic science around the turn of the 20th century [2,4].

In 1913, Osborn and another examiner, Elbridge W. Stein, met together to discuss topics dealing with handwriting examination and to initiate a program for the exchange of ideas and research. This was followed by a larger gathering in 1914 at Osborn's residence in Montclair, New Jersey. These meetings were by invitation only, and they gradually expanded and began to occur annually. Membership required annual attendance as well as active participation; attendees presented a paper each year that was then discussed by the group. In 1942, Osborn and 14 other prominent examiners formally organized as the American Society of Questioned Document Examiners, with Osborn as the first president [3,8].

Up until the 1920s and 30s, document examiners worked in the private sector, with many having started as penmanship instructors, though other professions included bankers, lithographers, engravers, court clerks and police officers [2–4]. The 1920s and 1930s saw the opening of the first federal, state and local crime laboratories in the United States. In the 1930s, two examiners were known to be doing questioned document work for the federal government in the Bureau of Standards and the Treasury Department [4]. The first major police questioned documents laboratory in the United States was opened in Chicago in 1938.

9.4.3 Ordway Hilton and the formation of the American Academy of Forensic Sciences (AAFS)

The First American Medico-legal Congress, which was the precursor to the Academy of Forensic Sciences, was addressed in January 1948 by George Swett, a document examiner with the US Postal Inspection Service. The steering committee meeting convened in October 1948 by Dr. R. B. H. Gradwohl at the Hotel Pierre, New York City, was attended by document examiners Ordway Hilton and Albert D. Osborn. In 1950, the American Academy of Forensic Sciences adopted a constitution and by-laws setting up seven sections, and James Clark Sellers, charter member and first vice-president of the ASQDE, became the first questioned documents chairman.

Presenters at the third Academy meeting in 1951 included questioned document examiners and, in 1952, section chairman Ordway Hilton began correspondence that influenced the development of an active QD Section within the AAFS. Largely due to the efforts of Hilton, assisted by secretary David J. Purtell, section membership number requirements were met and Questioned Documents held its first section meeting with a program of technical papers at the AAFS meeting in 1953.

At this time, forensic sciences were advancing as crime laboratories were growing. Initially, most members of the ASQDE were wary of meeting and sharing their collected knowledge with examiners outside of the society. However, Donald Doud, who was a well-respected private examiner and member of the ASQDE, worked along with Ordway Hilton to convince the society members to support the AAFS. Hilton, Purtell and Doud were the real pioneers who worked to form the QD Section. Thanks to these pioneers, government, as well as private FDEs from various areas of the country – including members of the ASQDE – eventually joined and supported the AAFS QD Section. Historically, there has been a very prolific group of FDEs who possess dual membership in both the AAFS and the ASQDE [4,9].

Ordway Hilton earned degrees in mathematics and statistics at Northwestern University, which had a scientific crime investigation laboratory within its law school. The Northwestern laboratory was developed toward the end of the 1920s. Northwestern Law Professor Emeritus Fred Inbau was instrumental in the transition of the Northwestern Laboratory to the City of Chicago for its police department in 1938. The same year, Ordway Hilton, who had pursued a career in questioned documents, was chosen to be the first questioned document examiner for the Chicago Police Department crime laboratory by Professor Inbau.

Hilton became highly regarded and internationally recognized for his achievements in the field of questioned documents. He was the author of a leading textbook and a monograph on pencil erasures, published over 90 articles for scientific and legal journals and lectured at law schools across the country. He was the president of the AAFS in 1959 and president of the ASQDE in 1960, and was awarded the honor of distinguished fellow by the Academy. In recognition of his outstanding contributions to the field of questioned documents, he was awarded the first the AAFS QD Section award, an award that is in his name, the 'Ordway Hilton Award'. Like his fellow QD Section pioneers, he was also a diplomate of the American Board of Forensic Document Examiners [10].

9.4.4 Questioned documents and the formation of the International Association of Forensic Sciences (IAFS)

The first International Association of Forensic Sciences Meeting, in conjunction with the International Meeting in Forensic Immunology, Medicine, Pathology and Toxicology, was held in London, England, in 1965. This meeting was attended by members of the AAFS QD Section as well as members of the ASQDE. From the inception of the IAFS, questioned documents has been one of the forensic disciplines represented at the Association's triennial meetings. The IAFS has fostered the international exchange of information in forensic sciences, including research and standard practices.

9.5 Key issues

9.5.1 Certification

In the 1970s, the need to identify qualified professionals in the forensic science disciplines lead to serious discussions regarding the establishment of certification boards within the United States. The Department of Justice became aware of the need to support the development of these professional boards.

The Department of Justice has funded, through the Law Enforcement Assistance Administration (LEAA), a program of certification in forensic sciences. This program is administered under the auspices of the AAFS. Certification boards in forensic toxicology, forensic odontology, forensic psychiatry, forensic anthropology and forensic document examination have already been constituted [11].

9.5.2 Certification and the American Board of Forensic Document Examiners, Inc. (ABFDE)

The AAFS Committee on Certification was formed in 1974, and Dr. Kurt M. Dubowski was selected to chair this important committee, which proposed and established certification programs in forensic sciences. In 1975, the Report of the Committee on Certification was presented to Dr. David A. Crown, President of the AAFS and Chief of the CIA's QD Laboratory. Following the report, the Forensic Science Foundation obtained an LEAA grant and helped to establish the American Board of Forensic Document Examiners, Inc. The ABFDE was incorporated in the District of Columbia in 1977 with John J. Harris, President, James J. Horan, Vice President, James H. Kelly, Secretary and Maureen Casey Owens, Treasurer.

The objectives of the Board were stated as being: 'to establish, enhance, and maintain standards of qualification for those who practice forensic document examination and to certify, as qualified specialists, those voluntary applicants who comply with the requirements of the Board' [9]. The basis of certification is each candidate's personal education, professional education, training and experience, and successful completion of a formal three-part examination process [9]. By 1980, more than one hundred experts had been certified by the ABFDE and granted diplomate status.

The Questioned Document Board certification is somewhat similar to the Board Certifications conferred in the medical field to various types of specialists. In other words, it is a formal recognition of expertise by a high quality professional board, though by no means a compulsory or legally required recognition for licensure or conduct of business [11].

9.5.3 Standardization

AAFS QD Task group on opinion terminology

Thomas V. McAlexander, Jan Beck, and Ronald Dick, the members of an AAFS QD Section task group on opinion terminology, presented a paper entitled 'Committee Recommendations: The Standardization of Handwriting Opinion Terminology' at the 1990 AAFS annual meeting, and this was published the following year as a letter in the *Journal of Forensic Sciences*. The terminology (a nine-level scale) was accepted by the QD Section of AAFS and by the ABFDE. The positive and negative terminology in the scale allows for the expression of various levels of certainty.

A need for 'qualified' conclusions exists because of possible limitations in the questioned and/or known materials that can preclude a definite conclusion. Restrictions affecting questioned documents such as copying processes, limited clarity, brevity and physical damage, as well as limitations in known documents such as insufficient comparability, limitations in amount, disguise and a lack of contemporaneous samples, are common factors that hamper examinations and which may lead to a 'qualified' conclusion.

The use of the term 'probable', 'probably' or 'probability' used in the standard is addressed by the authors: 'Probability in handwriting opinions is not a statistical measurement but a measure of the examiner's confidence, based on scientific principles and experienced judgment, that the opinion rendered is correct. This is true because probability relates to qualitative as well as quantitative processes' [12].

The acceptance of the standardized handwriting opinion terminology was a major accomplishment within the field. It required a great deal of interactive debate and compromise among the FDEs throughout the country. Prior to this standard terminology, the widespread diversity of expressions limited the possibility for a clear, uniform understanding of 'qualified' conclusion terms used in the field.

'Committee Recommendations: The Standardization of Handwriting Opinion Terminology' was the basis for ASTM E1658, Standard Terminology for

Expressing Conclusions of Forensic Document Examiners, which was drafted and passed by the ASTM E30.02 Questioned Document Subcommittee.

ASTM International

ASTM International (formerly the American Society for Testing and Materials) is one of the largest voluntary standards development organizations in the world. In 1970, ASTM E30 Committee on Forensic Science was established, with the goal of developing standards in the forensic sciences. The E30.02 Questioned Document Subcommittee was established to develop standards in forensic document examination. ASTM procedures accommodate the following processes:

- The drafting of standards.
- The balloting of the draft to the E30.02 Subcommittee (which considers the votes of the ballots at an annual meeting).
- Balloting of the draft to the E30 Committee.
- Final evaluation and vote by both the E30.02 Subcommittee and E30 Committee (at the following annual meeting).
- The request of the Committee on Standards for publication approval.

The E30.02 drafts have been developed by the diligent work of tasks groups. Scientific Working Group for Questioned Documents (SWGDOC) task groups deserve recognition for their efforts in developing the overwhelming majority of the drafts for the standards published to date:

- E444 Standard Descriptions Relating to the Scope of Work of Forensic Document Examiners
- E1422 Standard Guide for Test Methods for Forensic Writing Ink Comparison
- E1658 Standard Terminology for Expressing Conclusions of Forensic Document Examiners
- E1789 Standard Guide for Writing Ink Identification
- E2195 Standard Terminology Relating to the Examination of Questioned Documents
- E2285 Standard Guide for the Examination of Mechanical Checkwriter Impressions
- E2286 Standard Guide for the Examination of Dry Seal Impressions
- E2287 Standard Guide for the Examination of Fracture Patterns and Paper Fiber Impressions on Film Ribbons and Typed Text

- E2288 Standard Guide for Physical Match of Paper Cuts, Tears, and Perforations in Forensic Document Examinations.
- E2289 Standard Guide for Examination of Rubber Stamp Impressions
- E2290 Standard Guide for the Examination of Handwritten Items
- E2291 Standard Guide for Indentation Examinations
- E2325 Standard Guide for the Non-Destructive Examination of Paper
- E2331 Standard Guide for Examination of Altered Documents
- E2388 Standard Guide for Minimum Training Requirements of for Forensic Document Examiners
- E2389 Standard Guide for Examination of Documents Produced with Liquid Ink Jet Technology
- E2390 Standard Guide for Examination of Documents Produced with Toner Technology
- E2490 Standard Guide for Examination of Typewritten Items

Scientific Working Group for Questioned Documents

The Technical Working Group for Questioned Documents (TWGDOC) began in 1997 and was one of the first of the technical working groups in the forensic sciences. TWGDOC met in various Washington, DC area locations. This national group, composed of forensic document examiners from the private sector and from federal, state and county laboratories, was renamed the Scientific Working Group for Questioned Documents (SWGDOC) in 1999. After 1999, meeting space was offered to SWGDOC as well as other scientific working groups at the FBI Academy in Quantico, Virginia.

SWGDOC made the determination to publish through ASTM International, where draft documents are reviewed by a variety of forensic disciplines. SWGDOC sub-groups, usually composed of five to seven individuals, develop drafts or updates which are vetted through other sub-groups before submission, as drafts, to ASTM International E30.02 Questioned Documents Subcommittee.

In the beginning of 2012, SWGDOC reached the decision to no longer submit drafts to ASTM International Subcommittee E30.02 for balloting, but remained dedicated to the development of quality standards. A reorganization of SWGDOC (along with 19 other Scientific and Technical Working Groups) began with the goal of publishing documents that receive full public scrutiny and represent the consensus of a broad range of qualified government and private practice examiners as the best practices in the field.

The mission of SWGDOC is to assemble representatives from the forensic document examination community in order to:

- define the scope and practice areas of the profession;
- standardize operating procedures, protocols and terminology;
- consolidate and enhance the profession of forensic document examination;
- promote self-regulation, documentation, training, continuing education and research [13].

9.5.4 Forensic document examination and the courts

Forensic handwriting examination has been regarded as a valid and reliable expertise for over 100 years in US courts, through repeated admissibility, federal statute (1913 US Statute) and laws (*Frye* ruling [1923]) and the Federal Rules of Evidence (specifically Rule 901[b][3]).

In Daubert v. Merrell Dow Pharmaceuticals, Inc. 509 U.S. 579 (1993), the Supreme Court's decision suggested there should be additional criteria which may be applied to determine the reliability of an expertise. Daubert initially led to serious challenges to the validity of handwriting examination; to date, however, forensic handwriting examination has satisfied Daubert requirements in an overwhelming majority of federal and state district courts, and in every federal and state appellate court that has considered handwriting admissibility.

In Daubert, the court assigned judges the role of 'gatekeeper' over what scientific expert testimony was to be admitted for trial, and suggested several evaluation criteria, including testability and validity, peer review, known or potential error rates, existence of accepted methodology and general acceptance.

In United States v. Starzecpyzel, 880 F. Suppl. 1027 (S.D.N.Y.), 1995, the judge's ruling, based on Daubert criteria, was that Daubert did not apply to forensic handwriting examination because it was practical rather than scientific. The judge ruled that the expertise was admissible, but only as a technical skill.

In General Electric v. Joiner (1997) 78 F.3d 524, the judge's ability to reject opinion evidence was strengthened and, in *Kumho Tire v. Carmichael*, 526 U.S. 137, 119 S.Ct. 1167, 143 L.Ed.2d 238 (1999), the application of Daubert was extended to all forms of expert opinion testimony.

The decision in Starzecpyzel, the first post-Daubert forensic handwriting examination case, delivered a shocking blow to the questioned document profession. The lack of controlled studies became apparent and the need for change became obvious.

The American Board of Forensic Document Examiners (ABFDE) recognized the critical need for an organized response, and the Daubert Group, consisting of three FDEs, was formed. The group's primary functions are to track federal and state Daubert decisions and provide FDEs with condensed responses to each of the Daubert criteria. The effectiveness of the group is undisputed, and it has assisted FDEs and attorneys in over 30 Daubert challenges, all of them successful [14] (Daubert challenges were defeated and forensic handwriting testimony was allowed).

In the 1990s and into the new millennium, a great amount of research has been conducted, and 15 of the 18 current ASTM International E30.02 standard guides (in the field of questioned documents) were drafted and published in this period. Several research papers on the individuality of handwriting were also published in scientific journals. The published research projects conducted by Sargur Srihari, PhD, SUNY Distinguished Professor, Department of Computer Science and Engineering, and Director of the Center of Excellence for Document Analysis and Recognition, State University of New York, Buffalo, New York, demonstrate the individuality of handwriting, and Dr. Srihari's work is continuing. A series of six published research projects conducted by Moshe Kam, PhD, Distinguished Professor, Data Fusion Laboratory, Electrical and Computer Engineering Department, Drexel University, Philadelphia, Pennsylvania (between 1994 and 2008) demonstrate error rates of FDEs versus non-document examiners. Dr. Kam is continuing his advanced research in the area of forensic handwriting analysis. Today, there is ongoing research by FDEs as well as academics.

A more detailed discussion of questioned document research follows in the 'Research' section of this chapter.

9.6 Forensic document examination internationally

The field of forensic document examination has a long international history, and historical cases involving the acceptance of handwriting identification in the British and American courts, as well as early publications around the turn of the 20th century (1890s though 1910), have been discussed earlier in this chapter. A significant part of this profession's history was the formation of the American Society of Questioned Document Examiners (ASQDE) and the formation of the AAFS QD Section. These major organizations provide a formal setting for the exchange of information and research, while encouraging international participation. The British Academy of Forensic Sciences Society and the Canadian Society of Forensic Science have a history of encouraging international participation, and the IAFS fosters the international exchange of information in forensic sciences, including research and standard practices.

The field also has a history of exchanges of shared research, as new and evolving technologies have effected business machines and communications. One example is the long, complicated history of the evolution of typewriters. To provide an extremely brief synopsis of this, some of the changes have included:

- American companies who produced their own type began purchasing European type.
- A machine manufactured in Japan would be marketed under various American brand names.

- The single-element machine was equipped with interchangeable type elements and later dual pitch.
- The large-scale coping of IBM's most popular type style (Courier) occurred when numerous companies developed their own close-copies and copies.

There has been much exchange of research and data to deal with typewriter advances. The Interpol Typewriter Classification System was adopted and circulated in 1969, and later updated with supplemental materials. In 1972, the Haas Pica Atlas was introduced and later followed by three additional volumes (making it a more complete typewriter identification system). In 1973, Douglas Cromwell's 'A Method of Indicating the Manufacturer of Courier Style Type Fonts' (a flow chart for classification system for Courier type styles) was published. In 1993, Dr. Phillip D. Buffard presented the PC Based Typewriter Classification System for Courier Typewriters, and later digitally indexed the Haas Atlas materials. These data are used in numerous examinations worldwide. The origins of these shared data include Interpol (Europe), Germany, Great Britain and the United States.

Like the typewriter, evolving business technology is marketed internationally, and thus new questioned document challenges and problems have to be addressed. Related research is shared through international participation in meetings and publication in scientific and legal journals.

Companies market instrumentation used in forensic laboratories internationally. One example is Foster & Freeman, LDT, a prestigious English company that develops various specialized instruments that are used in questioned document laboratories worldwide (marketed in over 70 countries). The desire for information regarding research conducted with the use of these instruments is truly international.

There is a history of strong ties among American document examiners in both Canada and the USA. Many belong to organizations and/or attend meetings in both countries, and a number of Canadian FDEs are diplomats of ABFDE.

The European Network of Forensic Science Institutes (ENFSI) was established to share knowledge in the forensic sciences. The European Document Experts Working Group (EDEWG), a working group of ENFSI, is composed of three subgroups: ink dating, ink and toner analysis, and non-destructive examination of printing products. Formed in 1998, EDEWG has developed training and examination procedure standards which are similar to those published by the FDEs through ASTM International.

The European community's focus on standardization of methods and procedures for questioned document examination has advanced the international goal of seeking consistency among practitioners in the field. EDEWG holds biannual meetings with scientific sessions, workshops, poster sessions and sub-group sessions [15]. International participation is encouraged at these meetings. The European Network of Forensic Handwriting Experts (ENFHEX), also a working group of ENFSI, is working toward the development of the International Handwriting Information System (IHIS), a database of a collection of international hand printed copybook example references for Latin-based languages [16].

9.7 Research

9.7.1 Introduction

Prior to the late 1980s, the published works in the field of questioned documents were rather limited in comparison to some other disciplines. The early books of Osborn, Harrison, Hilton and Conway were the mainstays of the field. Much of the research done was not published in peer-reviewed journals but was disseminated via professional contacts and meetings, such as the meetings of the American Academy of Forensic Science, the American Society of Questioned Document Examiners and other regional organizations. This is not to say that there did not exist numerous research papers; these were published in various journals such as the *Journal of Forensic Sciences* and *Northwestern's Journal of Criminal Law, Criminology and Police Science* (1951–1972, continued by the *Journal of Criminal Law and Criminology*). However, some information was kept within the community itself. The training of a forensic document examiner was, and still is, comparable to an apprenticeship under the tutelage of an experienced examiner. Much of the material used in training was garnered from papers presented at the annual meetings and from the books listed above.

N. G. Galbraith [17] researched the trends in the field of questioned documents by reviewing publications in the *Journal of Forensic Sciences* from 1956 through 1978. The total number of articles numbered 106. These articles were broken down into 30 categories, which were placed in one of seven sections. The seven sections are as follows: Document Examiner, Document Examination, Means of Identification, Differences in Writing, Methods of Identification, Illustrations, and Other. Galbraith's research indicated that there were two major trends observed in the questioned documents field over the 23-year time period: ' . . . training for examiners and increased use of the scientific methods and equipment' [17].

Over the next ten years, 1979–1989, 111 articles and five technical reviews were published in the *Journal of Forensic Sciences* alone, and the scope and subject matter of the articles being published were increasing. Advances in instrumental methods, coupled with changing technology, expanded the range of subject matter researched by FDEs. Printed research papers included topics such as, 'Examination of Line Crossings by Scanning Electron Microscopy' [18], 'Electrophoretic Identification of Felt Tip Pen Inks' [19], 'The Application of Mass Spectrometry to the Study of Pencil Marks' [20], 'Examination of Ball Pen Ink by High Pressure Chromatography' [21], 'Visible and Infrared Luminescence in Documents: Excitation by Laser' [22], 'Microanalysis of Painted Manuscripts and of Colored Archeological Materials by Raman Laser Microprobe' [23], 'Ballpoint Ink Age Determination by Volatile Component Comparison – A Preliminary Study' [24], 'Applications of Pyrolysis Gas Chromatography/Mass Spectrometry to Toner Materials from Photocopiers' [25], 'Color Comparison in Questioned Document Examination Using Microspectrophotometry' [26] and 'Transmission and Reflectance Microspectrophotometry of Inks' [27].

We also see one of the first references to an image enhancement system [28]. However, the majority of the papers are practical applications of the questions arising from case-related material. Todd's [29] paper 'Do Experts Disagree?' attempted to provide statistical support to the supposition that ethical and competent FDEs, given the same materials to examine, will come to the same conclusion [29]. Even though research in the field was expanding, little had been done in the way of reliability studies. This was about to change.

9.7.2 Post-Daubert

In the late 1980s and early 1990s, a series of events occurred which had an explosive impact on the field of questioned documents. These events included a 1989 publication by Risinger, Denbeaux, and Saks [30], the rulings in *Daubert v. Merrell Dow Pharmaceuticals, Inc.* 509 U.S. 579 (1993), United States v. Starzecpyzel, 880 F.Supp. 1027 (S.D.N.Y.) 1995, General Electric Co., *et al.* v. Joiner *et al.* 1997 F.3rd 524 and Kumho Tire *et al.* v. Carmichael *et al.*, 526 U.S. 137, 119 S. Ct. 1167, 143 L. Ed.2d 238 (1999). Questions about the validity and reliability of many of the opinion evidence fields were being challenged, starting with attacks on handwriting identification.

In 1989, Risinger *et al.* published an article in the *University of Pennsylvania Law Review*, 'Exorcism of Ignorance as a Proxy for Rational Knowledge: The Lessons of Handwriting Identification Expertise', which challenged the validity of the premises upon which handwriting identification is based and the ability of an FDE to make a determination of authorship [30]. In this article, the authors conclude, based on the Forensic Science Foundation (FSF) test results from 1975 and 1984 to 1987, '. . . that no available evidence demonstrates the existence of handwriting expertise' [30].

This brought about an outcry from FDEs. Articles were written to refute the criticism raised by the law review article. In the article, 'The Principle of the Drunkard's Search as a Proxy for Scientific Analysis: The Misuse of Handwriting Test Data in a Law Journal Article', the authors refute the interpretations by Risinger *et al.* of the FSF studies of 1975 and 1984–1987 and, based on their analysis, conclude: 'If anything the data indicates that expert document examiners do provide valuable information to the court' [31].

Kam's study 'Proficiency of Professional Document Examiners in Writer Identification' points out the flaws in the findings by Risinger *et al.* and presents the results of a designed writer identification test which supports the premise that '. . . handwriting identification expertise indeed exists, and that the generally negative conclusions about this issue by Risinger *et al.* may have been premature' [32]. In addition, Andre Moenssens' University of Missouri-Kansas City Law Review article, 'Handwriting Identification Evidence in the Post – Daubert World' [33], addresses many of the flaws and deficiencies in the research and conclusions drawn in the Risinger *et al.* paper.

While the 'Exorcism' paper caused much controversy, it was the shifting legal landscape that inspired a new renaissance in research and publications in the questioned documents field. Prior to the ruling handed down by the Supreme Court in Daubert v. United States, 1993, the criteria for admissibility of scientific testimony was based on the 1923 Frye test of 'general acceptance', and in 1975 the Federal Rules of Evidence, rule 702, applied the 'general relevance' standard. Questioned document evidence had no problems being admitted under Frye and the Federal Rules of Evidence. However, numerous challenges had to be met to be in accord with the criteria set forth in the Daubert ruling [33,34].

In Daubert, the criterion for admissibility is 'general acceptance' based on testing the technique, peer view, error rates and standard operating procedures. To date, many challenges have been made concerning questioned document testimony. Some of the cases have resulted in allowing limited testimony of FDEs, but the majority of challenges have resulted in denying motions to exclude the testimony.

The issues raised by handwriting critics and the necessity to meet the court's interpretation of the criteria under Daubert brought about empirical, experimental research to resolve these matters. The studies, conducted by Dr. Moshe Kam, a researcher at Dexel University, and Dr. Sargur N. Srihari, Director, Center of Excellence for Document Analysis and Recognition, are considered by many to be the most noteworthy in addressing these topics. In his controlled study, 'Proficiency of Professional Document Examiners in Writer Identification' Dr. Kam's findings indicated that the professional document examiner is better at writer identification than the control group tested. In fact, the hypothesis that the professionals and non-professionals performed equally in identifying handwriting ' . . . was found via the Kruskal-Wallis test to have probability of less than 0.001' [32].

This was a pilot study which was followed in 1997 by the publication of 'Writer Identification by Professional Document Examiners' [35] – a comprehensive, controlled study that tested the proficiency of questioned document examiners in their ability to identify handwriting. The study found that the professionals did possess skills that non-professionals did not. In response to criticism that monetary incentives influenced the results of the non-professionals, Kam reported on a subsequent study which found ' . . . non-professionals performed the same regardless of incentive scheme, and exhibited markedly inferior performance . . . ' than professionals [36].

'Writer Identification Using Hand-Printed and Non-Hand-Printed Questioned Documents' [37] was a study in response to the inquires of federal district court judges on the ability of questioned document examiners to identify hand printing. Again, the skill and ability of questioned document examiners was tested and verified.

While Kam's research was geared toward verifying that FDEs possess a greater ability than non-professionals to identify handwriting, Srihari's research focuses on the consistency and individuality of a person's handwriting.

'Individuality of Handwriting' [34] by Srihari is a response to Daubert *et al.* and several United States court rulings which ensued concerning expert testimony and, specifically, handwriting testimony. This study was done to validate the individuality of handwriting by extracting features from scanned writing and analyzing the differences via computer algorithms. Based on the macro-features and micro-features in the handwriting, the study was '. . . able to establish with a 98% confidence that the writer can be identified' [34] and that, if statistically inferred to the US population, the computer system could validate handwriting '. . . individuality with a 96% confidence' [34].

The Srihari study was criticized in a commentary to the *Journal of Forensic Sciences* by Michael J. Saks [38], and this criticism was addressed point by point in the authors' response [39]. One of Saks's complaints was that samples were taken from a large, diverse group of writers, thereby making it easier to distinguish the handwriting by computer. 'The Determination of Authorship from a Homogenous Group of Writers' by Durina and Caligiuri [40] endeavored to address this as well as other issues raised by Saks.

Srihari's research 'On the Discriminability of the Handwriting of Twins' [41] expanded upon his previous study, comparing the handwriting of twins (identical and fraternal) and non- twins via an automatic handwriting system.

9.7.3 Current projects

Doctors Kam and Srihari are continuing to do research in the field of questioned documents. Dr. Kam is testing the proficiency of FDEs and laypersons in identifying and excluding specific writers of naturally written, as well as deliberately disguised, handwritten and hand-printed documents. His research also includes evaluating the effect of context (information provided to FDEs about document collection circumstance and other evidence) on the accuracy of the conclusions expressed by FDEs (i.e. context bias) (K. Singer, elec. comm. 2011).

Dr. Srihari is continuing research in methods of extracting letter combinations occurring in handwriting automatically to determine the frequency of the characteristics of those letter combinations in a large representative population in the United States (K. Singer, elec. comm. 2011).

The University of Central Florida was awarded an NIJ Grant to study the 'Frequency Occurrence of Handwriting and Hand-Printing Characteristics' [42].

Minnesota Department of Public Safety was awarded an NIJ Grant titled 'Development of Individual Handwriting Characteristics in 1,800 Students: Statistical Analysis and Likelihood Ratios That Emerge Over An Extended Period Of Time' (L. Hanson, elec. comm. 2011). All the writing samples (data) collected for this study will be analyzed to:

- identify and measure (statistically) individual writing habits in students' writing samples as they begin to deviate from the copybook style taught at IDS #832 (as students develop their own handwriting habits);

- measure the likelihoods ratios in which individual writing habits appear in the samples;
- measure the frequency in which these individual writing habits appear in combinations within each of the samples (L. Hanson, elec. comm. 2011).

Kentucky State University was awarded an NIJ Grant to study 'Validity, Reliability, Accuracy and Bias in Forensic Signature Identification' [42].

Current research in questioned documents is published in peer-reviewed journals such as *Journal of Forensic Sciences*, *Journal of the American Society of Questioned Document Examiners*, the *Journal of Forensic Identification*, *Forensic Science International*, the *Journal of the Forensic Science Society* and the *Canadian Society of Forensic Science Journal*, as well as others.

9.7.4 New developments and future directions

Newer developments in questioned document examination include computer-developed search, verification, identification and comparison systems such as the Forensic Information System for Handwriting (FISH), the Cedar Forensic Examination (CEDAR-Fox) system and the Forensic Language-Independent Automated System for Handwriting Identification (FLASH ID). These automated systems help to meet the criteria for admissibility of expert testimony in handwriting identification. In the future, these systems are certain to be improved upon and upgraded, although they are not a replacement for the forensic document examiner.

The utilization of digital technology is greatly enhancing the ability of the FDE to investigate changes to documents and to illustrate easily their findings to juries. The 'Information Age' has presented many new challenges to the forensic document examiner. Identity theft, authentication of official documents which can easily be replicated on a home computer, and just keeping up with advancing technology are all daunting tasks. However, the comparison and identification of handwriting and hand printing still comprises the majority of examinations performed by forensic document examiners.

Acknowledgments

We wish to acknowledge Robin K. Hunton, Maureen Casey Owens, Larry A. Olson, Alisa Skinner, Gerald M. LaPorte, Ted M. Burkes, Jeff Huber and Thomas W. Vastrick for their support and assistance.

References

1. ASTM, International. *ASTM Standard E444, Standard Guide for Scope of Work of Forensic Document Examiners*. West Conshohocken, PA: ASTM International; 2009.
2. Huber RA, Headrick AM. *Handwriting Identification Facts & Fundamentals*. New York: CRC Press; 1999; 3–8.
3. Levinson J. *Questioned Documents, A Lawyer's Handbook*. San Diego: Academic Press; 2001; 1–7.
4. Hilton O. History of Questioned Document Examination in the United States. *Journal of Forensic Sciences* 1979; 24 (4):890–897.
5. Louisiana Supreme Court. *Reports of Cases Argued and Determined by the Supreme Court of Louisiana, Volume 7, for the Year 1852*. New Orleans: T Rea;1854:565–6, http://books.google.com/books?id=AfsDAAAAYAAJ (accessed October 6, 2011).
6. Chabot C, Twisleton ETB. *The Handwriting of Junius, Professionally Investigated by Mr. Charles Chabot, Expert*. London: J Murray;1871, http://books.google.com/books?id=FXogAQAAMAAJ&dq=charles+chabot+junius&source=gbs_navlinks_s (accessed October 11, 2011).
7. ASQDE. Albert S. Osborn. Long Beach, CA: American Society of Questioned Document Examiners. www.asqde.org/about/presidents/osborn_as.html (accessed October 11, 2011).
8. ASQDE. *The Origin of the ASQDE*. Long Beach, CA: American Society of Questioned Document Examiners, www.asqde.org/about/history.html (accessed October 11, 2011).
9. ABFDE. *Background, Functions and Purposes of the ABFDE*. Houston: American Board of Forensic Document Examiners, www.abfde.org/htdocs/AboutABFDE/BackgroundFunctionsPurposes.pdf (accessed October 7, 2011).
10. Harris JJ. A Tribute to Ordway Hilton. *ABFDE News* 1998 July; 9(3):1, 17.
11. Crown D. Questioned Document Examination. In: Wecht CH, editor. *Forensic Sciences Law/Science Civil/Criminal*, Vol 3. New York: Matthew Bender & Co, Inc.; 1989; 1–39.
12. McAlexander TV, Beck J, Dick R. The Standardization of Handwriting Opinion Terminology. *Journal of Forensic Sciences* 1991; 36(2):313.
13. SWGDOC. Scientific Working Group for Forensic Document Examination [updated 2011 Nov 15]. *About SWGDOC* [updated 2011 Sep 22]. www.swgdoc.org/about_us.htm (accessed October 7, 2011).
14. Singer K. The Daubert Era. In: Kelly S, Lindblom BS, editors. *Scientific Examination of Questioned Documents*. Boca Raton: CRC Taylor & Francis; 2006; 37–42.
15. ENFSI. The Hague: European Network of Forensic Science Institutes; c1999–2009. European Document Experts Working Group (EDEWG). Document [updated 2011 Nov 21]. www.enfsi.eu/page.php?uid=55 (accessed October 18, 2011).
16. ENFSI. The Hague: European Network of Forensic Science Institutes; c1999–2009. European Network of Forensic Handwriting Experts (ENFHEX). *Handwriting* [updated 2011 Nov 21]. www.enfsi.eu/page.php?uid=64 (accessed October 18, 2011).

17. Galbraith G. Trends in the Field of Questioned Document Examination. *Journal of Forensic Sciences* 1980; 25(1):132–140.
18. Waeschle PA. Examination of Line Crossings by Scanning Electron Microscopy. *Journal of Forensic Sciences* 1979; 24(3):569–78.
19. Moon HW. Electrophoretic Identification of Felt Tip Pen Inks. *Journal of Forensic Sciences* 1980; 25(1):146–149.
20. Zoro JA, Totty RN. The Application of Mass Spectrometry to the Study of Pencil Marks. *Journal of Forensic Sciences* 1980; 25(3):675–678.
21. Lyter AHIII. Examination of Ball Pen Ink by High Pressure Liquid Chromatography. *Journal of Forensic Sciences* 1982; 27(1):154–160.
22. Dalrymple BE. Visible and Infrared Luminescence in Documents: Excitation by Laser. *Journal of Forensic Sciences* 1983 Jul; 28(3):692–696.
23. Guineau B. Microanalysis of Painted Manuscripts and of Colored Archeological Materials by Raman Laser Microprobe. *Journal of Forensic Sciences* 1984; 29 (2):471–485.
24. Stewart LF. Ballpoint Ink Age Determination by Volatile Component Comparison—A Preliminary Study. *Journal of Forensic Sciences* 1985; 30(2):405–411.
25. Levy EJ, Wampler TP. Applications of Pyrolysis Gas Chromatography/Mass Spectrometry to Toner Materials from Photocopiers. *Journal of Forensic Sciences* 1986; 31(1):258–271.
26. Olson LA. Color Comparison in Questioned Document Examination Using Microspectrophotometry. *Journal of Forensic Sciences* 1986; 31(4):1330–1340.
27. Zeichner A, Levin N, Klein A, Novoselsky Y. Transmission and Reflectance Microspectrophotometry of Inks. *Journal of Forensic Sciences* 1988; 33 (2):1171–1184.
28. Schuetzner E. Examination of Sequence of Strokes with an Image Enhancement System. *Journal of Forensic Sciences* 1988; 33(1):244–248.
29. Todd I. Do Experts Frequently Disagree? *Journal of Forensic Sciences* 1973; 18(4): 455–459.
30. Risinger D, Denbeaux M, Saks MJ. Exorcism of Ignorance as a Proxy for Rational Knowledge: The Lessons of Handwriting Identification Expertise. *University of Pennsylvania Law Review* 1989; 137:731–791.
31. Galbraith O, Galbraith CS, Galbraith NG. The Principle of the 'Drunkard's Search' as a Proxy for Scientific Analysis: The Misuse of Handwriting Test Data in a Law Journal Article. *International Journal of Forensic Document Examiners* 1995 Jan/Mar; 1(1):7–17.
32. Kam M, Wetstein J, Corm R. Proficiency of Professional Document Examiners in Writer Identification. *Journal of Forensic Sciences* 1994; 39(1):5–14.
33. Moenssens AA. Handwriting Identification Evidence in the Post Daubert World. *UMKC Law Review* 1997; 66(251):251–343.
34. Srihari SN. The Individuality of Handwriting. *Journal of Forensic Sciences* 2002; 47(4):856–872.
35. Kam M, Fielding G, Conn R. Writer Identification by Professional Document Examiners. *Journal of Forensic Sciences* 1997; 42(5):778–786.
36. Kam M, Fielding G, Conn R. Monetary Incentives on Performance of Nonprofessionals in Document-Examination Proficiency Tests. *Journal of Forensic Sciences* 1998; 43(5):1000–1004.

37. Kam M, Lin E. Writer Identification Using Hand-printed and Non-Hand-Printed Questioned Documents. *Journal of Forensic Sciences* 2003; 48(6): 1391–1395.
38. Saks M. Commentary on: The Individuality of Handwriting, Srihari *et al. Journal of Forensic Sciences* 2003; 48(4):916–918.
39. Srihari SN. Author's Response: The Individuality of Handwriting, Srihari *et al. Journal of Forensic Sciences* 2003; 48(4):919–920.
40. Durina M, Caligiuri MP. The Determination of Authorship from a Homogenous Group of Writers. *Journal of the American Society of Forensic Document Examiners* 2009 Dec; 12:77–90.
41. Srihari S, Huang C, Srinivasan H. On the Discriminability of the Handwriting of Twins. *Journal of Forensic Sciences* 2008; 53(2):431–446.
42. NIJ. *Forensic Science Research and Development Awards 1992–2011*. Washington DC: US Government, www.nij.gov/topics/forensics/forensic-awards.htm (accessed December 1, 2011).

Further reading

Ames DT. *Ames on Forgery.* San Francisco, CA: Ames-Rollinson; 1901.

Browning BL. *Analysis of Paper.* 2nd ed. Berkeley, CA: M. Dekker; 1977.

Brunelle RL, Crawford KR. *Advances in the Forensic Analysis and Dating of Writing Ink*. Springfield, IL: Charles C. Thomas; 2003.

Brunelle RL, Reed RW. *Forensic Examination of Ink and Paper.* Springfield, IL: C. C. Thomas; 1984.

Conway JVP. *Evidential Documents*. Springfield, IL: C. C. Thomas; 1959.

Ellen D. *Scientific Examination of Documents, Methods and Techniques*. 3rd ed. Boca Raton, FL: Taylor & Francis; 2006.

Harrison WR. *Suspect Documents, Their Scientific Examination*. New York, NY: Praeger; 1958.

Hilton O. *Scientific Examination of Questioned Documents*. Rev ed. New York, NY: Elsevier Science publishing Co., Inc; 1982.

Huber RA, Headrick AM. *Handwriting Identification: Facts and Fundamentals*. Boca Raton, FL: CRC Press; 1999.

Moenssens AA, Henderson CE, Portwood SG. *Scientific Evidence in Civil and Criminal Cases*. 5th ed. New York, NY: Foundation Press Thompson/West; 2007.

Morris R. *Forensic Handwriting Identification, Fundamental Concepts and Principles*. San Diego, CA: Academic Press; 2000.

Osborn AS. *Questioned Documents*. 2nd ed. Albany, NY: Boyd Printing Company; 1929.

Osborn AS. *The Problem of Proof.* 2nd ed. Newark, NJ: Essex Press; 1926.

Seaman Kelly J. *Forensic Examination of Rubber Stamps*. Springfield, IL: C. C. Thomas; 2002.

Seaman Kelly J, Lindblom BS. *Scientific Examination of Questioned Documents*, 2nd ed. Boca Raton, FL: CRC/Taylor & Francis; 2006.

Questioned document research is published in the following scientific journals:

- *Journal of Forensic Sciences*
- *Journal of the American Society of Questioned Document Examiners*
- *Journal of Forensic Identification*
- *Canadian Society of Forensic Science Journal (CSFSJ)*
- *The International Journal of Forensic Document Examiners*
- *Forensic Science International*
- *Forensic Science Review*

10

Digital evolution: history, challenges and future directions for the digital and multimedia sciences section

David W. Baker[1], Samuel I. Brothers[2], Zeno J. Geradts[3], Douglas S. Lacey[4], Kara L. Nance[5], Daniel J. Ryan[6], John E. Sammons[7] and Peter Stephenson[8]

[1]*MITRE Corporation, McLean, Virginia, USA*
[2]*United States Customs and Border Protection, Springfield, Virginia, USA*
[3]*Digital Technology and Biometrics Section, Netherlands Forensic Institute, The Hague, Netherlands*
[4]*DL Technology LLC, Fredericksburg, Virginia, USA*
[5]*Computer Science Department, University of Alaska Fairbanks, Fairbanks, Alaska, USA*
[6]*Information Resources Management College, National Defense University*
[7]*Marshall University, Huntington, West Virginia, USA*
[8]*Norwich University, Northfield, Vermont, USA*

10.1 Introduction to digital and multimedia forensics

In criminal and civil litigation, digital evidence presents itself in an almost limitless number of scenarios. Due to the ubiquity of technology such as laptop computers, cell phones, digital cameras (both still and video) and pocket recorders, digital evidence is routinely recovered in relation to criminal acts such as homicide, robbery, burglary, larceny, illicit drug trafficking and many others. Additionally, forensic analysis is a necessity in 'digitally centered crimes' such as computer network intrusions, identity theft with associated fraud, and much of the possession, distribution and production of child pornography.

Digital evidence plays a major role in civil cases as well. Parties to a civil action are often required to exchange electronically stored information during the course of

Forensic Science: Current Issues, Future Directions, First Edition. Edited by Douglas H. Ubelaker.

litigation. In this process, known as Electronic Discovery (or eDiscovery), digital forensic protocols, procedures and technology are often used.

Digital and multimedia evidence is also heavily relied upon in the operations of both governmental intelligence agencies and the military. Digital forensics is used routinely to identify, track and monitor terrorists and their activities. It features heavily in battlefield operations as well, with soldiers seizing electronic devices in order to exploit the information they contain to gain a tactical and or strategic advantage. In addition, cyber warfare and cyber terrorism pose a significant threat to many nations.

Administrative proceedings also utilize digital forensics capabilities. Administrative proceedings typically revolve around the violation of organizational policies such as those governing the acceptable use of computing resources. These actions typically result in suspensions and terminations, rather than criminal charges.

10.2 Definition of digital and multimedia sciences

The digital and multimedia sciences (DMS) area cover several different, but related, technical disciplines. For our purposes, DMS comprises forensic analysis of:

- computers and computing devices of all types (including standalone and networked devices, as well as software which runs on the devices);
- storage devices or media utilized by such devices;
- network communications between such devices; analysis of images (analog and digital);
- analysis of audio (analog and digital);
- analysis of video (analog and digital).

Analysis of audio, images and video in this context primarily refers to analysis for authenticity, photogrammetry, comparison and enhancement.

Computer forensics involves analysis of devices and data from a multitude of computing environments. These environments include standalone computers, networks, mainframes, 'cloud' environments and small-scale digital devices.

A standalone machine is a single computer not connected to another computer or network. In comparison to the other environments, standalone computers are the most contained and have been the focus of what most consider traditional computer forensics. Typical examination involves imaging (making an exact digital copy) of the storage media associated with the computer. Images of media are typically validated using a cryptographic hashing algorithm to verify the integrity and accuracy of the image. The image is then used for the analysis, without using the original media for the analysis.

There are a number of tools used to examine or analyze the media, and various of these can determine current contents, deleted data, timelines of activity and other types of data, all of which may have evidentiary value. Computer forensics practitioners may have to analyze a vast array of storage media and technology, including magnetic (hard disks and tape), optical (CD, DVD and Blu-ray) and flash (USB and solid state drives). Data may also be stored in a variety of different file systems and/or file formats, requiring extensive knowledge of various ways in which data is stored. Some data is extracted directly from the memory of digital devices, requiring yet another set of skills in understanding the ways various devices' operating systems process instructions and data, and understanding the structures used by those operating systems. Additionally, many standalone computer systems will access the internet, using a variety of methods via an internet service provider.

Examination of systems can also identify internet activity by the system and its users, including web browsing, email, instant messaging and other typical internet application usage. Examination may also identify malicious software that has infected the system, which users of the system may or may not be aware of. With the advent of malicious software, or malware, forensics examinations may require skills in reverse engineering to understand the effect that malware may have on computers, networks and data.

Of course, use of the internet does begin to cross the boundary from a standalone system analysis to a network investigation. Any two or more computers connected together form a network. Networks can be as small as two computers, or as large as thousands of computers. Mainframe systems centralize the computing power and resources and are typically found in very large organizations as part of their internal network resources. The internet is a network of networks, providing a means for sharing information and capabilities across the globe.

Regardless of the type of network, once systems participate in a network, data about the communications between systems, which may include content, is no longer restricted to a single system. This adds to the complexity of investigations but also provides more sources of information that may be evidence. It can also affect timelines about activity, by having independent time sources from the different systems.

Logs from various network devices can confirm or refute that a system performed certain actions. Communications between systems consist of layers, commonly called the network and transport layers, from the OSI model [1]. Devices that enable networking often have log data which provides details about such communication. Systems offering services such as databases, web pages or email accounts are called servers. These often maintain logs of activity which are used primarily for making sure that things work properly, but these same logs can often provide valuable evidence about activity by users and client systems connecting to the server.

As network services and networks become more powerful and more complex, it is possible for network services to begin to offer the types of capabilities previously seen only on dedicated systems, which brings us from basic network investigations to investigations in the 'cloud'. The cloud takes computing resources that were once located on individual machines and delivers them as services to users that are connected to a network, which could include the internet. The cloud offers multiple types of services, including Hardware as a Service (HaaS), Software as a Service (SaaS), Infrastructure as a Service (IaaS) and Platform as a Service (PaaS).

Mobile device forensics is a sub-discipline of digital forensics and relates to the recovery of information from a given mobile device under forensically sound conditions. The methods used for mobile device forensics vary greatly between devices, due to the great deal of diversity and the proprietary nature of hardware interfaces and hardware configurations. To make matters even more complex, many of the operating systems used are undocumented and proprietary. Mobile device forensics also includes the discipline of global positioning system (GPS) forensics.

Both cell phone and GPS forensics require a vast array of tools, as no one single tool is able to obtain data from all devices that an examiner may encounter. Traditionally, a cell phone contains the following data types: contacts, call logs and text messages. However, smart phones (e.g. the Apple iPhone and HTC Droid) are more like mini personal computers, as they contain many more data types, including internet browser history, digital camera pictures, videos and GPS data, to name just a few.

GPS devices traditionally hold the following three types of information: way points (documenting where a given device has been over a period of time), track points (documenting specific points on the Earth) and track logs (a collection of track points). GPS devices may contain additional data types, as well (e.g. pictures and videos). GPS forensic practitioners are again required to use a wide array of tools, as there is no one single tool that can obtain information from all GPS devices. This is a growing field, and many cell phone forensic tool vendors have added data extraction support for GPS devices in the last 2–3 years.

Image and video forensics involves the analysis of devices and data from a variety of cameras. The number of mobile phones containing cameras, as well as hand-held video cameras and closed circuit television/monitoring (CCTV) systems, has grown very rapidly in the last decade. Additionally, the capability to network these devices and provide a means for sharing video and images has also developed rapidly, for example with services such as YouTube and Flickr. In many investigations, video and images are considered as potential evidence, and a number of questions arise about the usefulness of the images or video. Some questions are related to enhancement of the image, such as whether there any possibility of improving the quality of an image. Others seek to determine if the images have sufficient detail in order to perform some sort of identification of the content of the images, such as:

- Is the person in the image the same person as the suspect?
- How tall is/are the person(s) in the image?
- Can you conduct a facial comparison?
- Can you perform a posture comparison?

- Can you identify the clothing with a comparison?
- Is the object (car, bag, wall, etc.) in the image the same as an object from the suspect?

Sometimes, the analysis seeks to determine if certain images or video had been made with a specific camera or imaging device. Other analysis seeks to determine the authenticity of a particular image, or investigate whether the images have been tampered with. Image and video analysis can also combine inputs and data from several cameras and recreate a three-dimensional visualization of the route or activities of persons/cars/objects within a scene. Finally, image and video analysts are often provided with damaged media, damaged data files or data files of an unknown type, and asked to determine whether the images or video might be repaired or converted into a form such that the content is available for review.

Audio forensics involves the analysis and/or processing of audio recordings, which can come from a variety of sources, including undercover law enforcement recorders, dashboard video/audio recording systems, 911 and other telephonic recordings systems, cockpit voice recorders from commercial and private aircraft, broadcast recordings and recordings produced by private individuals.

There are four primary types of examination conducted in the audio forensics field: enhancement, authenticity, signal analysis and voice comparison:

- In an enhancement examination, a recording is digitally filtered in an effort to reduce noise which is interfering with the desired signal (typically voice information) [2,3].
- An authenticity examination seeks to determine whether or not a recording is original (or a clone of the original recording, in the case of a digital recording), continuous, unaltered or, in some cases, consistent with having been produced on the alleged recording device and in the manner represented by the person(s) who made the recording [4–6].
- Signal analysis examinations encompass detailed time- and frequency-domain studies of non-voice signals contained in a recording, such as gunshots, telephone signaling, jet engines and the like, in order to glean information about specific audio event(s) [7–9].
- In voice comparison examinations, a questioned, unknown voice in an evidentiary recording is compared with known, verbatim samples of an individual's voice, to provide an opinion on whether or not that person voiced the evidentiary recording [10,11].

Since forensic audio examinations rely heavily on knowledge of various scientific principles, including acoustics, electronics and digital signal processing theory, practitioners in the field often have bachelor of science degrees in fields such as engineering, physical sciences or computer engineering.

Just because a conversation is recorded digitally does not guarantee that it will be of better quality than an analog recording of the same conversation. Digital recordings typically employ compression, which, while allowing for longer recordings to be made, sacrifices quality and can render low-level voice information unintelligible, even after attempts at digital filtering. The finer details of the recording may be lost in the artifacts of the compression schemes utilized by most commonly available digital audio recorders.

10.3 History of digital and multimedia sciences

The history of digital and multimedia forensics brings together three threads: the use of digital notation, binary expressions and instantiation in information technology; the laws of evidence applicable to scientific, technical or specialized evidence; and substantive Cyberlaw (or law pertaining to computers, networks, digital devices and communications between such devices). The threads join when digital evidence is used to assist courts in determining facts at issue in criminal and civil cases.

10.3.1 Binary expressions and the evolution of information technology

Much of the world is continuous in nature – inherently analog. But when we count things, we move away from the continuous to the discrete, with early counting using fingers or digits (from the Latin 'digitus' for finger). We enter the digital world. Early on in history, dots, slash marks, loops or circles on clay tablets stood for fingers [12]. Some early number systems used 5, 12, 20 or 60 as bases, rather than 10 or 2 (12, p. 11). The Mayans were using zero as a number as early as 36 BCE, but it was likely known even earlier to the Olmecs (12, p. 43). The Indians were using a blank space for zero as early as the 4th century BCE, and the Chinese understood both negative numbers and zero since at least the 4th century BCE.

Decimal numbering systems, much like those we use today, were developed in India by about 100–200 CE. By 130 CE, Ptolemy was using zero as a standalone number, not just as a placeholder. Place-value notation together with zero was in use by the Jains in India by 458 CE. The Hindu-Arabic numerals we use today appeared about 500 CE, but they were not used in the West until the 11th Century, when they were introduced in Europe through Spain by the Moors.

The Indian scholar Pingala (circa 5th–2nd centuries BCE), about whom little is known, developed a mathematical approach to prosody, and his works give us the first known description of a binary number system [13]. In 1605, Francis Bacon discussed a system whereby letters of the alphabet could be reduced to sequences of binary digits [14] and, in 1679, in his *Explication de l'Arithmétique Binaire*,

Gottfried Wilhelm von Leibniz (1646–1716) used modern binary notation of 1 and 0 to explore binary arithmetic [15].

In 1854, George Boole published a landmark paper, *An Investigation of the Laws of Thought on Which are Founded the Mathematical Theories of Logic and Probabilities*, that gave us Boolean algebra [16]. In 1937, Claude Shannon produced his master's thesis at MIT, *A Symbolic Analysis of Relay and Switching Circuits*, that implemented Boolean algebra and binary arithmetic using electronic relays and switches [17]. Shannon would go on to publish *A Mathematical Theory of Communication* in 1948 [18] and *Communication Theory of Secrecy Systems* in 1949 [19], establishing information theories that were essential in enabling telecommunications to move from analog to digital during the 1960s and thereafter.

The use of technology to facilitate numerical operations kept pace with the development of binary mathematics. Technology appeared very early in the form of clay tablets or wooden planks covered with wax, on which numbers could be recorded with a stylus. Counting and arithmetic were accelerated by the invention of the abacus. In 1614, John Napier discovered logarithms, making it possible to convert problems in multiplication and division into problems in addition and subtraction [20], and leading to the invention of the slide rule by William Oughtred [21].

Slide rules were still in use in the mid-20th century until they were made largely obsolete in the early 1970s by electronic calculators. Before the appearance of electronic calculators, however, there were generations of mechanical calculators. Blaise Pascal (1623–1662) invented in 1642 a geared mechanism that could calculate with eight-digit numbers and carry powers of ten[22]. In 1671, Gottfried Wilhelm von Leibnitz (1646–1716) invented a stepped-drum machine to calculate mechanically [23]. Joseph Marie Jacquard (1752–1834) built a loom in 1801 that used binary logic, following a fixed program on punched cards operating in real time to produce complex woven patterns [24]. And, of course, Charles Babbage (1792–1871) designed, but was never able to build, steam-driven mechanical calculators, including a 'difference engine' for calculating mathematical tables and an 'analytical engine' that would have been the world's first general purpose computer had it been completed [25].

Augusta Ada Byron King, Countess of Lovelace, daughter of the poet George Gordon Lord Byron and Anne Isabella Milbanke, met Charles Babbage when she was 18, whereafter she, an untrained but gifted mathematician, became fascinated with Babbage's machines and began to write instructions for their use, thereby becoming the world's first computer programmer [26].

Herman Hollerith (1860–1929) used the punched-card approach of Jacquard and, in 1890, developed a punched-card tabulating machine [27]. Electrical signals could be conducted only when holes in the punched cards allowed contact. Electromechanical tabulators counted the signals that got through, reading, counting and classifying the cards. Later models could also sort the cards and incorporated automatic card punching. This was the beginning of modern data processing. The Tabulating Machine Company he founded to distribute Hollerith's machines merged with other companies and eventually was purchased by its general manager, Thomas J. Watson, who renamed it the International Business Machines Corporation, or IBM, in 1911.

In November 1937, George Stibitz, an associate of Claude Shannon at Bell Labs, built a relay-based computer which calculated using binary addition (28, pp. 193–256). On January 8, 1940, he completed a machine that could calculate using complex numbers. Dr. Howard H. Akin of Harvard University had IBM build the IBM Automatic Sequence Controlled Calculator, called the Mark 1, for the university. It was delivered in February of 1944 and used not only electric switches and relays, but also shafts and clutches, so the Mark 1 was an electro-mechanical computer. It could store 72 numbers of 23 decimal digits each and perform three additions or subtractions each second. Evaluating a logarithm or trigonometric function took over a minute. There was no conditional branch instruction so, technically speaking the Mark 1 did not employ a Von Neumann architecture (28, pp. 134–190).

ENIAC (the Electronic Numerical Integrator and Computer) was built at the University of Pennsylvania's Moore School of Electrical Engineering for the United States Army. It was dedicated on February 15, 1946 and transferred to the Aberdeen proving ground in Maryland, where it remained in operation until 1955. ENIAC was built using vacuum tubes, crystal diodes, relays, resistors and capacitors, and so was a fully electronic device. Its programs could branch, so it was more versatile than the Mark 1. It could perform up to 385 multiplications, or 40 divisions, or three square root operations, per second (28, pp. 257–301).

Following the Second World War, digital technology began to advance exponentially. Vacuum tubes were replaced with transistors. Punched cards gave way to magnetic tape, and magnetic cores replaced rotating magnetic drums for internal storage after 1953. Machine-language coding of sequences of instructions that could be used directly by the computer was replaced by assembly language coding, which was in turn replaced by higher-level languages that the computer compiled into the sequence of machine code instructions comprising the program to be performed. FORTRAN (FORmula TRANslator) appeared in 1955, LISP (LISt Processor) in 1958, ALGOL (ALGOrithmic Language) in 1958, COBOL (COmmon Business Oriented Language) in 1959, and BASIC (Beginner's All-purpose Symbolic Instruction Code) in 1964. Use of higher-level languages greatly simplified programming of digital devices and created the programming profession we know today [29].

The transistor was invented in 1947 and quickly replaced vacuum tubes as the key components of digital logic circuits [30]. Then, in 1958, Jack Kilby of Texas Instruments created the first integrated circuit, which carved many circuit components into a single block of semiconductor material, making it possible to reduce the size of the components and increase the speed at which the circuits could operate [31]. Kilby received the Nobel Prize in physics in 2000 for his invention. Gordon Moore, one of the founders of Intel, famously stated in April of 1965 that the number of transistors that could be inexpensively incorporated into an integrated circuit would approximately double every two years, a trend that has held more or less constant for fifty years [32]. Today, transistors are about 90 nanometers in size, and the most advanced integrated circuits contain several hundred millions of components.

The earliest computers did not have operating systems to manage the resources of a computer efficiently as it performed its programmed tasks. In the 1950s, computers ran only a single program at a time. To cut down on the wasted time needed to set up each job, in 1958, General Motors Research Laboratory developed a program for its IBM 701 mainframe to allow multiple programs, each on a deck of punched cards, with each deck separated by control cards, to be batched into a sequence of jobs in a single deck. This is considered to have been the first operating system [33].

In the 1960s, programmers improved the operating system to permit different programs to concurrently use different computer resources – input and output channels and devices, central processing units, storage devices and so forth – a technology called multiprogramming. This greatly enhanced the utilization of components and overall throughput. Further improvements allowed the computer to better control the timing of shifts in control among the several programs in a multiprogramming environment, and so timesharing was invented. Timesharing reduced program turnaround from hours or days to minutes [34].

The 1960s also saw the development of large-scale database management systems. In 1961, President Kennedy charged the nation to send a man to the moon and return safely. Rockwell won the Apollo contract and, in 1965, partnered with IBM to create an automated system to manage the huge bills of material for the project. The space program would have been much more difficult or impossible without the Information Control System and Data Language/Interface (later renamed the Information Management System/360) database management they created. The system began operations in 1968 [35].

The 1970s saw the introduction of TCP/IP (Transmission Control Protocol/Internet Protocol) in military and university computing environments [36]. Xerox's Palo Alto Research Center developed the Ethernet standard, and Local Area Networks (LANs) appeared. The UNIX multitasking operating system was developed by Bell Labs in 1969 and was working well by the early 1970s [37]. Dennis Richie developed the *C* programming language to facilitate writing programs, and translating UNIX into *C* made it the first operating system that was machine-independent [38].

The 1970s also saw the development of the microprocessor, which enabled the development of personal computers. The integrated circuits of the early 1960s contained a half-dozen components per chip. By 1970, Moore's Law made it possible to make chips containing thousands of active components. In 1971, Intel brought the 4-bit Intel 4004 microprocessor to market. In 1973, the 4004 was replaced with the 8-bit Intel 8008, and the 8-bit standard – still in use in computer architectures and other digital devices – was born [39].

In January of 1975, the Altair 8800, the very first microprocessor-based computer, was brought to market as a hobbyist's kit [40]. It used an Intel 8080 chip, had no display, no keyboard and too little memory to do much. It was programmed in binary code, using small switches on its front panel. It was not much, but it was the start of something really big. Bill Gates and Paul Allen decided to develop BASIC for the Altair and formed a partnership called Micro-Soft, later Microsoft. They licensed the software, rather than selling it outright, setting the pattern for software still in use today [41].

The first operating system for the 8080 chip was written in 1976 by Gary Kildall, who settled on the floppy disk as a mass storage device for storing programs and data. Files were stored as fragments in whatever space was available on the disk, so the operating systems for personal computers were required to find free space of the disk, store the program or data, retrieve them later and reconstruct the original file. Kildall called the specialized code to accomplish these tasks the BIOS (Basic Input/Output System). By 1977, many microprocessor-based computers routinely used 8-inch floppy disks for storage [42].

In 1976, Steve Jobs and Steve Wozniak founded Apple Computer, to focus on creating microprocessor-based computers for consumers who were not experts in computers and programming. The Apple I reached market in 1976, using an ordinary television set as its display. In 1977, the Apple II replaced the Apple I, boasting excellent color graphics [43]. Initially, Apple II used a cassette tape for storage, but switched to 5.25-inch floppy disks by the end of 1977. In 1979, Apple marketed the first spreadsheet software and word processing software, producing a truly consumer-friendly work station [44].

In 1980, IBM decided to enter the microprocessor-based computer market and chose Microsoft to develop its operating system (34, pp. 25–30). Microsoft, not having a suitable operating system ready, bought the QDOS (Quick and Dirty Operating System) from Seattle Computer Products and renamed it MS-DOS, which it licensed to IBM. In the summer of 1991, IBM began shipping personal computers with 64 kilobytes of memory and a floppy disk unit, which were sold by Sears and ComputerLand for US$ 2,880 (34, p. 27).

The IBM personal computer quickly dominated the workstation market. Many clones appeared, and so widespread became these devices that a software industry quickly arose to provide applications for business and personal use. At the same time, an industry producing peripherals evolved. In 1983, the impact of the personal computer on society was so profound that *Time* magazine awarded the PC its annual Man of the Year award (34, p. 28).

Almost every IBM PC or clone thereof used MS-DOS. However, MS-DOS was not user-friendly, requiring the user to interact with the operating system through a command line with a relatively inflexible instruction set (34, p. 29). In 1984, Apple responded by introducing its Macintosh line of personal computers, which were much more user-friendly. The Macintosh used a graphical user interface (GUI) with pull-down menus, icons and a mouse, enabling point-and-click activities in place of command line programming (34, p. 30).

Microsoft recognized the advantages of Apple's approach and, in 1985, released *Windows* 1.0, based on the Macintosh user interface (34, p. 32). When Intel 386 and

486 processors became available in the late 1980s, providing the speed needed to make the GUI sufficiently responsive, Microsoft released *Windows* 2.0 and, in May of 1990, *Windows* 3.0, and added *Excel* and *Word* to its product line (34, p. 33). In 1995, the greatly improved *Windows 95* was a big success, and later versions of *Windows* and its descendants have continued to allow Microsoft to dominate the market (34, p. 34).

The 1980s saw the dynamic growth of local and wide-area networks and widespread deployment of distributed computing and client-server architectures. The File Transfer Protocol (FTP) and other protocols had been developed to facilitate the exchange of computer files across networks. The original specification for FTP appeared in 1971 [45], even before TCP/IP existed, but it provided the basis for what would, when combined with hypertext, become the World Wide Web (WWW).

In March of 1989, Tim Berners-Lee of CERN proposed 'a large hypertext database with typed links' [46]. By 1990, he had combined the HyperText Transfer Protocol (HTTP), the HyperText Markup Language (HTML), a web browser called *WorldWideWeb*, HTTP-based server software, a web server and the first web pages in a functioning, if miniature, web. On August 6, 1991, the WWW became a publicly available service on the internet [47]. In late 1992, work began on the *Mosaic* web browser at the University of Illinois. It was released in February of 1993 [48]. Its developers later founded Netscape and *Mosaic* became *Netscape Navigator* [49].

The use of digital technology continued to accelerate. By the mid-1990s, growth of the WWW had expanded beyond its early roots in the academic and research communities, and the commercial world was beginning to recognize its promise for commerce as well as collaborative efforts in product development. Google, eBay and Amazon appeared and prospered. The early years of the 21st century saw the appearance of YouTube and the development of social networking on sites like MySpace and Facebook. The capacity increases made streaming video practical, and led to Really Simple Syndication, or RSS, widgets and video embedding.

Internet ubiquity has made it possible and practical to network a wide variety of digital devices. Today it is not just workstations and servers that have IP addresses. There are millions of digital devices that are now networked. Printers and facsimile machines contain computers and digital memory. Cell phones and e-book readers incorporate computers, as do cars, cameras, watches, sewing machines, thermostats, washers and dryers, and even some refrigerators. Sensors detect environmental data and report via networks. Supervisory Control and Data Acquisition (SCADA) systems control manufacturing processes and delivery of electricity, water and other essentials.

All of these digital devices make our everyday lives richer and more productive, but they may also contain digital evidence in the form of pictures, text files, email, activity logs and other digital records of the lives and actions of all of us. Before we take a look at how digital evidence pervades our homes, our offices and all of our lives, we need to review how the laws of evidence evolved.

10.3.2 The laws of evidence and digital evidence

Digital and multimedia evidence is, first of all, evidence. Its use forensically is, therefore, controlled by the laws of evidence. In order to resolve disputes and dispense justice, modern legal systems rely on testimony concerning events and evidence, including testimony by expert witnesses when the court needs help understanding how scientific, technical or specialized evidence, such as digital evidence, should be understood and interpreted. This was not always the case.

For most of history, the outcome of trials was based upon a determination by God – *judicium Dei* – controlling the outcome of trial by ordeal or by battle [50]. In trials by ordeal, the court imposed a physical challenge and assumed that, since God certainly knew the truth of what had happened, He would ensure that the accused would pass the test if innocent and fail if guilty. Similarly in trial by battle, an innocent defendant would prevail while a guilty defendant would lose. When disputants were allowed to choose champions presumably more capable than the disputants themselves, it made no difference, since it was assumed that God would see justice done. We hear faint echoes of such trials today in our courts, where we have attorneys as champions committed to 'zealous advocacy' for their clients.

Amazingly, trial by battle was not abolished in England until 1818, following the famous case Ashford v. Thornton [51]. Nevertheless, as early as the 12th century, trial by ordeal was no longer seen as a rational path to justice, and the legal system gave way to the Romano-canon inquisition process on the European continent, and to trial by jury in England (52, p. 3). Practices were developed for testimony under oath and for the use of documentary evidence (52, pp. 202, 223).

After 1500 CE, we see the beginnings of a theory of admissibility of evidence. The 18th century established a right to cross-examination and, after 1790, the use of written *Nisi Prius* reports describing the facts and rulings of cases, and the logic for those rulings. This solidified the doctrine of *stare decisis*, the doctrine requiring that later cases with similar facts be resolved in the same way as earlier cases were resolved [53].

The role of witnesses evolved along with the role of juries (52, pp. 186–198). During the early evolution of the English system of jurisprudence, everyone involved in the community in which a legal dispute arose could be counted upon to know the disputants or the defendant, and they probably had a good idea as to what the outcome should be before the trial even started. However, as populations grew, it was no longer true that jurors would necessarily know the disputants or the accused. Witnesses were needed, to tell the jury what happened and to explain evidence. Testimony was given under oath and there were penalties for perjury (52, p. 188).

As regards experts with scientific, technical or other specialized knowledge, testimony by expert witnesses would rarely have been needed in medieval times. No expert was needed in a dispute over damage caused by a fire resulting from spontaneous combustion of a haystack. Everyone knew how to stack hay, and that spontaneous combustion would occur if proper methods were not used [54]. But with the rise of industrialization, advances in science and technology, as well as increasingly complex feats of engineering, the average juror began to need not only lay testimony as to the events and evidence involved in the dispute or crime, but expert help in understanding how to interpret the events or evidence when such interpretation required scientific, technical or other specialized knowledge. The law of evidence had to evolve to meet this new reality and so, eventually, forensic science was born.

The Europeans that settled colonial America brought their notions of jurisprudence with them, both from England and from the continent. In what would become the United States and Canada, except in Louisiana [55], the English common law system of jurisprudence came to dominate. There were few legal professionals at that time, so the colonists often made do with vigilante justice. Such legal codes as existed in any formal way were strongly influenced by the religions practiced in the different colonies. Judges were civic or religious leaders, and their tribunals were simpler and less formal that those in Europe. Disputes were resolved by decree and, at least in criminal matters, magistrates decided if there had been a crime, conducted trials in which an elected District Attorney prosecuted, but in which defense attorneys were rarely present, and made final judgment. Since the same judge who had decided that a crime had been committed and that a trial was necessary was the judge who decided guilt or innocence, it should come as no surprise that 'guilty' verdicts were more common than 'not guilty' verdicts [56].

In the colonies, and later in the United States, the law of evidence was developed in case law well into the 20th century. Particularly vexing were problems where cases had to be decided based on scientific, technical or specialized evidence. The courts, wrestling with this problem, have devised standards for admissibility of scientific or technical evidence. Prior to 1975 and the enactment of the Federal Rules of Evidence, the most commonly used standard came to us from the case Frye v. United States [57].

James Alphonzo Frye was accused of shooting and killing a wealthy black physician, Dr. Robert W. Brown. He escaped, but eight months later he committed an armed robbery and was apprehended. Under questioning, he admitted that he had perpetrated the robbery, and he also confessed to killing Dr. Brown. Frye's lawyer advised Frye to recant his confession, and Frye did so. At trial, the defense sought to have an expert testify that he had tested Frye's story by taking Frye's blood pressure after each question posed to him by the defense attorney when the defense was preparing its case for trial – a sort of crude polygraph test. The judge refused to allow the testimony and Frye was convicted [58]. On appeal, in a two-page opinion that referenced not a single precedent, the court framed the admissibility standard as follows:

Just when a scientific principle or discovery crosses the line between the experimental and demonstrable stages is difficult to define. Somewhere in this twilight zone the evidential force of the principle must be recognized, and while courts will go a long way in admitting expert testimony deduced from a well-recognized scientific principle or discovery, the thing from which the deduction is made must be sufficiently established to have gained general acceptance in the particular field in which it belongs (57, p. 1014).

The Frye standard would control the admissibility of scientific, technical or specialized evidence in federal courts until the case of Daubert v. Merrill Dow Pharmaceuticals [59] was decided by the Supreme Court in 1993 – although Frye continues to control admissibility in some State courts even today.

In 1934, Congress passed the Rules Enabling Act [60], giving the judicial branch the power to develop and promulgate rules of judicial procedure. Based on the Rules Enabling Act, the courts then developed the Federal Rules of Civil Procedure [61] and the Federal Rules of Criminal Procedure [62]. Chief Justice Warren, in 1965, proposed to follow a similar path in developing the Federal Rules of Evidence (FRE). The United States Supreme Court circulated drafts of the FRE in 1969, 1971 and 1972. However, Congress exercised its power under the Rules Enabling Act to suspend implementation of the Federal Rules of Evidence until it could study them further. Eventually, after modifying the Court's proposed rules in several places, Congress itself enacted the FRE as the Act to Establish Rules of Evidence for Certain Courts and Proceedings, Pub. Law 93–595, 88 Stat. 1926 [63].

FRE Rule 701 tells us:

> *'If the witness is not testifying as an expert, the witness' testimony in the form of opinions or inferences is limited to those opinions or inferences which are (a) rationally based on the perception of the witness and (b) helpful to a clear understanding of the witness' testimony or the determination of a fact in issue and (c) not based on scientific, technical, or other specialized knowledge within the scope of Rule 702 [63].'*

FRE Rule 702 specifies that:

> *'If scientific, technical, or other specialized knowledge will assist the trier of fact to understand the evidence or to determine a fact in issue, a witness qualified as an expert by knowledge, skill, experience, training, or education, may testify thereto in the form of an opinion or otherwise, if (1) the testimony is based upon sufficient facts or data, (2) the testimony is the product of reliable principles and methods and (3) the witness has applied the principles and methods reliably to the facts of the case [63].'*

Digital evidence frequently requires the testimony of experts regarding the nature or source of the evidence, how the evidence was collected, transported, stored and analyzed, and how the evidence should be understood in relation to other evidence relevant to the case at trial. Thus, understanding of these rules is critical when digital evidence is involved in federal courtrooms.

The same year James Frye died, and 20 years after the opinion in Frye v. United States set the standard for admissibility of novel scientific evidence, Richardson-Merrill pharmaceuticals proposed a combination drug to reduce morning sickness among pregnant women [64]. Beginning in 1956, Bendectin was marketed in several countries, including the United States, and it rapidly became the leading drug prescribed for morning sickness and a very profitable product for Merrill. By the mid-1970s, questions were being asked concerning the safety of the drug. At that time, the United States Food and Drug Administration had more than 90 drug evaluation reports noting birth defects in children whose mothers had taken Bendectin. Shortly afterwards, the first lawsuit was filled, alleging that Bendectin was teratogenic.

In a series of lawsuits, perhaps numbering in the hundreds, across the United States, Merrill-Dow was sued by other parents of children who had allegedly suffered birth defects when their mothers took Bendectin. Plaintiffs had to prove beyond a preponderance of the evidence that Bendectin caused the birth defects, and that Merrill could have discovered this fact and could have warned consumers of the risk of teratogenic effects. Expert testimony concerning scientific evidence was proffered. Some plaintiffs won; some lost. Eventually, the parents of Jason Daubert filed suit against Merrell-Dow in Federal District Court in Los Angeles, alleging that Jason's mother had used Bendectin and that, as a consequence, Jason was afflicted with limb-reduction birth defects similar to those of other plaintiffs' children across the nation. The trial court granted Merrell-Dow's motion for summary judgment, finding the proffered scientific evidence was not sufficiently conclusive regarding causality to be admissible.

The Supreme Court's opinion in Daubert held that, at least in the Federal courts, the Frye standard had been superseded by the enactment of the Federal Rules of Evidence by Congress in 1975 (59, p. 587). For the court, the phrase 'scientific knowledge' replaced 'general acceptance' as the key to admissibility. Scientific knowledge sets the standard for reliability. Scientific certainty is not required, but a basis in scientific principles and methods is required: '*in order to qualify as "scientific knowledge", an inference or assertion must be derived by the scientific method.*'

The reasoned application of the scientific principle and methods must be sound as well. '*Proposed testimony must be supported by appropriate validation – i.e., "good grounds", based on what is known.*' (59, p. 590) The restrictive Frye standard that was used to exclude the evidence at the trial level was replaced with Federal Rule of Evidence 702. This is the standard that controls admissibility of scientific, technical or specialized evidence, including digital and multimedia evidence, in all federal courts and many state courts today.

10.3.3 Digital and multimedia evidence comes of age

By the last quarter of the 20th century, it had become apparent that the widespread and pervasive use of information systems based on digital technologies required the development of a substantive law regulating the use of such systems. The commercial sector discovered the internet and began conducting commercial transactions online, necessitating modifications to the Uniform Commercial Code [65] and the creation of the Uniform Electronic Transactions Act [66] and the less successful Uniform Computer Information Transactions Act [67]. The Electronic Signatures in Global and National Commerce Act [68] at the federal level ensured the validity and enforceability of contracts entered into electronically. Internationally, in the 21st century, the United Nations Convention on the Use of Electronic Communications in International Contracts [69] has proved very successful in enabling and regulating electronic commerce.

Privacy has become a significant concern as information systems based on digital technologies are increasingly used to create, store, process and communicate valuable information. Issues concerning privacy as a constitutional right stem from the 4th, 5th, 6th and 14th Amendments. Three federal statutes deter and punish invasions of privacy involving computers and the internet. The Wiretap Act [70], the Stored Communications Act [71] and the Pen Register Statute [72] were enacted to create a statutory form of 4th Amendment applicable to cyberspace. Many state statutes also protect privacy, and tort law governs injurious human acts that invade privacy [73]. Expert testimony regarding digital and multimedia evidence appears frequently in such cases.

Criminal activity also appears in cyberspace late in the 20th century. Beginning in 1970, over a period of some three years, the chief teller at the Park Avenue branch of the Union Dime Savings Bank in New York City embezzled over US$ 1.5 million using the bank's computers [74]. In 1981, a hacker called 'Captain Zap' (real name Ian Murphy) broke into AT&T computers and changed the billing clock [75]. Lacking laws to effectively control hacking, the United States passed the Computer Fraud and Abuse Act (CFAA) in 1986 [76].

An early CFAA case arose on November 2, 1988, when Robert Tappan Morris released a worm into the internet that eventually infected one-tenth of its computers [77]. He was convicted and was fined and required to perform community service [78]. The case led to the founding of the first Computer Emergency Response Team (CERT) at Carnegie-Mellon University to deal with such incidents in the future [79]. In 1995, Kevin Mitnick was charged with stealing sensitive data and credit card information, hacking into the California motor vehicles data base, and taking control of telephone systems in New York and California [80]. Also in 1995, Vladimir Levin hacked into Citibank and stole US$ 10 million on line [81]. In addition to many other incidents, the end of the 20th century saw the destructive spread of the 'ILoveYou' and 'Melissa' viruses [82,83].

The 21st century has seen an exponential increase in the number of cases involving digital and multimedia evidence. Crimes against property are evidenced in cyberspace as crimes against intellectual property in the form of copyrights, patents, trademarks and service marks, trade dress and trade secrets. Two types of cybercrime can be categorized as crimes against persons: first, threats and harassment; and second, the invasion of privacy crimes discussed earlier.

Vice crimes – those involving pornography, narcotics, prostitution and gambling – criminalize behavior that many believe to be immoral. The internet significantly increases the ability of users to access websites that facilitate the commission of vice crimes, and at the same time makes it more difficult to enforce vice crime statutes effectively. Especially troubling is the exploitation of children in creating obscene images and videos. Child pornography cases involve the use of the internet to send, receive and/or possess images of minors engaged in explicit sexual activity. Other cases involve stalking of minors by adults for actual or attempted sexual activity. Law enforcement and the legal system have rushed to keep up.

In 1984, the Federal Bureau of Investigation founded the FBI Magnetic Media Program, which would later become the Computer Analysis and Response Team (CART) [84]. 1993 saw the First International Conference on Computer Evidence. In 1995, the International Organization on Computer Evidence (IOCE) was formed to develop standards for digital evidence. In 1998, the Scientific Working Group on Digital Evidence (SWG-DE) was formed to enable United States participation in standards development, and the Association of Chief Police Officers (ACPO), the Forensic Computing Group (FCG) and the European Network of Forensic Science Institutes agreed to participate [85]. The ASTM Committee E30 on Forensic Sciences created a sub-committee on Digital Forensics [86]. There has also been the creation of the Scientific Working Group on Imaging Technology (SWGIT) [87], the Audio Engineering Society [88] and the Forensic Video Analysis Certification Study Committee of the International Association of Identification [89].

SWG-DE has published documents relevant to this field, notably *Best Practices for Mobile Phone Examinations*, which outlines many practice procedures. NIST has also completed quite a bit of work with mobile devices with their Computer Forensic Tool Testing program to include test assertions, test plans and tool specifications for many different mobile devices [90].

Analysis of analog video and images for forensic purposes has been used since the early 19th century, as well as photogrammetric methods. Early in the 1970s, the first attempts were made to remove background noise from fingerprint images [91]. Methods for digital enhancement of images in forensic context have been described since 1985 [92]. Methods for image restoration have been used in court from 1991, for example with fingerprints, documents and video images (and have been challenged in Frye and Daubert hearings), and now are often included in standard equipment and software along with audit trails [93]. In the past, there was an SPIE working group, Digital Investigation, active in the field.

Audio forensics has its roots in the early 1960s, when the FBI and other organizations began enhancing and authenticating audio recordings submitted by law enforcement and other government agencies [94]. Initially, the number of requests was relatively small, especially for authenticity examinations. However, following the analysis of the infamous 18.5 minute gap on one of the recordings associated with the Watergate scandal in the early 1970s, public awareness of the field grew and resulted in an increase in examination requests (94, p. 86).

Another factor in the growth of the field in the 1970s and 1980s was the development and implementation of digital signal processing tools, such as audio filters and Fast Fourier Transform analyzers (94, p. 90). These tools provided for more efficient and effective filtering/analyses which had not been possible with traditional analog tools. Since the 1980s, these tools have become increasingly more complex and novel in their approach to addressing intelligibility problems of the past, and overcoming hurdles in certain types of analyses.

Traditionally, voice comparison examinations involved aural and spectrographic evaluations of known, verbatim voice samples with unknown voice samples in an evidentiary recording. In 1991, the Voice Identification and Acoustic Analysis Subcommittee of the International Association for Identification published the peer-reviewed standards for the aural-spectrographic method of voice comparison [95]. Since that time, significant research has been conducted into automatic or semi-automatic comparisons of voice samples using sophisticated algorithms. To date, automatic and semi-automatic systems are not widely accepted in the United States, though many European countries are utilizing the technology.

Signal analysis examinations have progressed from simple examinations, such as decoding dual-tone multi-frequency (DTMF) telephone signaling to determine an outgoing telephone number, to complicated analyses of recorded gunshot sounds to discriminate between different weapons and shooter locations within the original environment. Other types of recorded signals examined in these analyses include, but are not limited to, jet engine sounds (to determine the thrust percentage of an aircraft), aluminum baseball bat 'pings' and alarm tones from Personal Alert Safety Systems (PASS) utilized by firefighting personnel. Probably the most well-known signal analysis examination involved the refutation of the existence of a 'second shooter' in the 1963 assassination of President John F. Kennedy, as based upon a 1978 examination of two audio recordings of Dallas Police Department radio traffic in the vicinity of the scene [96].

10.4 A Brief History of AAFS section

During the first decade of the 21st century, efforts began to create a Digital and Multimedia Science Section within the American Academy of Forensic Sciences. Practitioners of several of the disciplines related to digital and multimedia sciences were members of the AAFS but were spread across multiple sections within the AAFS, including Criminalistics, Engineering Sciences and the General Section. A community of interest group meeting of these practitioners was held at the 2002 meeting of the AAFS, with the intent to determine if there was sufficient interest to form a section dedicated to their disciplines. Subsequently, a presentation to the AAFS Board of Directors in 2002 outlined the then-current status of digital forensics. The Board suggested that potential members involved in digital forensics could join the General Section of the Academy and, when enough members had joined, a separate section might be created.

Tireless efforts ensued by Carrie Whitcomb, Zeno Geradts, David Baker, Doug Lacey, Mark Pollitt and others, eventually leading to the creation of the Digital & Multimedia Sciences Section of the Academy at the 2008 annual meeting of the Academy in Washington, DC. The section was formed with Carrie Whitcomb selected as the director, Zeno Geradts as the chair and David Baker as the secretary, with an initial membership of 42 members. The section has grown to 94 members in its three-year existence, representing all of the disciplines within the section. The section's disciplines and areas of interest are aligned with the American Society of Crime Laboratory Directors Laboratory Accreditation Board (ASCLD-LAB) and joint Scientific Working Group on Digital Evidence (SWGDE)/Scientific Working Group on Imaging Technology (SWGIT) definition of digital and multimedia forensics.

10.5 Key issues

Digital and multimedia sciences, as forensic sciences, are still very young and face a number of key issues, including rapid technological innovation, increasing quantity of digital evidence, widespread use of new digital devices, advanced data hiding/data protection techniques and remote accessibility of devices and data which may contain evidence. Additionally, some of the issues facing practitioners in this field are the same as in others, in that adequate training, certification and accreditation must keep pace with the changing landscape of both the underlying technologies and the evolving legal landscape.

The rapid evolution and the blistering pace of technological change pose major challenges to our practitioners. Anti-forensic techniques in the digital forensics area are now surfacing in alarming numbers. Anti-forensic technology is used to hide or destroy data – for example, data can be hidden using encryption, which scrambles data in such a way as to render it unreadable to unauthorized individuals. Steganography, the hiding of a file within a file, is another effective anti-forensic technique – sophisticated enough to make detection and recovery decryption without the hidden file a practical impossibility.

Even the adoption of newer technologies that utilize different mechanisms for operation of storage devices can make recovery of data difficult, if not impossible. For example, the use of wear-leveling technologies in solid state storage devices, intended to extend the life of the device, have a negative impact on the length of time that deleted data might be recoverable and used as evidence.

Not all of the challenges facing DMS are directly connected to technology. By and large, the legal system has not been able to keep pace with the rapid evolution and explosive growth of technology.

For some devices, no tool may exist for data extraction or the data port may be locked, prohibiting digital extraction. Data can then obtained by digitally recording (via digital camera) the information as seen on the screen of the device. This type of analysis should only be done after ensuring that the device is prohibited from sending or receiving any type of radio signals. Many mobile devices store information on NAND/NOR chips. Recovery of the data directly off these chips is generally possible but can result in the destruction of the device or even the chips themselves in the removal process. Once removed, the wear-leveling algorithm must be reverse-engineered and then the data can be viewed and 'mounted' as a complete file system. Additional data may be recovered that has been previously deleted, using traditional digital forensic techniques. This technique is relatively new within mobile device forensics and derives many skills, techniques and methods from the fields of solid state or 'flash' drive data recovery/forensics.

The 'cloud' presents another obstacle from a legal perspective. In the cloud, data of probative value can be stored not just in other US states, but in foreign nations around the world. These nations may have radically different rules, requirements and procedures regarding the discovery of digital evidence within their jurisdictions. These vast differences create a major barrier to obtaining the electronic evidence in question.

The so-called 'CSI effect' has caused the overestimation of possibilities for image enhancement [97]. If the image is of limited resolution, the limitation is inherent in the pixels. The quality of an image is limited by the resolution of the system capturing and storing the image. There are commonly used possibilities for enhancing an image by modifying contrast and intensity. The best quality should be derived from a 1:1 copy of the original. Often, digital systems store images in different resolutions, and the examiner must be aware that, in exporting images from a CCTV system, reduction of quality is possible because of recompression of the image.

Advanced enhancement must be done with great care, since such techniques can alter the image and can result in a wrong interpretation of the image. The examiner should also be aware of other factors (such as damage to the system, lighting conditions, etc.) and the influence of compression on the image, as well as how digital image formats are displayed and possibilities for recovering the best quality images from a digital system. The SWGIT (Scientific Working Group of Imaging Technology) guidelines describe various enhancements methods that are often used [98]:

1. **Basic**
   - Brightness and contrast adjustment, including dodging and burning
   - Resizing (file interpolation)
   - Cropping
   - Positive to negative inversion
   - Image rotation/inversion
   - Conversion to grayscale
   - White balance
   - Color balancing and/or color correction
   - Basic image sharpening and blurring (pixel averaging)
   - Stabilizing a video (removing the movement of a person or object in the video so that it is easier to interpret)
   - De-interlacing.

2. **Advanced**
   - Frame averaging
   - Fourier Analysis (including the use of FFT)
   - De-blur
   - Noise reduction
   - Image restoration
   - Color channel selection and subtraction
   - Perspective control and/or geometric correction
   - Advanced sharpening tools, such as unsharp mask.

In the comparison of persons and objects, it is important to have the same conditions in the images being compared, preferably images from the same camera system. Knowledge of photogrammetric methods is important, as described in section 16 of the SWGIT guidelines. For facial comparison, it is important to have knowledge of the face and aging, as is described by the Scientific Working group for Facial Identification [99]. For the complex examination of images or video for signs of tampering or authentication, the metadata of the files should be examined, in addition to considering continuity issues (such as edges from cut-and-paste operations), also described in section 14 of the SWGIT guidelines.

The repair of analog video is often a case of repairing video tapes and then using the right players to display the video. For the repair of digital video, the process often requires analysis of the digital video data file with hex editors or other available tools, such as the open source tool *Defraser* [100]. Extensive knowledge of the file formats and their implementation is necessary, especially if the files are fragmented. Knowledge of video encoding/decoding processes is often necessary, too [101].

For examining whether a set of images or a video have been produced with a specific camera, several methods can be used. In older cameras, defective pixels might be found, whereas newer cameras have Photo Response Non Uniformity (PRNU) patterns which might be useful for comparison [102,103]. It is important in all of these methods to use the same type and brand of camera to validate whether the pattern is really random. Also, metadata can be used in combination with such analyses. In complicated cases, it can be helpful to make a 3-D visualization of the crime scene. This kind of 3-D visualization can aid in identifying different scenarios as to what could have happened, and it may eliminate some scenarios as being impossible. However, visualizations must be used with great care, lest they influence juries and judges based on the scenarios rather than the facts [104].

The Scientific Working Group on Digital Evidence (SWGDE) has, over the past several years, published documents which are aimed at providing the field with information regarding training and qualifications of forensic audio practitioners. These documents include the *Guidelines & Recommendations for Training,* published in 2010 and the *Core Competencies for Forensic Audio Technicians*, currently in draft form and released for public comment [105,106].

As in all forensic disciplines, training is a vital part of quality assurance. Defining appropriate quality standards can be a challenge in fields like computer forensics, where the principles of evidence collection are not governed by the stable laws of physics and chemistry. Rather, they are governed by a combination of electronic hardware and logical software that, together, define virtual realities which are continuously updated and which require examiners to develop new tools and attain new skills in order to remain competent. To assist in this regard, SWGDE is working with NIST to standardize the validation of computer forensics tools and methods, so that laboratories can better leverage the efforts of others in their practices and procedures.

Training in DMS is especially important, given the rapid rate at which technology changes. Procedures, tools and techniques can quickly become antiquated. The Forensic Science Education Programs Accreditation Commission (FEPAC) offers accreditation for both undergraduate and graduate degree programs. FEPAC accredited baccalaureate degrees in digital and multimedia forensics requirements include a solid curriculum with foundations in a variety of both forensic-focused and technology-focused courses. The curriculum includes courses in general forensic science, computing and information science, as well as specialized courses in computer and network forensics.

Technology-focused courses include topics such as computer organization and architecture, programming theory, data structures, file and operating systems and more. General forensics courses cover legal issues, criminal investigations and procedure, quality assurance, testimony and ethics. The curriculum for accredited graduate programs includes crime scene investigation, physical evidence concepts, quality assurance, drug chemistry/toxicology, pattern evidence, etc. In addition to specialized courses covering digital forensics, graduate students are also required to complete independent research projects.

10.6 International perspectives

The field of digital forensics is coming of age across the globe. As a relatively new field of endeavor, it has grown and evolved along with the technologies it addresses. There are a number of international organizations, including standards organizations, which are essential in a field as dynamic as information technology and digital processing. One of the organizations, the International Federation for Information Processing or IFIP, has been in existence since 1960. The IFIP created a technical committee on security and protection in information systems, under which the digital forensics working group was formed [107].

The Forum for Incident Response and Security Teams (or FIRST) began in 1990, in response to the growing number of computer network incidents requiring effective ways to share information about such events across national boundaries [108].

The International Standards Organization has a technical committee for information technology standards (SC27), under which are working groups and standards for digital forensics practitioners, all under the ISO/IEC 27000 series. Many of these are very active and include inputs from other standards organizations such as NIST, the British Standards Institute, the Information Security Forum (formerly the European Security Forum, the Institute of Electrical and Electronic Engineers (IEEE), Bundesamt fur Sicherheit in der Informationstechnik (or BSI, the German information security standards organization), the ASTM International (formerly the American Society for Testing and Materials) and others. The European Network of Forensic Institutes (ENFSI) formed a working group to address forensic IT in 1997 and holds annual conferences and organizes proficiency tests as well as offering a best practice guide.

Organizations such as SWGDE, formed in 1998, have been working with many of these international organizations as the field of digital forensics grows and matures. Evidence of global participation in this field is easily seen by the number of technical conferences for research in the area of digital forensics. The FIRST conference has been well attended internationally for 23 years. The DFRWS has been attracting international attendance since its first meeting in 2001. The IFIP WG 11.9 Digital Forensics conference has been popular, too, since 2004. The European Conference on Computer Network Defense has been held since 2005, and a relatively new conference, the International ICST Conference on Digital Forensics & Cyber Crime, celebrated their third conference in 2011. This is by no means a complete or full listing, but simply demonstrative of the international flavor of the research and work in the digital forensics disciplines. And of course, the AAFS has had presentations on digital forensics issues since the late 1980s.

Image and video processing is widely used throughout the world. Within Europe, ENFSI created the Digital Imaging Working Group, which is active in the field of research and proficiency tests [109]. Most forensic labs have departments that are active in this field. Interpol also has covered this field of interest since 1995, in their three-year Forensic Review [110]. The EESAG is an Australian-New Zealand Group that organizes proficiency tests for video and audio processing [111]. The Arbeitsgtruppe Identifikation nach Bildern, or AGIB, is a working group in Germany on facial comparison [112]. Also, in the United States, the group LEVA is focused on video processing and training [113]. The International Association of Identification is also very active in the field, with guidelines in collaboration with the scientific working group [114].

Forensic audio examiners from around the world face the same challenges when conducting examinations and presenting their results in a courtroom setting. Groups such as SWGDE in the United States and ENFSI in Europe each contain subgroups involved with forensic audio analysis, aimed at providing 'best practice' documents and other informative material to the field as a whole. For example, in 2008, SWGDE published its *Best Practices for Forensic Audio* and, in 2009, the Forensic Speech and Audio Analysis Working Group of ENFSI published its *Best Practice Guidelines forElectrical Network Frequency (ENF) Analysis in Forensic Authentication of Digital Evidence* [115,116]. Examination methodologies and case reports have also been published in peer-reviewed scientific publications such as the *Journal of Forensic Sciences*, the *Journal of the Audio Engineering Society*, the *Journal of Forensic Identification* and *Forensic Science Communications*.

10.7 Research and future directions

10.7.1 Expansion to new devices

Digital forensics is already concerned with a much wider array of digital devices than just computers, including servers, workstations and laptops. However, the diversity of digital devices that are likely to be within the scope of an investigation in the future is growing rapidly and shows no sign of slowing, as computer systems become fundamental to almost all aspects of modern life. Mobile devices (e.g. smart phones and tablets) are certainly increasingly common, but there is also a wide variety of other digital systems that will likely play a part in investigations in the future. Examples include:

- VoIP systems are replacing home phones.
- Enterprise phone systems, home and office automation systems manage the environments in which we live and work
- Electronic access systems track our movements through the physical world.
- Cars and home appliances have increasingly complex automation systems.
- Medical devices include onboard diagnostic and monitoring components.

In addition, the range of devices in any specific category is growing rapidly. For example, there are multiple mobile device operating systems, many of which have multiple versions active in the marketplace at any time, and some of which feature vendor-specific customizations. This wide range of operating systems runs on an even wider range of hardware devices. As a result, a given investigator is less likely to be familiar with the range of devices involved in a given investigation. Even an expert in any given sub-area (e.g. GPS device forensics) is likely frequently to encounter new devices. Therefore, an investigator may need to spend a significant amount of time in each investigation determining what devices are likely to have relevant data and developing a process by which such data can be reliably recovered.

10.7.2 The cloud and virtualization

At a superficial level, the cloud can seem to be similar previous iterations of distributed systems, such as grid computing. However, it really is a fundamentally different environment, both in the scale and flexibility of the computing resources, the widespread availability of fast networking, even in mobile environments, and the consumption of cloud resources by all segments of the IT community – from enterprise systems to home users. There are numerous challenges that the cloud and its associated enabling technology – virtualization – brings to the DF community:

- The ephemeral nature of cloud resources means that once a cloud system is un-deployed or released, there may no longer be any remaining significant artifacts of that system to examine.
- The shared nature of the cloud means that it may be challenging to attribute artifacts reliably to a particular cloud resource (e.g. virtual machine).
- If the virtualization layer is compromised (either by an external attacker or an insider), it may result in the modification of other virtual machines running within that virtualization layer. This, in turn, could result in the appearance of activities apparently initiated by a given cloud resource having, in fact, been initiated by the virtualization layer or someone that controls it.

10.7.3 Control systems

Control systems, or SCADA systems, have long been used to provide computer control of physical devices, from small environmental control systems to large-scale industrial facilities. While such systems are not new, their integration and connection (albeit sometimes indirectly) to local and wide area network systems (such as the internet) have made them much more likely to be exploited and, as such, much more likely to be a relevant component of an investigation. These systems typically include user interface components (often Windows-based workstations) that manage a wide range of devices that interface with the physical systems. While the user interface systems, or at least the underlying operating systems if not the specific applications, are likely to be familiar to a wide range of digital forensics investigators, the controller devices and the protocols used to communicate with them are far less likely to be so. As with mobile devices, there exists a vast range of such systems.

10.7.4 New challenges for traditional approaches

Existing and well understood approaches to digital forensics are also facing challenges that will likely change the way in which investigations are conducted in the future. One of the obvious challenges is the overwhelming number of devices and the amount of data associated with investigations. The proliferation of massive storage devices on even the lowest-end computer system, or the massive volumes of network data sent across the many types of network connections available today, makes the search for evidence an increasingly time-consuming process that overwhelms the resources and time constraints of almost any investigative agency. To address this issue, it is likely that the data acquisition process will have to be more selective in the future, to ensure that the analyst's time is spent on the body of data most likely to yield results.

Traditional examination of hard drives, in which the system is placed in a quiescent state and the media is copied using write-blocking devices, has provided investigators with access to the current contents of a computer's file system, in addition to large volumes of previously deleted files in slack space and unallocated clusters. However, the widespread use of cryptography and solid state devices is likely to change that process. Solid state drives, for example, can be instructed by the operating system to prepare unallocated blocks preemptively for subsequent write operation, meaning that deleted files are more likely to be inaccessible to an investigator.

There is widespread support for encryption in a range of modern operating systems and is likely to be more widely used in the future. For example, a simple checkbox setting to encrypt entire disks (or portions thereof) is common in Windows, Mac and Linux systems today. Once encrypted, the contents of the media are essentially inaccessible to the investigator without the decryption key. In some very limited cases, it may be possible to recover the key from RAM, and it is possible that the key might be recovered through non-technical means (such as finding a written or separately stored copy of the key). Also, some jurisdictions have laws than can compel a suspect to reveal their decryption key. In the absence of such means, however, modern encryption, with carefully chosen keys, can easily render media inaccessible to any publicly known efforts to recover the plaintext data.

10.7.5 Live analysis challenges

Live analysis offers a method by which an investigator can examine or record the state of a running system. It provides access to the dynamic state of the system, including data that may never be written to disk. Such data may include current network connections, plaintext versions of otherwise encrypted data, and process and system memory. It may also provide insight into the state of a system that examination of the media alone would not provide. While this is becoming a mainstream operation today, it is likely to become more critical in the future as static analysis has to address the challenges of cryptography and solid state devices. Some techniques for live memory acquisition and analysis exist today, but we are likely to see much more effort put into expanding these capabilities in the future, in addition to understanding what impact live analysis efforts have on the state of the system being examined.

10.7.6 Digital forensics for first responders

As digital forensics becomes more important to a wider range of investigators and scenarios, it will become more important to provide a wider range of people with the ability to acquire, analyze and act upon digital evidence. For example:

- First responders need to have the ability to identify and preserve digital evidence.
- Non-technical investigators in a law enforcement agency could be given the ability to perform some initial analysis of digital evidence.
- Soldiers in the field need to be able prioritize digital evidence and quickly move from digital data to actionable intelligence in a battlefield environment.

Clearly, more advanced analysis will remain the purview of highly experienced analysts, but tools and processes for data identification, preservation, acquisition, and even analysis, can be developed and distributed to a much wider group. Some tools have already been developed for acquisition, but we are likely to see more advanced tools developed to allow non-analysts to perform some of the other digital forensics functions, too.

10.7.7 Cybercrime assessment

The notion of crime and crime scene assessment is well established in physical space [117]. In crime scene assessment, the investigator examines the crime scene and draws conclusions, which he or she then imposes upon the description of the likely perpetrator. The bulk of the work in this regard has been done in the area of violent crime, such as rape or murder.

However, there is evidence to suggest that the same techniques can be applied also to non-violent crime in cyberspace [118]. Looking forward, it is clear that more effective means of analyzing cybercrime are needed. Traditional computer and network forensics techniques become less and less effective as file systems explode in size, become stretched across networks (even the internet, in some cases), cloud systems become the norm and networks become increasingly complex. For investigators of cyber events – whether those events are crime or intelligence based – improvements in tools are as important as keeping pace with the technologies that these investigators face on a daily basis.

The result would be a crime scene profile – a concept more typical of computing systems than of the physical world – that may be applied to the computer or network under investigation. Analysis of the characteristics of the crime scene in context with the sub-types may cause clues as to where to search for additional evidence (e.g. the 'cyber trail' on the computer, elsewhere in the local network, etc.) to become manifest. Such clues suggest a pattern of behavior that may be applied to a human perpetrator. Using the clues, evidence and behavior patterns can lead to a shortlist of a few possible offenders.

In computing, we use the term 'profile' somewhat differently from the way it is used by psychologists. In computing jargon, a profile refers to a description of the computing platform under examination. This may include such factors as memory, storage, processor and processor speed, network operating system, etc. When we add the results of the crime scene assessment, we can add that information – usually associated with one or more of the sub-types – to the profile of the computer/network, resulting in a cybercrime profile that is specific to a given crime scene under examination. Because evidence and clues may stretch across cyberspace, we may need to consider the network as well.

Such approaches to cybercrime, and identification of cyber criminals that take the human factor into account, have promise. They have promise, simply because people commit crimes – computers do not. A pure focus on technology will not solve the problem. It is the combination of people, products and processes that gets anything done – for good or ill – in cyberspace. With that in mind, we can expect more than cursory examination of techniques such as this, or those of Rogers [119], Bocij [120], McQuade [121] and Shindler and Cross [122], in the near future. Cyber forensics – no longer simply digital forensics – has become sufficiently complicated that traditional tools no longer suffice by themselves.

10.7.8 Envisioning image and video advances

The number of papers in this field is growing rapidly, as can be seen from the Interpol review [123]. Most important developments are reported in the areas of:

1. Digital image and video technology.
2. Facial image comparison.
3. Photogrammetry, crime scene recording and 3-D-modeling.

In the digital image and video technology, the processing of large amounts of video data has become a huge challenge, while developments of new methods are not being made as quickly. However, many new methods have been reported on detection of image manipulation and identification of cameras. Validation of these methods in forensic environments is very important, since often methods might work in a lab situation but not in large-scale real environments.

New methods and technology are being developed in the field of facial comparison, since validation of the methods remains important. The introduction of 3-D-scanners for face models has resulted in a variety of new field for research and development. This can also be implemented in other field of biometric comparison, such as gait analysis.

In the field of photogrammetry and crime scene recording, introduction of software that can handle large amount of point data is expected to reduce time needed for the modeling process. Also, new hand-held scanners and developments in the gaming industry will stimulate the developments in the field. In all of these areas, however, validation of techniques and measurement errors remains important.

10.7.9 Future tracks in audio forensics

Recent research efforts in the forensic audio field have focused primarily on the development of automated or semi-automated speaker recognition systems, and of new techniques to supplement the established methods for the authentication of digital audio recordings. With regard to the automated or semi-automated speaker recognition systems, the research is aimed at developing a system which is content-independent (i.e. not reliant on verbatim samples), channel-independent (i.e. not reliant on identical transmission lines – telephone recording vs. room recording), and accurate for recordings made under less than ideal investigative conditions, with the rate of false positive matches minimized.

As new digital audio recording formats are developed and utilized by recorder manufacturers, supplementary techniques to authenticate these recordings also need to be established. Research into these techniques often takes the form of case reports, as each format typically must be dealt with on a case-by-case basis. The analysis of embedded metadata within digital file formats has become increasingly important, as non-audio data can often provide crucial information regarding a recording's continuity and integrity. Further development in these research areas is critical to the continued success of examiners' techniques in the field.

As in the forensic video field, fewer and fewer analog audio recordings are being submitted for processing and analysis. Digital recordings clearly dominate the field, and new processing techniques must continually be developed to keep up with the wide array of physical and file formats. Further research into these various formats, and into automated or semi-automated speaker recognition systems, may lead to major breakthroughs into examinations of recordings and recorded voices. At the very least, advancements will continue to be made.

Individuals interested in entering this field must possess a strong technical background, preferably through formal education and training, with emphases on digital audio theory, acoustics and, more and more, computer science.

References

1. Zimmerman H. OS1 Reference Model–The ISO Model of Architecture for Open Systems Interconnection. *IEEE Transactions on Communications* 1980 April; Com-28(4):425–432.
2. Koenig BE. Enhancement of forensic audio recordings. *Journal of the Audio Engineering Society* 1988; 36(11):884–894.
3. Koenig BE, Lacey DS, Killion SA. Forensic enhancement of digital audio recordings. *Journal of the Audio Engineering Society* 2007; 55(5):352–371.
4. Bolt RH, Cooper FS, Flanagan JL, McKnight JG, Stockham TG, Jr., Weiss MR. *Report on a technical investigation conducted for the US District Court for the District of Columbia by the Advisory Panel on White House Tapes*. Washington, DC: US Government Printing Office; 1974.

5. Koenig BE. Authentication of forensic audio recordings. *Journal of the Audio Engineering Society* 1990; 38(1/2):3–33.
6. Koenig BE, Lacey DS. Forensic authentication of digital audio recordings. *Journal of the Audio Engineering Society* 2009; 57(9):662–695.
7. Koenig BE, Hoffman SM, Nakasone H, Beck SD. Signal Convolution of Recorded Free-Field Gunshot Sounds. *Journal of the Audio Engineering Society* 1998; 46(7/8):634–653.
8. Marr KW, Koenig BE. Fundamental Frequency Analysis of a Metal Baseball Bat. *Forensic Science Communications* 2007; 9(1).
9. Beck SD, Nakasone H, Marr KW. Variations in recorded acoustic gunshot waveforms generated by small firearms. *Journal of The Acoustical Society Of America* 2011; 129(4):1748–1759.
10. Koenig BE. Spectrographic Voice Identification. *Crime Laboratory Digest* 1986; 13(4):105–118.
11. Campbell JP, Shen W, Campbell WM, Schwartz R, Bonastre JF, Matrouf D. Forensic Speaker Recognition. *IEEE Signal Processing Magazine* 2009; 26(2):95–103.
12. Cajori F. *A History of Mathematical Notations*. Mineloa, NY: Dover Publications; 1993.
13. vanNooten B. Binary Numbers in Indian Antiquity. *Journal of Indian Studies* 1993; 21:31–50.
14. Bacon F. *The Advancement of Learning*. London: Cassell & Company; 1893.
15. Gerhardt CI,editor. *Die mathematische schriften von Gottfried Wilheim Leibniz* [The mathematical writings of Gottfried Wilhem Leibniz]. 1703; VII:223–7. http://www.leibniz-translations.com/binary.htm (accessed December 8, 2011).
16. Boole G. *An Investigation of the Laws of Thought, on Which Are Founded the Mathematical Theories of Logic and Probabilities*. Cambridge: MacMillan & Co.; 1854.
17. Shannon C. A Symbolic Analysis of Relay and Switching Circuits. *Transactions American Institute of Electrical Engineers* 1938; 57:713–723.
18. Shannon C. A Mathematical Theory of Communication. *Bell System Technical Journal* 1948 Jul and Oct; 27:379–423, 623–656.
19. Shannon C. Communication Theory of Secrecy Systems. *Bell System Technical Journal* 1949; 28:656–715.
20. Napier M. *Memoirs of John Napier of Merchiston: his lineage, life, and times, with a history of the invention of logarithms*. Edinburgh: William Blackwood; 1904.
21. The Oughtred Society. *Slide Rule History*. www.oughtred.org/history.shtml (accessed December 8, 2011).
22. Ketelaars N.Pascal's Calculator. *AIMe Magazine* 2001/2, 3–5. www.win.tue.nl/aime/Files/dec2001_pascal.pdf (accessed December 8, 2011).
23. Chase GC. *History of Mechanical Computing Machinery. Proceedings of the 1952 ACM national meeting (Pittsburgh)* New York, NY: Association for Computing Machinery, 1952; 1–28.
24. Delve J. Joseph Marie Jacquard: Inventor of the Jacquard Loom. *IEEE Annals of the History of Computing* 2007 Oct–Dec; 29(4):98–102.
25. Hook DH, Norman JM. *Origins of cyberspace: a library on the history of computing, networking*. Novato, CA: historyofscience.com; 2001.

26. Gleick J. *The Information: A History, a Theory, a Flood*. New York: Random House; 2011; 78–124.
27. Randell B,editor. *The Origins of Digital Computers*, Selected Papers. 3rd ed. Springer-Verlag; 1982.
28. Ceruzzi PE. *The Prehistory of the Digital Computer, 1935-1945: A Cross-Cultural Study* [dissertation]. Lawrence, KS: University of Kansas; 1981.
29. Wexelblat RL,editor. *History of Programming Languages*. New York: Academic Press; 1981.
30. Riordan M, Hoddeson L. *Crystal Fire*. New York & London: W.W. Norton & Company Limited; 1998.
31. Reid TR. *The Chip: How Two Americans Invented the Microchip and Launched a Revolution*. New York: Random House; 2001.
32. Moore GE. Cramming More Components onto Integrated Circuits. *Electronics* 1965 April; 38(8):114–117.
33. Bashe CJ, Johnson LR, Palmer JH, Pugh EW. *IBM's Early Computers*. Cambridge: MIT Press; 1986.
34. Moumina A. *History of Operating Systems*, 2001. www.computinghistorymuseum.org/teaching/papers/research/history_of_operating_system_Moumina.pdf (accessed December 8, 2011).
35. International Business Machines. *History of IMS: Beginnings at NASA*. http://publib.boulder.ibm.com/infocenter/zos/basics/index.jsp?topic=/com.ibm.imsintro.doc.intro/ip0ind0011003710.htm (accessed December 8, 2011).
36. Cerf VG, Kahn RE. A protocol for packet network interconnection. *IEEE Transactions on Communication Technology* 1974 May; COM- 22(5):627–641.
37. Raymond E.Origins and History of UNIX, 1969–1995. *The Art of UNIX Programming*, 2003. http://www.faqs.org/docs/artu/ch02s01.html (accessed December 8, 2011).
38. Raymond E. Early History of C: *The Art of UNIX Programming*, 2003. http://www.faqs.org/docs/artu/c_evolution.html#id2998213 (accessed December 8, 2011).
39. Dixon J.History of Microprocessors. *Computer History: Tracing the History of the Computer*, 2005. http://www.computernostalgia.net/articles/HistoryofMicroprocessors.htm (accessed December 8, 2011).
40. Greenia M. *History of Computing: An Encyclopedia of the People and Machines that Made Computer History*. Elverta, CA: Lexikon Services; 2001. http://www.computermuseum.li/Testpage/Altair-8800-1974.htm (accessed December 8, 2011).
41. Poole K. William H. *Gates*, 2002. http://www.voteview.com/gates.htm (accessed December 8, 2011).
42. Morgan D. *Short History of the Micro Computer*: Gary Kildall, 2002. http://www.fortunecity.com/marina/reach/435/kildall.htm (accessed December 8, 2011).
43. Weyhrich S. *The Apple II History 1969–1977*, 1991. http://apple2history.org/appendix/ahb/ahb1/(accessed December 8, 2011).
44. Weyhrich S. *The Apple II History 1978–1981*, 1991. http://apple2history.org/appendix/ahb/ahb2/(accessed December 8, 2011).
45. Bhushan A. *A File Transfer Protocol, Request For Comments 114*. Internet Engineering Task Force: 1971. http://tools.ietf.org/html/rfc114 (accessed December 8, 2011).

46. Berners-Lee T. *Information Management: A Proposal*, 1990. http://www.nic.funet.fi/index/FUNET/history/internet/w3c/proposal.html (accessed December 8, 2011).
47. Peter I. *History of the World Wide Web*, 2004. http://www.nethistory.info/History%20of%20the%20Internet/web.html (accessed December 8, 2011).
48. Andreessen M. *NCSA Mosaic Technical Summary*, Feb. 20, 1993. http://web.archive.org/web/20000919210952/http://www.cbl.leeds.ac.uk/WWW/ps/mosaic.orig.ps (accessed December 8, 2011).
49. Stewart W. *The Internet: Netscape-The First Commercial Web browser*, 2007. http://www.livinginternet.com/w/wi_netscape.htm (accessed December 8, 2011).
50. Kirsch JP. *Ordeals In The Catholic Encyclopedia*. New York: Robert Appleton Company, 1911. http://www.newadvent.org/cathen/11276b.htm (accessed October 29, 2011 from New Advent).
51. Megarry R,Garner B,editors. *A New Miscellany at Law. Yet Another Diversion for Lawyers and Others*. Clark, New Jersey: The Lawbook Exchange, Ltd; 2005.
52. Shapiro BJ. Beyond Reasonable Doubt and Probable Cause: Historical Perspectives on the Anglo-American Law of Evidence. Berkeley: University of California Press, 1991.
53. Langbein JH. Historical Foundations of the Law of Evidence: A View from the Ryder Sources. *96 Columbia Law Review* 1168 (1996).
54. Vaughan v.Menlove, Common Pleas, 3 Bing. N. C. 468 (1837).
55. Fernandez MF. *From Chaos to Continuity: The Evolution of Louisiana's Judicial System*, 1712–1862. Baton Rouge: Louisiana State University Press; 2001.
56. Friedman LM. *History of American Law. Review.* ed. New York: Touchstone, 1986.
57. Frye v.United States, 293 Fed. 1013(DC Cir. 1923).
58. Starrs JE. A Still-Life Watercolor: Frye v. United States. *Journal of Forensic Sciences* 1982; 27(3):684–694.
59. Daubert v. MerrillDow Pharmaceuticals, 509 U. S. 579; 113 S. Ct. 2786; 125 L. Ed. 2d 469 (1993).
60. Ch. 651 Pub. Law 73–415, 48 Stat. 1064, enacted June 19, 1934, 28 U.S.C. Ï 2072.
61. *Federal Rules of Civil Procedure* (as amended to December 1, 2010). http://www.law.cornell.edu/rules/frcp/(accessed December 8, 2011).
62. *Federal Rules of Criminal Procedure* (as amended to December 1, 2011). http://www.law.cornell.edu/rules/frcrmp/(accessed December 8, 2011).
63. Federal Evidence Review: *Legislative History Overview Resource Page*, 2011. http://federalevidence.com/node/638 (accessed December 8, 2011).
64. Mekdeci B. *Bendectin – How a Commonly Used Drug Caused Birth Defects.* http://www.birthdefects.org/research/bendectin_1.php (accessed December 8, 2011).
65. *Uniform Commercial Code.* http://www.law.cornell.edu/ucc/ucc.table.html (accessed December 8, 2011).
66. *Uniform Electronic Transactions Act* (1999). Drafted by the National Conference Commissioners on Uniform State Laws. http://www.law.upenn.edu/bll/archives/ulc/ecom/ueta_final.pdf (accessed December 8, 2011).
67. *Uniform Computer Information Transactions Act.* Drafted by the National Conference of Commissioners on Uniform State Laws. http://www.law.upenn.edu/bll/archives/ulc/ucita/ucita200.pdf (accessed December 8, 2011).

68. 15 USC 7001, Public Law 106–229, *Electronic Signatures in Global and National Commerce Act*. http://www.gpo.gov/fdsys/pkg/PLAW-106publ229/html/PLAW-106publ229.htm (accessed December 8, 2011).
69. United Nations Convention on the Use of Electronic Communications in International Contracts. http://www.uncitral.org/pdf/english/texts/electcom/06-57452_Ebook.pdf (accessed December 8, 2011).
70. 18 USC 2511, Interception and Disclosure of Wire, Oral or Electronic Communications Prohibited. http://www.law.cornell.edu/uscode/usc_sec_18_00002511----000-.html (accessed December 8, 2011).
71. 18 USC 2701, Unlawful Access to Stored Communications. http://www.law.cornell.edu/uscode/18/2701.html (accessed December 8, 2011).
72. 18 USC 3121, General Prohibition on Pen Register and Trap and Trace Device Use, Exception. http://www.law.cornell.edu/uscode/18/usc_sec_18_00003121----000-.html (accessed December 8, 2011).
73. DeVore P, Nelson M, Killeen M. Legal Guide 1 (Part III Defamation and Privacy). Seattle, WA: *Seattle Times*, 1996. http://seattletimes.nwsource.com/legalguide/3_2_B.html (accessed December 8, 2011).
74. *Computer Crime*. Answers.com, 2011. http://www.answers.com/topic/computer-crime (accessed December 8, 2011).
75. Murphy, IA *Security Threat Profile*™. http://attrition.org/errata/charlatan/ian_murphy/threat_profile/(accessed December 8, 2011).
76. 18 USC 1030, Fraud and Related Activity in Connection with Computers. http://www.law.cornell.edu/uscode/18/1030.html (accessed December 8, 2011).
77. Spafford EH. *The Internet Worm Program: An Analysis*, Purdue Technical Report CSD-TR-823. West Lafayette, IN: Purdue University, 1988.
78. U.S. v. Morris, 928 F.2d 504 (2nd Cir. 1991), http://www.rbs2.com/morris2.htm (accessed December 8, 2011).
79. Killcrece G, Kossakowski K, Ruefle R, Zajicek M. *State of the Practice of Computer Security Incident Response Teams (CSIRTs)* Technical Report CMU/SEI-2003-TR-001. Pittsburgh, PA: Carnegie Mellon University, 2003; 17-19. http://www.sei.cmu.edu/reports/03tr001.pdf (accessed December 8, 2011).
80. Shimomura T, Markoff J. *Takedown: The Pursuit and Capture of Kevin Mitnick, America's Most Wanted Computer Outlaw – By the Man Who Did It*. NY: Hyperion Books, 1996.
81. Bhattacharjee S. *Internet Fraud Case Study – Vladimir Levin*. http://www.cab.org.in/Lists/Knowledge%20Bank/Attachments/64/InternetFraud-VL.pdf (accessed December 8, 2011).
82. Sprinkel S. *Global Internet Regulation: The Residual Effects of the ILoveYou Computer Virus and the Draft Convention on Cyber-Crime. Suffolk Transnational Law Review* 2002; 25 (3): 491–514.
83. Garber L. Melissa Virus Creates a New Type of Threat. *Computer*. 1999; 32 (6): 16–19.
84. Kacoyannakis C. *Regional Computer Forensic Laboratory (RCFL)* National Program Office (NPO), September 8, 2004. http://jcots.dls.virginia.gov/2005%20Content/pdf/FBI-RCFL.pdf (accessed December 8, 2011).
85. Noblett MG, Pollitt MM, Presley LA.Recovering and Examining Computer Forensic Evidence. *Forensic Science Communications* 2000; 2 (4). http://www.

fbi.gov/about-us/lab/forensic-science-communications/fsc/oct2000/computer.htm (accessed December 8, 2011).

86. Committee E30 on Forensic Sciences. ASTM International, Denver, CO. http://www.astm.org/COMMIT/SUBCOMMIT/E30.htm (accessed December 8, 2011).
87. Scientific Working Group on Imaging Technology (SWIGIT). http://www.theiai.org/guidelines/swgit/(accessed December 8, 2011).
88. The Audio Engineering Society. http://www.aes.org/(accessed December 8, 2011).
89. The International Association for Identification. http://www.theiai.org/(accessed December 8, 2011).
90. US National Institute of Standards and Technology, Information Technology Laboratory, *Computer Forensics Tool Testing Program*. http://www.cftt.nist.gov/(accessed December 8, 2011).
91. Crime-photo.com. *History of Digital Imaging in Law Enforcement From the early 1970's to present*. http://www.crime-photo.com/HistoryOfDI.pdf (accessed December 8, 2011).
92. Sawney DJ. A way forward – with computers? *Journal of the Forensic Science Society* 1987 September; 27 (5):349–352.
93. State v. Hayden, 90 Wn. App. 100, 104, 950 P.2d 1024(1998). http://caselaw.findlaw.com/wa-court-of-appeals/1189825.html (accessed December 8, 2011).
94. Maher RC. Audio Forensic Examination: Authenticity, Enhancement, and Interpretation. *IEEE Signal Processing Magazine* 2009 March; 26(2):84–94.
95. IAI Voice Identification and Acoustic Analysis, Subcommittee. Voice Comparison Standards. *Journal of Forensic Identification* 1991; 41(5):373–392.
96. Koenig BE. Acoustic Gunshot Analysis – The Kennedy Assassination and Beyond. *FBI Law Enforcement Bulletin* 1983; 52(11):1–9 and 52(12):1–9.
97. Shelton DE. *The "CSI Effect": Does It Really Exist?* https://www.ncjrs.gov/pdffiles1/nij/221501.pdf (accessed December 8, 2011).
98. Scientific Working Group on Imaging Technology. *Best Practices for Documenting Image Enhancement*. http://www.theiai.org/guidelines/swgit/guidelines/section_11_v1-3_20100115.pdf (accessed December 8, 2011).
99. Scientific Working Group on Facial Identification. http://www.fiswg.org/ (accessed December 8, 2011).
100. Netherlands Forensic Institute. *NFI Defraser*. www.sourceforge.net/defraser (accessed December 8, 2011).
101. Poisel R, Tjoa S. *Forensics Investigations of Multimedia Data: A Review of the State-of-the-Art. Proceedings of the Sixth International Conference on IT Security Incident Management and IT Forensics;* Stuttgart, Germany: Institute of Electrical and Electronics Engineers (IEEE); 2011; 48–61.
102. Chen M, Fridrich J, Goljan M, Lukáš J. Determining Image Origin and Integrity Using Sensor Noise. *IEEE Transactions on Information Security and Forensics* 2008 March; 1(1):74–90.
103. Geradts Z, VanHouten W. Source video camera identification for multiply compressed videos originating from YouTube. *Journal of Digital Investigation* 2009 Jul; 6(1–2):48–60.
104. Schofield D. Playing with evidence: Using video games in the courtroom. *Entertainment Computing* 2011; 2(1):47–58.

105. Scientific Working Group on Digital Evidence. *Guidelines & Recommendations for Training* (v2.0), 2010, http://www.swgde.org/documents/current-documents/2010-01-15%20SWGDESWGIT%20Guidelines%20%20Recommendations%20for%20Training%20v2.0.pdf (accessed October 8, 2011).
106. Scientific Working Group on Digital Evidence. *Core Competencies for Forensic Audio Technicians* (v1, Draft 2), 2010, http://www.swgde.org/documents/released-for-public-comment/2010-09-16_SWGDE_Core_Comptencies_ForAudio_Tech_v1_DRAFT2.pdf (accessed October 8, 2011).
107. International Federation for Information Processing Technical Committee 11: *Security and Protection in Information Processing Systems*. http://ifiptc.org/?tc=tc11 (accessed December 8, 2011).
108. Forum of Incident Response and Security Teams (FIRST). http://www.first.org/about/history/(accessed December 8, 2011).
109. European Network of Forensic Science Institutes (ENFSI). http://www.enfsi.eu (accessed December 8, 2011).
110. Daeid N, Houck M. *Interpol's Forensic Science Review*. Boca Raton, FL: CRC Press; 2010.
111. Bijhold J, Ruifrok A, Jessen M, Geradts Z, Ehrhardt S Alberink I. *International Criminal Police Organization (Interpol). Electronic Evidence – Forensic Audio and Visual Evidence – A Review:* 2004-2007. https://www.interpol.int/Public/Forensic/IFSS/meeting15/Papers03.pdf (accessed December 8, 2011).
112. Arbeitsgruppe Identifikation nach Bildern (Working Group for Identification by Pictures). http://www.bildidentifikation.de/startseite.html (accessed December 8, 2011).
113. Law Enforcement and Emergency Services Video Association (LEVA). http://www.leva.org (accessed December 8, 2011).
114. International Association for Identification. http://www.theiai.org/ (accessed December 8, 2011).
115. Scientific Working Group on Digital Evidence. Best Practices for Forensic Audio (v1.0), 2008. http://www.swgde.org/documents/current-documents/2008-01-31%20SWGDE%20Best%20Practices%20for%20Forensic%20Audio%20v1.0.pdf (accessed October 8, 2011).
116. ENFSI Working Group for Forensic Speech & Audio Analysis. Best Practice Guidelines for ENF Analysis in Forensic Authentication of Digital Evidence. 2009. http://www.enfsi.eu/uploads/files/ENFSI-FSAAWG-BPM-ENF-001-20090615.pdf (accessed October 8, 2011).
117. Keppel RD, Walter R. Profiling Killers: A Revised Classification Model for Understanding Sexual Murder. *International Journal of Offender Therapy and Comparative Criminology* 1999; 43(4):417–437.
118. Stephenson P, Walter R. *Toward Cyber Crime Assessment: Cyberstalking*. Albany: University of Albany – Annual Symposium on Information Assurance; 2011.
119. Rogers M, Brinson A, Robinson A. A cyber forensics ontology: Creating a new approach to studying cyber forensics. *Journal of Digital Investigation* 2006; 3S: S37–S43.
120. Bocij P. *Cyberstalking – Harassment in the Internet Age and How to Protect Your Family*. Westport: Praeger Publishers; 2004.
121. McQuade SC. *Understanding and Managing Cybercrime*. Boston, MA: Allyn & Bacon; 2005.

122. Shindler DL, Cross M. *Scene of the Cybercrime*. 2nd ed. Rockland, MA: Syngress; 2008.
123. Bijhold J, Ruifrok A, Geradts Z. *Forensic Visual Evidence 2007–1020: A Review*. Netherlands Forensic Institute 2010. http://forensic.to/webhome/enfsidiwg/interpol2010.pdf (accessed December 8, 2011).

List of relevant legal cases

- *Ashford v. Thornton*, 1B. & Ald. 405 on an appeal of murder (1818).
- *Blackwell v. Wyeth*, 408 Md. 575, 971 A.2nd 235 (2009).
- *Burral v. State*, 724 A.2d 65 (holding that Maryland has not abandoned Frye in favor of standards set forth in Fed. R. Evid. 702) (Md. 1999).
- *Conaway v. Deane*, 401 Md. 219, 932 A.2d 571 (2007).
- *Daubert v. Merrell Dow Pharmaceuticals*, 727 F. Supp. 570, 574, S.D.Cal (1989).
- *Daubert v. Merrell Dow Pharmaceuticals*, 951 F.2d 1128, 1130, 9th Cir. (1991).
- *Daubert v. Merrill Dow Pharmaceuticals*, 509 U.S. 579; 113 S. Ct. 2786; 125 L. Ed. 2d 469; 1993 U.S. LEXIS 4408 (1993).
- *Daubert v. Merrell-Dow Pharmaceuticals*, 43 F. 3d 1311, 1314, Ninth Cir. Ct. App. (1995).
- *Erie Railroad v. Tompkins*, 304 U.S. 64 (1938).
- *Frye v. United States*, 293 Fed. 1013, D. C. Cir. (1923).
- *Joiner v. General Electric*, 864 F. Supp. 1310; 1994 U. S. Dist. LEXIS 13326 (1994).
- *Joiner v. General Electric*, 78 F.3d 524; 1996 U. S. App. LEXIS 5590 (1996).
- *General Electric Co. v. Joiner*, 522 U. S. 136 (1997).
- *Giant Food v. Booker*, 152 Md. App. 66, 831 A.2d 481 (2003), cert. denied 378 Md. 614, 837 A.2d 926 (2003).
- *Hurtado v. California*, 110 US 516 (1884).
- *Hutton v. State*, 663 A.2d 1289, 1295 n.10 (Md. 1995) (applying the Frye standard and referencing the committee note appended to Md. R. Evid. 5–702).
- *Carmichael v. Samyang Tires, Inc.*, 923 F. Supp. 1514, S.D. Ala. (1996).
- *Carmichael et al. v. Samyang et al.*, 131 F.3d 1433; 1997 U.S. App. LEXIS 35981, 11 Cir. (1997).

- *Kumho Tire Co. v. Carmichael*, 526 U. S. 137 (1999).
- *Lisenba v. California*, 314 US 219, 236 (1941).
- *Montgomery Mutual Insurance Co. v. Chesson*, 399 Md. 314, 923 A.2d 939 (2007).
- *Oxendine v. Merrell Dow Pharm., Inc.*, 506 A.2d 1100 (D.C. Ct. App. 1986), on remand, 563 A.2d 330 (D.C. Ct. App. 1989), cert denied 493 U.S. 1074 (1990), on remand, 593 A.2d 1023 (D.C. Ct. App. 1991), on remand, 649 A.2d 825 (D.C. Ct. App. 1994), on remand, Civ. No. 82–1245, 1996 WL 680992, D. C. Super. Ct. (Oct. 24, 1996).
- *Palko v. Connecticut*, 302 US 319 (1937).
- *Reed v. State*, 283 Md. 374, 391 A.2d 364, 97 A.L.R.3d 201 (1978).
- *Richardson v. Richardson-Merrell, Inc.*, 649 F. Supp. 799; 506 A.2d 1100 (D.D.C. 1986) (citing and following independent FDA advisory panel finding nothing to implicate Bendectin exposure as cause of increased incidence of birth defects), aff'd, 857 F.2d 823 (D.C. Cir. 1988), cert. denied, 493 U.S. 882 (1989).
- *Smith v. State*, 388 Md. 468, 880 A.2d 288 (Md., 2005) ("Maryland has not rejected the Daubert standard, leaving to case-by-case development whether and to what extent Daubert may apply here. See Committee Note to Md. Rule 5-702." 388 Md. 468 at footnote 12.).
- *Wood v. Toyota Motor Corp.*, 134 Md. App. 512, 760 A.2nd 315, cert. denied, 364 Md. 141, 803 A.2nd 1034 (2002).

Further reading

Alder K. *The Lie Detectors: The History of an American Obsession*. New York: Free Press; 2007.

Anson S, Bunting S. *Mastering Windows Network Forensics and Investigation*. Berkley, CA: Sybex, Inc.; 2007.

Bonifant NE. Blackwell v. Wyeth: It's Our Courtroom and We'll Frye (Only) If We Want To – The Maryland Court of Appeals's Unstated Adoption of Daubert. 69 *Maryland Law Review* 719 (2010).

Carrier B. *File System Forensic Analysis*. Boston, MA: Addison-Wesley; 2005.

Carvey H. *Windows Forensics and Incident Recovery*. Boston, MA: Addison Wesley Professional; 2004.

Carvey H. *Windows Forensic Analysis and DVD Toolkit*. 2nd ed. Burlington, MA: Syngress (Elsevier); 2009.

Casey E. *Digital Evidence and Computer Crime*. 2nd ed. Burlington, MA: Academic Press; 2004.

Casey E. *Handbook of Digital Forensics and Investigation*. Burlington, MA: Academic Press; 2009.

Cheng EK, Yoon AH. Does Frye or Daubert Matter? A Study of Scientific Admissibility Standards. 91 *Virginia Law Review* (2005).

Chiesa R, Ducci S, Ciappi S. *Profiling Hackers*. Boca Raton, FL: CRC Press; 2009.

Clark F, Diliberto K. *Investigating Computer Crime*. Boca Raton, FL: CRC Press; 1996.

Thornton A, Cooper J. *A Report of the Proceedings against Abraham Thornton*. London: Heathcote and Foden; 1818.

Cranor L, Garfinkel S. *Security and Usability*. Sebastopol, CA: O'Reilly; 2005.

Crowley P,Kleiman D,technical editor. *CD and DVD Forensics*. Burlington, MA: Syngress; 2006.

Dixon L, Gill B. *Changes in the Standards for Admitting Expert Evidence in Federal Civil Cases Since the Daubert Decision*. Santa Monica, CA: RAND Institute for Civil Justice; 2001.

Farmer D, Venema W. *Forensic Discovery*. Boston, MA: Addison-Wesley; 2009.

Fedotov NN. Форензика – компьютерная криминалистика *(Forensics – Computer Forensics)*. Moscow, Russian Federation: The Legal World, 2007.

Garfinkel S. *Database Nation; The Death of Privacy in the 21st Century*. Sebastopol, CA: O'Reilly; 2000.

Garfinkel S, Rosenberg B. *RFID: Applications, Security and Privacy*. Boston, MA: Addison-Wesley; 2005.

Geradts ZJMH,Franke K,Veenman CJ,editors. Computational Forensics Third International Workshop, IWCF 2009. *Proceedings Series: Lecture Notes in Computer Science Subseries: Image Processing, Computer Vision, Pattern Recognition, and Graphics,* Vol. 5718; 2009 Aug 13–14; The Hague, The Netherlands: Springer-Verlag, 2009.

Geschonneck A. *Computer-Forensik*. 2nd ed. (German language) Berlin, Germany: Dpunkt Verlag, 2006.

Ghirardini A, Faggioli G. *Computer Forensics*. 1st ed. (Italian language) Milan, Italy: Apogeo, 2007.

Groscup J, Penrod S, Studebaker C, Huss M, O'Neil, K. The Effects of Daubert on the Admissibility of Expert Testimony in State and Federal Criminal Cases. *Psychology, Public Policy & Law* 2002; 8 (4): 339–372.

Hoog A, *Strzempka K. iPhone and iOS Forensics: Investigation, Analysis and Mobile Security for Apple iPhone, iPad and iOS Devices*. Burlington, MA: Syngress; 2011.

Hoog A. *Android Forensics: Investigation, Analysis and Mobile Security for Google Android*. Burlington, MA: Syngress; 2011.

Howard RC, Jr., Case Study: Kumho Tire v. Carmichael: Kumho Tire v. Carmichael – Facts and Background. 9 *Kansas Journal of Law & Public Policy* 134 (1999).

Howell BT, Thomas DS, editors. *State Trials*. London: Routledge & K. Paul; 1972.

Huber PW. *Galileo's revenge: Junk science in the courtroom*. New York: Basic Books (Original work published 1991); 2001.

Jaishankar K, editor. *Cyber Criminology – Exploring Internet Crimes and Criminal Behavior*. Boca Raton, FL: CRC Press; 2011.

Katzenbeisser S, Petitcolas F. *Information Hiding: Techniques For Steganography and Digital Watermarking*. Norwood, MA: Artech House; 2000.

Kipper G. *Investigator's Guide to Steganography.* Boca Raton, FL: Auerbach Publications; 2003.

Krafka C, Dunn MA, Johnson MT, Cecil JS, Miletich D. Judge and Attorney Experiences, Practices, and Concerns Regarding Expert Testimony in Federal Civil Trials. *Psychology, Public Policy & Law* 2002; 8. (3):309–332.

Kubasiak RR,Morrissey S,Varsalone J,technical editors. *Mac OS X, iPod, and iPhone Forensic Analysis DVD Toolkit.* Burlington, MA: Syngress; 2008.

Levy AD, Hornstein AD, Weissenberger G. *Maryland Evidence: 2011 Courtroom Manual.* LexisNexis; 2011.

Lillquist RE. A Comment on the Admissibility of Forensic Evidence. *Seton Hall Law Review* 1189–1204 (2003).

Litchfield D, Anley C, Heasman J, Grindlay B. *The Database Hacker's Handbook – Defending Database Servers.* Indianapolis, IN: Wiley; 2005.

Mandia K, Prosise C, Pepe M. *Incident Response & Computer Forensics.* 2nd ed. Emeryville, CA: McGraw-Hill/Osborne; 2003.

Marcella AJ, Jr., Greenfield RS. *Cyber Forensics: A Field Manual for Collecting, Examining, and Preserving Evidence of Computer Crimes.* Boca Raton, FL: Auerbach Publications; 2002.

Megarry R.Garner BA, editor. *A New Miscellany at Law: Yet Another Diversion for Lawyers and Others.* Clark, New Jersey: The Lawbook Exchange, Ltd.; 2005.

Moreno JA. Eyes Wide Shut: Hidden Problems and Future Consequences of the Fact-Based Validity Standard. *Seton Hall Law Review* 2004; 34 (1):89–103.

Morrissey S. *iOS Forensic Analysis for iPhone, iPad and iPod Touch.* New York, NY: Apress; 2010.

NIJ – NCJ, 187736. *Electronic Crime Scene Investigation: A Guide for First Responders.* Washington, DC: National Institute of Justice; 2001.

NIJ – NCJ 199408. *Forensic Examination of Digital Evidence: A Guide for Law Enforcement.* Washington, DC: National Institute of Justice; 2004.

Risinger DM. Navigating Expert Reliability: Are Criminal Standards of Certainty Being Left on the Dock? *Albany Law Review* 2000; 64 (1):99–152.

Rodrigues de Freitas A. Perícia Forense Aplicada à Informática *(Applied Computer Forensics).* 1st ed. Rio de Janeiro, Brazil: Brasport, 2006.

Roisman AZ. The Implications of G. E. v. Joiner for Admissibility of Expert Testimony. *Vermont Journal of Environmental Law* 1998-1999; 1.

Spafford E, Garfinkel S, Schwartz A. *Practical UNIX and Internet Security.* 3rd ed. Sebastopol, CA: O'Reilly; 2003.

Stephenson P. *Investigating Computer-Related Crime.* Boca Raton, FL: CRC Press; 1999.

Stoll C. *The Cuckoo's Egg – Tracking a Spy Through the Maze of Computer Espionage.* New York, NY: Doubleday; 1989.

Vickers L. Daubert, Critique and Interpretation: What Empirical Studies Tell Us About the Application of Daubert, 40 *University of San Francisco Law Review* 109, 109–110 (2005).

Vorder Bruegge RW, Evison MP,editors. *Computer-aided Forensic Facial Comparison.* Boca Raton, FL: CRC Press; 2009.

Wigmore JH. *Select Cases on the Law of Evidence*. 2nd ed. Boston: Little, Brown, and Company; 1904.

Woodall WO. The Trial of Abraham Thornton for the Murder of Mary Ashford, at the Warwick Summer Assizes, Before Mr. Justice Holroyd on August 8th, 1817. In: *A Collection of Celebrated Trials, Civil and Criminal Vol. I*. London, UK: Shaw and Sons; 1878; 1–74.

Zdziarski J. *iPhone Forensics*. Sebastopol, CA: O'Reilly; 2008.

11

Global thinking and methodologies in evidence-based forensic engineering science

Laura L. Liptai[1], Adam Aleksander[2], Scott Grainger[3], Sarah Hainsworth[4], Ryan Loomba[5] and Jan Unarski[6]

[1]*BioMedical Forensics, Moraga, California, USA*
[2]*Aleksander and Associates P.A., Boise, Idaho, USA*
[3]*Grainger Consulting, Inc., Mesa, Arizona, USA*
[4]*University of Leicester, United Kingdom*
[5]*Nanomix, Emeryville, California, USA*
[6]*Institute of Forensic Research, Krakow, Poland*

11.1 Introduction

In the forensic analysis of engineering system failures, the forensic engineering sciences are uniquely situated to offer insights and to engineer a safer world. Engineering, as a discipline, encompasses a very broad range of subject matter studied by a common iterative problem-solving approach. The main branches of the engineering sciences include biomedical, civil, electrical, environmental, materials, mechanical, metallurgical, physics, and others. There is also a wide bandwidth of subtypes, depending upon the often esoteric problems to be solved.

All of these types of engineering hold in common an approach to scientific knowledge: building scientifically grounded tools, or 'engines', which analyze and solve problems [1]. One of the basic tools in science that is utilized in applied engineering is geometry, which, when coupled with its modern counterpart, mechanics, has seeded most of the growth of human civilization. However, no matter how complex, all systems – both natural and man-made – must still follow a simple design principle: *design evolves towards ease of movement.*

Forensic engineering is the application of engineering methods to forensic problems, and it is similarly diverse in expertise and application; however, the

Forensic Science: Current Issues, Future Directions, First Edition. Edited by Douglas H. Ubelaker.

diverse analyses of forensic engineering sciences are bound together by a common paradigm. This particular paradigm is especially important in the current forensic climate, because the tool-building, tool-testing engineering methods lend themselves very naturally to sound engineering methodologies. Forensic sciences, with more traditional pattern-based forensics, are increasingly utilizing long-standing engineering science methods to establish pathways for growth and innovation [2].

This work will not cover all the ways in which engineering scientists have applied their skills to forensic problem-solving, but the standard engineering principles that govern the world in which we live apply over a vast bandwidth of specialties. Our governing goals in the engineering sciences include promoting forensic engineering education and research, and improving the practice and elevating the standards of forensic engineering sciences worldwide.

This chapter encompasses engineering sciences such as those represented in the American Academy of Forensic Sciences (AAFS) and the Society of Forensic Engineers and Scientists (SFES). We report current progress, training, certification and accreditation programs and also research opportunities. Perhaps the most complete reference, including nearly a decade of forensic engineering case studies and research results, is *Forensic Engineering Sciences*, which is part of the American Academy of Forensic Sciences Reference Series [3].

11.2 The growth of forensic engineering

Over the past 70 years, forensic engineering has grown with the surge of industrial development and motorized vehicles. Today, with the foundation of forensic engineering analytical methodologies, there is an increasing need for more esoterically skilled individuals. The modern forensic engineer is a specialized individual who is grounded in experience and embraces educational growth.

One of the earliest examples of employing forensic engineering to investigate accidents was the fall of the Dee Bridge in Chester, England. The bridge was strengthened by wrought iron bars and approved for use in 1846. However, in 1847, a train fell through the bridge, resulting in five deaths and nine injuries. The builder of the bridge was accused of negligence. An expert, Captain Simmons of the Royal Engineers, was called to conduct a study. This marks one of the first instances of an expert forensic engineer. Captain Simmons conducted a series of tests and determined that the design was flawed [4].

Decades later, on September 20, 1893, Frank Duryea drove a gasoline-powered contraption 200 feet down the Main Street of Chicopee Falls, Massachusetts, beginning the automobile revolution in the United States. By 1970, over 100 million vehicles were registered in the United States. In 1980, the US census reported that there were 17.9 million motor vehicle accidents [5], including 52,000 fatalities (National Highway Traffic Safety Administration, NHTSA). By 2008, vehicular fatalities had dropped to 37,423, 31% or 11,711 of fatalities traced to alcohol-impairment. By 2010, seat belt use for occupants of passenger vehicles was 85%, according to the National Occupant Protection Use Survey [6].

The definition of a forensic engineering *expert* has developed and changed drastically over the past 70 years. In the distant past, to select an expert, attorneys would find a local who knew something about the subject matter, regardless of his or her expertise in forensics. In the late 1930s and early 1940s, engineers specializing in forensics were rare. The Accident Investigation Manual was first published in the 1940s by the Northwestern University Traffic Institute, and it offered a structured approach for police officers to develop accident investigation tools and techniques [7]. Law enforcement officers dramatically improved the documentation of important evidence at the scene [8].

In 1979, the book *Accident Reconstruction* was published [9], marking the first publication about accident reconstruction written from a scientific and engineering prospective. Dr. James Collins, Dr. John L. Habberstad, Dr. Robert Liptai, Dr. J. Michael Stephenson and Dr. Richard N. Stuart, engineers and scientists working at Lawrence Livermore National Laboratory, applied physical laws and modern science to interpret physical evidence and failure, to analyze automobile and motorcycle collisions as well as to evaluate pedestrian incidents. This notably established the application of science and engineering expertise to forensics. During this time, Hugh H. Hurt Jr., a safety engineer, started to investigate motorcycle crashes. His meticulous documentation and analysis showed that helmets that met official specifications drastically reduced head injuries [10]. During this period, the consensus of a forensic engineering method emerged.

Beginning in 2011, incident-specific research accessibility improved. The American Academy of Forensic Science's book *Forensic Engineering Sciences* published a decade of proceedings [3], exponentially improving the ease of historical case study and research result access. In 2012, the National Academy of Forensic Engineers also offered a collection of the organization's forensic engineering papers for the first time by specialty.

11.3 Methodology

Due to the complex, real-world, multiple-variable nature of engineering forensic problems, it is rare to isolate or control for numerous simultaneous forensic variables as required in scientific null hypothesis testing. Methodologically, forensic engineering is most often the application of existing science. Therefore, the forensic engineering method is a variant of the scientific method that utilizes controlled variable null hypothesis testing [11]. Engineering forensics requires analysis of an incident that has previously occurred with multiple simultaneous real-world variables. It is normally not practical, or even possible, to isolate individual variables using the classic scientific method in pre-versus-post, test-versus-control experimental protocols. Forensic problems most often use existing science

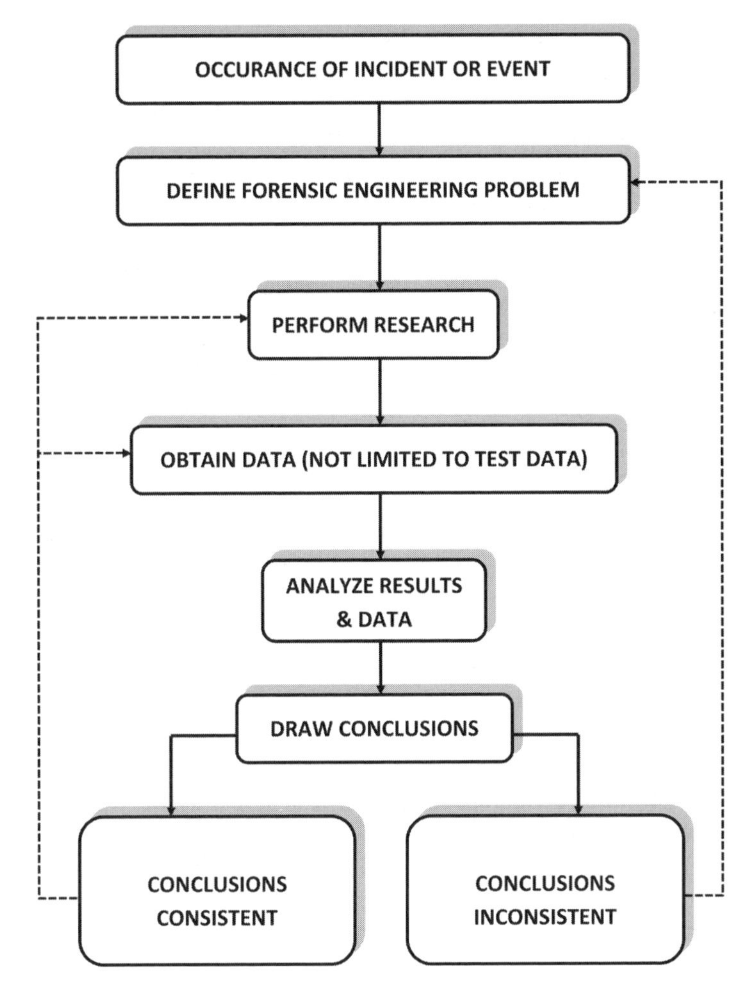

Figure 11.1 The forensic engineering method.

(Newton's Laws, for example), applying proven principles to a specific incident or failure. It is noteworthy that forensic engineering findings and methodologies may not be published, often due to the lack of uniqueness and/or narrow application, esoteric and incident-specific nature of the incident.

The long-standing and widely utilized forensic engineering method is displayed in Figure 11.1.

11.4 Education and qualifications

In the United States, forensic engineers start out their education with a high school diploma and then receive technical instruction from an accredited four-year university. At university, students usually major in engineering or applied science. Depending on a student's interest, one may choose to major in a number of different subjects such as civil engineering, mechanical engineering or applied physics. During university schooling, students may choose to take internship positions at companies to obtain work experience. After obtaining a bachelor's or master's degree, a student may choose to start to work in industry or continue school to obtain their PhD.

In the United States, there is an official board certification specifically in forensic engineering sciences. The International Board of Forensic Engineering Sciences (IBFES) administers the exams annually. There is a written and oral portion in front of a panel that must include a judge; this exam is similar to doctoral oral exams. This program is developed by the American Academy of Forensic Sciences, with support from the US Department of Justice. The IBFES executes strict criteria, culminating in peer reviews as well as oral and written examinations for domestic and international candidates.

Forensic engineers who testify in court generally have advanced degrees and many years' experience. Some engineers have a professional engineering license (PE), which indicates they have completed engineering understanding in their traditional engineering field. However, the professional engineering licensure does not test forensic engineering; therefore, a PE is not required as a prerequisite to testify according to the Reference Manual on Scientific Evidence [2].

In the United States, forensic engineers are retained by governmental agencies and private parties. Oftentimes, an expert witness will be retained by the plaintiff's lawyer and a separate expert will be retained by the defense's lawyer. Similarly, the prosecution and defense in criminal matters each have their own experts. Forensic engineers are not allowed by law to work on a contingency basis, because this could undermine the neutrality of their opinion. Some experts are engaged in their field of study in industry or academia and choose to consult as an expert witness part-time. Other experts practice forensic engineering on a full-time basis, alone or as a part of a firm of engineers. The forensic engineer's hourly rate ranges from $100–$500 USD, depending on the type of work and experience of the expert.

11.5 Forensic engineering in a changing world

The future of forensic engineering relies in part on the improvements in technology pertaining to forensics. Faster and accessible computing power will allow for a greater quantity of calculations, e.g. more detailed simulations. One field of forensics that relies on modern technology is natural language engineering. The development of electronic communications such as the internet, email and smart phones has provided an opportunity for natural language engineering to be applied forensically. Two issues which exploit fundamental methods of natural language engineering are speaker recognition and author identification.

Speaker recognition is the use of a machine to recognize a person from a spoken phrase. Generally, these machines operate to identify a specific person or to verify a person's claimed identity. Automatic speaker recognition systems, which can be implemented on a modest personal computer, have been operated with 98.9% accuracy [12].

In addition to speaker recognition, author identification is another branch of forensic engineering which utilizes complex technology attempt to identify or rule-out a person's identity using kinematic factors such as handwriting to supplement the works of questioned document and other forensic experts. One engineering approach to author identification and recognition is utilizing neural network analysis, or a software simulation of the human brain. A neural network can learn; it receives an input and then produces an output based on the weights attached to the links connecting the nodes of the network. In order for the weights to change or strengthen, both nodes must have a positive output simultaneously. A study using neural networks for author identification was successful in classifying the works of Shakespeare and Marlowe accurately [13].

Some established methods in the field of criminalistics rely on pattern and impression evidence. Fingerprints, palm prints, shoeprints, ballistics and certain types of handwriting identification have traditionally relied on the experience level of the experts. The first use of the Daubert criteria in a criminal prosecution involved the reliability of handwriting identification [14] and the reliability of fingerprint identification [15]. The resulting legal challenges have provided an engineering (as well as digital and multimedia sciences) opportunity for new computational methods based on image processing, and to bring traditional pattern-based problems into the biometric paradigm of facial recognition and iris recognition.

Many motor vehicle manufacturers have taken cruise control to the next level, by implementing various degrees of automation while driving. These may include adaptive cruise control, which can sense cars in front and adjust a vehicle's speed accordingly. Similarly, these sensors can be used to detect if a crash is imminent and will then apply the brakes, to bring the car to a complete stop if necessary. Future improvements include integration of onboard sensors with GPS and other resources, which would allow a vehicle to adapt to changing weather and road conditions. The Lexus LS 600h currently has automated parking capabilities; a user chooses a space, and then the system takes over steering and acceleration to park the car. Full automation is still in development and looks to be a promising feature in future vehicles [16].

Airbags are commonplace safety features found in currently manufactured automobiles. In contrast, motorcycles do not come standard with airbags, as the technology is in early phases. Specifically in frontal collisions, airbags absorb and disperse a rider's kinetic energy, which would reduce rider separation velocity on a motorcycle. Crash sensors are mounted in the front wheel assembly and activate airbags located just below the front displays and gauges. In preliminary testing, airbag deployment during an impact decreased the head injury criterion significantly. The motorcycle airbag has proven to be effective in helping to reduce the injury severity in frontal collisions, by absorbing energy and slowing the rider's velocity. More research and testing are needed before airbags can be a standard feature on motorcycles [17].

11.6 International perspectives on forensic engineering

Countries across the globe have developed the application of forensic engineering to best match the needs of their systems. These country-specific systems often include different requirements for education and certification, as well as unique adaptations for local laws. Despite these differences, the forensic engineering method (Figure 11.1), a derivative of the universally accepted scientific method, is accepted generally as well as applicable to the vast bandwidth of forensic applications worldwide. Below are sections which give examples of how forensic engineering problems are solved in other countries. In learning from each other and challenging our thinking we become better forensic engineering scientists.

11.6.1 The United Kingdom

In the UK, there is no formalized, prescribed route for training to practice as a forensic engineer. Most forensic engineers have developed their expertise over a considerable time following their initial education, which usually includes obtaining a bachelor's or a master's degree. Thus, the initial qualification is either a BSc or BEng degree. Master's qualifications have become more prevalent over the last 15 years, and many engineers now have either an MSc or MEng degree. They differ slightly, in that an MSc is usually obtained following a one-year further period of study in a specialist or more advanced areas. The MEng degree is an integrated master's, where a student studies a four-year course which leads to the master's qualification. Thus, an MSc holder usually has a BSc or BEng qualification as well as the MSc, whereas an MEng holder usually has only the MEng qualification. Some (few) experts also hold a doctorate – PhD or DPhil.

There are a few universities that offer forensic engineering at the undergraduate level as a course, and more that offer master's qualifications. Many of the courses that offer forensic engineering still have a high chemistry or biological sciences emphasis.

Many engineers in the UK join a professional engineering institution and subsequently gain Chartered Engineer status, which reflects a period of further training and usually some responsibility for project management. Chartered Engineer (CEng) is a title conferred by the Engineering Council, which is responsible for the UK-wide standards across all engineering specialties. It is not, however, formally regulated in the way the PE designation is regulated in the USA. A similar professional qualification of Chartered Scientist (CSci) can also be obtained for those with a science bias in their work, and this is overseen by the Science Council.

At the same time as the award of CEng, it is usual for someone to gain professional member status in their relevant engineering institution. There is a wide range of engineering institutions in the UK – 36 licensed institutions covering the whole spectrum of engineering. It is not mandatory to have the title CEng to act as an engineer in the UK, and many 'engineers' do not have CEng status. Forensic engineering is not regulated and there is no institution that promotes the profession. There is an Expert Witness Institute that promotes the professional training, education and representation of experts across all disciplines, including science.

Engineers in the UK act as experts usually on a consultancy basis. There are a number of firms offering forensic engineering services; also, expert witnesses may come from universities or firms whose primary activity is not forensic engineering. The company or individual will be instructed either by a law firm, by insurers, by a company or by the police services or government in the case of major accidents. In rare cases, a private individual may commission an investigation. The type of work that is undertaken is typically root-cause engineering failure analysis, vehicle or accident reconstruction and investigation, patent work and buildings-related services related to working environment. Consulting engineers are typically experts in mechanical, materials, civil, construction, transport or automotive engineering.

Research in forensic engineering is largely undertaken by either government laboratories, large engineering companies who offer forensic services, or universities. An example of a government laboratory would be the Transport Research Laboratory, which has an Incident Investigation and Reconstruction group, or the Air Accident Investigation Branch (AAIB).

However, forensic engineering science in the UK in 2011 was in a period of considerable flux. In May 2010, a coalition government was formed after none of the major political parties secured an overall majority. The new government (comprising the Conservative and Liberal Democrat parties) took office in a period of financial difficulties in which the UK overall had a large budget deficit. Their agenda was to cut the costs of public services and to abolish government quangos where possible. As a result, the Forensic Science Services (the government-owned provider of forensic science and engineering services), which at the time was making considerable losses (of the order of £2 million ($3 million) per month) were targeted for phased closure in 2011 [18]. The Forensic Science service was considered to be one of the UK's leading providers of forensic science services, having pioneered many of the breakthroughs in forensic science (such as the use of DNA evidence in the criminal justice system). They did not have a remit for research *per se*, but were important in supporting research through a number of routes.

Along with the closure of the Forensic Science Services, the Home Office Scientific Development Branch was reviewed and is now operating under the new remit of the Home Office Centre for Applied Science and Technology (which includes engineering). The other body that was a part of the change imposed by government was the National Policing Improvement Agency (NPIA), which is also targeted for closure. The NPIA works closely with the Association of Chief Police Officers to determine what the research priorities are from a police perspective; it forms an important link in connecting forensic experts to the various police authorities in cases where expert analysis can inform case-related inquiries. Simultaneously, cuts to the overall police budget have led to many police services reorganizing their scientific services and making decisions as to which cases are worthy of investigation based on the probability of science or engineering helping to secure a conviction. Research is also supported by the Serious Organized Crime Agency (SOCA).

Many of the government agencies rely on working in collaboration with universities so that funding can be obtained by the UK Research Councils, which provide government funding to research activity across all academic disciplines, not just engineering. Given the considerable changes taking place, a review of current forensic science provision in the UK was commissioned to better understand the forensic landscape, headed by Professor Bernard Silverman, the Chief Scientific Adviser to the UK Home Office [19]. The changes have not been without controversy: the House of Commons Science and Technology Select Committee reported that:

> *'In making its decision to close the FSS, the government failed to give enough consideration to the impact on forensic science research and development, the capacity of private providers to absorb the FSS's 60% market share and the wider implications for the criminal justice system. These considerations appear to have been hastily overlooked in favour of the financial bottom line.'* [20]

The future of forensic engineering research in the UK is likely to be driven by interested individuals in universities, working with organizations such as the Home Office, the National Policing Improvement Agency (or its successor), SOCA, or the specialist forensic providers developing their own research portfolios. The priority areas for the NPIA are in cost-effectiveness and critical national services. It is likely that the UK Research Councils will closely monitor applications for funding to ensure that they tie in with these priority areas.

There is a challenge in that applications to research councils are reviewed by academics who are often not forensically literate, and thus there will be a period of time where educating reviewers will need to take place. Much of the research will be around making turnaround of analysis faster and more robust; for example, engineering has a role to play in refining techniques that forensic scientists use for DNA analysis.

Another area where engineers and physicists are contributing is in developing hand-held detectors for the rapid analysis of residues at crime scenes – for example, is a white powder a drug or a biohazard? An example of where this would have been helpful was in the 7th July terrorist bombings in London, where there was delay in getting fire crews into the London Underground until the scene had been analyzed to ensure their safety.

Other areas where engineering has provided solutions is in developing software to allow multi-detector computer tomography imaging of bodies to be presented to juries, where gruesome pathology photographs would be likely to cause upset and trauma.

There are many examples where engineers are working in combination with others across discipline areas to drive developments. Despite the challenging economic times, there are many positive developments that are influencing the way in which accidents and crime are analyzed by forensic engineers in the UK.

11.6.2 The Philippines and Bangladesh

The Philippines has a population of approximately 94 million, and Bangladesh 150 million. The urban population of the Philippines is approximately 49%, while that of Bangladesh is 28%. The capital city of Bangladesh is Dhaka, arguably one of the fastest growing cities in the world with a population of 15 million, including some 3.4 million slum dwellers. Bangladesh is also one of the most densely populated countries, with the attendant problems of poverty and resource management issues. The Philippines is spread over some 7,000 islands, with major population concentrations in the northern island of Luzon, Manila (20 million) and a dozen or so major cities in the Visayas and Mindanao regions.

We begin by understanding the effects of large population growth in third world countries and the resulting strain on education, transportation, economic development and the legal systems. Simply put, the perception of readers from developed countries (Europe, Canada, the United States, etc.) must be tempered by the realization that forensic engineering practice is limited and is often unavailable. Forensic sciences in the criminal domain are more likely to be funded by local or national agencies in these countries. Both countries suffer from 'brain drain' – the systematic departure of qualified students and graduate professionals for better-paying job opportunities abroad. And this topic would not be complete without the mention of the word *corruption*, as it can be an important influence.

The Philippines

In the Philippines, the term 'forensic engineering' appears to be non-existent, although the concept is used in the legal system. Those engineers who practice before the courts are registered professional engineers. Philippine engineers can obtain professional engineering registration in a number of disciplines, including chemical, civil, computer, electrical, electronics, industrial, information technology and mechanical engineering, but not in forensic engineering. The Philippine Professional Regulation Commission (PRC) supervises and controls state-administered board examinations for engineers and other professional groups.

There is, at this time, no coursework offered in the Philippines in forensic engineering, nor is there an association of forensic engineers, nor do the courts identify them as forensic engineers. However, although there are relatively few practicing engineers in this area, they are involved in forensic engineering activities, such as accident reconstruction, product failure investigations, structural failure analysis and related investigations.

Under the Philippine setting, professionals performing forensic engineering activities are employed either in a private sector or in the public/government sector. In the private sector, they either work for a company engaged in providing engineering services or are retained by a private company. In the government sector, forensic engineering services are found among various government agencies, such as, among others, Bureau of Fire Protection, National Bureau of Investigation, Department of Transportation and Communication and the Philippine National Police.

All of these agencies/departments work hand in hand to support and help one another. For example, in cases of fire incidents, the Bureau of Fire Protection works with the Philippine National Police and the National Bureau of Investigation to gather evidence, determine the cause of fire and resolve issues involving said fire incident. Also, in cases where forensic engineering is required, the Philippine National Police has Scene of the Crime Operation (SOCO) teams who perform the vital task of gathering forensic evidence. These pieces of evidence will be directed to the Philippine National Police Crime Laboratory for evaluation and further investigation.

If and when a court action is involved, as regards forensic engineering matters, both the plaintiff/complainant and respondent/defendant may retain their respective 'forensic' engineers. Both sides may present their respective engineers as witnesses; however, the parties must prove that said engineers are experts on the matter. There is no definite standard of determining the degree of skill or knowledge that a witness must possess in order to testify as an expert. It is sufficient that the following factors be present:

- Training and education.
- Particular first-hand familiarity with the facts of the case.
- Presentation of authorities or standards upon which his or her opinion is based.

The question of whether the professional engineer is properly qualified to give an expert opinion rests with the discretion of the trial court.

Forensic engineering, as a concept, is very much interwoven into the Philippine legal system. Considering that it is an intricate matter, the Philippine legal system relies on those engineer experts who may not be forensic engineers by title but perform certain tasks of forensic engineers by practice.

As regards matters involving compensation, there is no standard salary that can be considered. Experience and skills are the factors considered as to what the salary is. There is a hierarchy of salary grade in the government sector, but, again, experience and skills are given importance. As regards whether these individuals work full-time or part-time, it is definitely full-time in the government sector, but in the private sector some may be full-time while others may be part-time.

As has already been mentioned, forensic engineering as a separate and distinct profession in the Philippines seems non-existent. As such, it is possible that Philippine engineers engaged in performing the activities covered by forensic engineering use materials produced by Filipino experts as well as international articles and journals.

If the Philippines is to truly have forensic engineering not only in concept but as a separate and independent field, it can definitely do so. The National Engineering Center (NEC) is an extension arm of the University of the Philippines College of Engineering. It was established in 1978 and mandated to strengthen and ensure a steady supply of technical manpower in the various fields of engineering and to develop technologies which utilize indigenous resources and are appropriate to the needs of local industries. Furthermore, it provides research, consultancy and continuing education services in engineering and related fields. The best thing about it is that the NEC serves industry, government agencies and academic institutions in need of engineering solutions. What the Philippines needs to do in order to truly have a field for forensic engineering is to have it pushed both academically and legally.

The Philippines legal system is an amalgam of Spanish, American and Philippine heritage, with some variants of Islamic Sharia Law in parts of Mindanao. There have been long-standing issues in the practical application of the civil and criminal codes and the timely progress of cases through the courts. In the Philippines, the courts have great discretion in the progress of a trial. It is not unusual for cases to linger for years – sometimes ten years or more. In a system where there are no juries, judges are in essence the sole triers of fact. The same pressures that affect the criminal cases, including prosecutors, police, and politicians, also affect civil cases. Although the Philippines is making some progress in the 'justice' of the legal system, there is a long way to go to assure a level playing field, even for technical experts to present their analysis of a case.

Bangladesh

Bangladesh suffers from many of the same pressures of a developing economy as does the Philippines. In addition, Bangladesh lies in a great delta, prone to severe flooding, cyclones and ensuing human disasters.

The country has emerged from a century of turmoil, having at times been Bengal, then a British colony, then partitioned into East Pakistan and then, through a revolt in 1971, becoming a parliamentary democracy in 1991. It is easy to see that many of the vital legal and socio-economic institutions necessary to support forensic engineering have suffered from developmental delays. The Bangladesh legal system has roots in British common law, but is heavily influenced by the official state religion, Islam, particularly in civil law. The system has an overwhelming backlog of court cases, affecting dispute settlement and foreign investment. Jury trials were abandoned in 1959.

Currently, the emphasis has been for the national and local government agencies to promote the development of forensic science units to support the needs of the criminal justice system. These operate under the CID (Criminal Investigation Department) branch of the police force, while the forensic chemical laboratories are located at multiple sites.

The Bangladesh Professional Engineers Registration Board (BPERB) administers the issuance of registrations, many of which go to graduates of the oldest engineering institution, Bangladesh University of Engineering and Technology (BUET), which traces its roots to British India in 1876.

A journey to Dhaka and spending a few weeks traveling the roads of Bangladesh quickly brings one face to face with the reality of life and death in this country. Transportation, including roads, rails, barges, boats, bridges, buses, trucks cars, rickshaws and pedestrians, all engage in a daily dance of death.

Mass deaths are common through overturned boats, and truck/bus collisions are daily occurrences (e.g. the July 2011 death of 42 teens when a truck overturned). The niceties of forensic engineering courtroom testimony (e.g. in North America and Europe) are far removed from the challenge of providing rudimentary services in the most densely populated country on earth. This is an environment with two-way traffic, potholes, water, no licenses, no traffic controls, no lights, no signs, no enforcement and no culpability except, of course, on paper.

There are very few practitioners of forensic engineering in Bangladesh. One such entity is the BUET Accident Research Institute, which is a government-funded safety research center. It provides a few personnel to assist in the investigation of major transportation mishaps related to roads, railways and waterways. More recently, the Institute itself (along with others) was called before the High Court demanding an explanation for the failure of authorities to assure road safety.

Similar chaos abounds in the steel, garment and construction industries, where accidents, environmental disasters and building collapses are common. As in transportation, the government is ineffective in the face of massive growth pressures.

One of the reasons forensic engineering practice evolved in the US, Canada and to a lesser extent Europe was the potential for timely recovery of damages from the negligent party or their insurance provider. This includes the establishment of foundation regulations such as the OSHA, EPA, and CPSC Acts in the US, enforcement by (more or less) impartial agencies, and a legal process including attorneys, experts and juries to render (more or less) fair decisions.

The discussion of tort reform in the US and limits on awards pales in importance compared to the lack of opportunity in most cases for reasonable compensation in the Philippines or in Bangladesh, or for that matter many third world or developing countries.

Until such systems are established in the Philippines and/or in Bangladesh, the true need for forensic engineering practitioners is minimal.

11.6.3 United States: fire protection engineering

Forensic fire protection engineering practice includes the following areas of expertise:

- Fire suppression systems (pipe, tubing, controls, pumps, nozzles, sprinkler heads).

- Fire alarm and notification systems (cable, panels, controls, activation, detection and notification devices).
- Fire separations (fire walls, floor/ceiling systems, roofs, draft stops, fire stopping systems).
- Structural fireproofing (cementitious, intumescent and constructed).
- Pipe (chemical and biological attack, connections and support).
- Fire and building code application and analysis.

There is no training specific to forensic fire protection engineering. A bachelor's degree in civil engineering provides much of the basic engineering and physics involved with pipe, fluid flow, structural statics and dynamics and materials science that are necessary. The engineering degree also provides a basic ability to apply the scientific method for solving problems. A college degree is important in the recipe for a successful forensic practice, but it must be combined with field experience.

Similar to that of obtaining a professional license which requires a minimum of four years of applied experience under the supervision of a licensee, a successful forensic engineer needs field experience. The major difference is that there is no specific license for forensic engineering, any more than there is a college degree for it. Success is market-driven, not controlled by a scholarly test. In addition to education, experience in construction, renovation, demolition, project management, inspection and rejection of equipment and facilities provides even broader scope for forensic practice.

Not all forensic engineering is driven by litigation. A small but active portion of work may involve the investigation of failures or the investigation of new materials and systems, for the express purpose of correcting or avoiding a failure under specific condition when the failure is not known.

The resources available to a forensic engineer are seemingly endless. The basic sources take the form of technical and professional associations, which provide opportunities to exchange thoughts, ideas and methodologies with peers, whether through conventions, meetings, seminars or blogs. They are also a very good source of papers and articles on many subjects of interest. The electronic age has made the exchange of ideas much easier and faster, all of which helps the forensic engineer during an investigation. Much of the older material is replaced by new data when using internet sources.

Computer simulations can be of great value, but the old adage of 'garbage in – garbage out' still applies. Is the software designed to do the analysis of interest? Is the engineer sufficiently familiar with the software to know when the software is not applicable? Were the data properly entered and documented? Too often, simulations are misused – for instance, the use of lots of color graphics in a routine that may be eye-catching to watch as it pretends to describe fire travel through a three-dimensional space on the video screen, when in fact the 'simulation' is nothing of the kind and has no basis in the engineering and science of fire travel.

The future is bright for forensic fire protection engineering. There are many areas where research is warranted, and one of particular interest is the use of epoxy-bound fibers to improve the strength of concrete and steel structural members. The fibers used are either carbon or fiberglass, and the strength of both types is highly temperature-dependent, with yield points at temperatures far below that of what is typically experienced under fire conditions. This dependency drives a need for the products to be protected from fire using compatible fireproofing materials. Investigations into this area indicate that much of the fireproofing work applied to fiber reinforcement has not been properly tested, which may lead to failures under fire conditions.

11.6.4 Poland

In Poland, two groups of people can provide their expert services to the courts pertaining to car accidents and car mechanics. Court experts are appointed by presidents of district law courts, who do not have detailed regulations in this matter. In practice, however, a national standard has been established which requires the candidate to hold a university degree in mechanical engineering specialized in automobile technology or a similar area. An MSc degree is obtained following a five-year course of studies. Nowadays, it is also frequently required to complete postgraduate studies in the field of accident reconstruction at a technology university. Moreover, candidates are required to substantiate, in writing, his or her qualifications for the job, such as professional experience and employment in this field.

The second group of experts is comprised of individuals associated with organizations such as the Association of Experts in Automobile Technology and Road Traffic, the Polish Automobile Federation or the Association of Transport Engineers and Technicians. They can provide services to the public and also can be accepted by the courts. These organizations have their own requirements similar to those mentioned above, and may include a two-year training as an assistant to a certified expert.

New regulations introduced recently require experts to hold a university degree, driving license categories A, B and C, have two years training and to have completed certifying procedure to be registered as experts for the Ministry of Transport. Among court experts, there are sometimes university staffs who, if not registered as court experts, are hired as '*ad hoc*' experts. Additionally, the Ministry of Justice runs the Institute of Forensic Research – Department of Accident Analysis, which can be used for difficult or controversial matters.

Court experts are usually certified by their professional organizations, e.g. expert accountants, expert psychologists, etc. The role of such organizations is to guarantee their experts' competence and high quality qualifications, as well as observance of deontology by them. This is accepted by law courts.

Recently, court experts in automobile technology and road traffic have been encouraged to undergo the certifying procedure. To this aim, the Polish

Accreditation Centre (PCA) has been appointed to operate three certifying institutions. PCA has set up a commission for establishing norms and certification minima to guarantee similar requirements and standards of experts. The Ministry of Transport maintains an official list of certified experts, and this procedure is also recommended to appoint new court experts. Certification has to be renewed every three years.

A court expert is only permitted to offer his or her service to the police, to public prosecutors or to courts. Non-court experts offer their expert opinions on private commission more often, but they also may give court testimony, depending on the judge's decision.

Experts are contracted mainly by police, prosecutors or courts. For the majority of court experts, working as an expert is an extra job, but for some it is their main source of income. Most experts work as independents, but some join together in engineers' offices (three to six persons). Expert reconstructionists often cooperate with experts in forensic medicine, if such is the decision of the organization commissioning the opinion. A significant number of expert opinions is produced by branches of the Association of Experts in Automobile Technology and Road Traffic or the Polish Automobile Federation, operating in district cities.

The payment for an expert opinion is granted by the commissioning institution (law court, prosecutor's office, the police) and is in the range of $10–$15 USD per hour. The costs of an expert analysis are in the range of $150–$500 USD. Presently, in criminal trial proceedings, agreements are possible without a trial, when the defendant submits voluntarily to the penalty. Also, in civil cases, agreement proceedings are possible between the litigants, which reduce the need for experts' opinions. The high costs of computer software, literature and other expenses do not make the experts' job highly profitable.

In this area, the Institute of Forensic Research works as the main source of technical information for experts. It organizes regular international conferences on subjects, and also works to publish books and monthly journals for experts. The Institute also cooperates in the postgraduate studies with the University of Technology in Kraków. This activity is supported by the earlier mentioned organizations and other universities (Warsaw, Lublin). Court experts can belong to the Polish Association of Courts Accidents Investigators. Many are members of EVU (European Association for Accident Research and Analysis). The Institute of Forensic Research and Central Forensic Laboratory of the Polish Police are also members of ENFSI (European Network of Forensic Science Institutes).

One of the popular techniques used in car reconstruction is simulation. There are two famous software programs in use: *PCCRASH* and domestic *V-SIM*. Reconstruction experts often cooperate with specialists of forensic medicine.

Some major institutions provide research and publish the results at principal conferences and journals (SAE, EVU, ITAI, Verkehsunfall and Fahrzeugtechnik). The future in car accident reconstruction sciences lies in solving the problem of 'clean accidents', i.e. accidents that do not leave traces on the road. Europe is still waiting to adopt the widespread use of EDRs (Event Data Recorders), as well as vehicular video cameras, which will be of help in further developing simulation techniques that will combine with the more precise methods of biomechanics.

Summary

An engineering forensic analysis is a multifaceted endeavor, requiring engineering knowledge and education as well as supplemental experience, training and skill in forensics. As seen here, the processes vary dramatically from country to country and specialty to specialty, sometimes even within the same country. Understanding these differences challenges our critical thinking as forensic engineering scientists and engineers. Forensic engineering scientists from all over the world share the common goal of uncompromised data collection and systematically considered, iteratively derived and objectively balanced conclusions. We offer unique insights into a safer world.

Acknowledgements

The following authors provided international perspectives: Adam Aleksander, PhD (Bangladesh, The Philippines), Jan Unarski, PhD (Poland), Sarah Hainsworth, PhD (United Kingdom) and Scott Grainger, PE (Fire Protection: United States).

References

1. Bejan A. *Advanced Engineering Thermodynamics*. Hoboken, NJ: John Wiley & Sons; 2006.
2. Committee on the Development of the Third Edition of the Reference Manual on Scientific Evidence. *Reference Manual on Scientific Evidence*. 3rd ed. Washington, DC: National Academy; 2011.
3. Liptai L, Michaels, A. Forensic Engineering Sciences: American Academy of Forensic Sciences Reference Series-A Decade of Research and Case Study Proceedings. AAFS Press/Amazon, 2013.
4. Lewis PR. *Disaster on the Dee: Robert Stephenson's Nemesis of 1847*. Stroud, UK: Tempus; 2007.
5. *Motor Vehicle Accidents, 2011*. http://www.census.gov/compendia/statab/2011/tables/11s1102.pdf (accessed October 14, 2011).
6. *FARS Data Tables*, 2011. http://www-fars.nhtsa.dot.gov/Main/index.aspx (accessed September 30, 2011).
7. Baker JS. *Traffic Accident Investigator's Manual for Police*. Evanston, IL: Traffic Institute of Northwestern University; 1957.
8. Renfroe DA. *The Role of Expert Witnesses in Accident Reconstruction Cases: Leading Experts on Utilizing Expert Witness Testimony, Understanding Technical and Scientific Evidence, and Preparing for Trial*. Boston, MA: Aspatore; 2009.

9. Collins JC. *Accident Reconstruction*. Springfield, IL: Thomas; 1979.
10. Martin D, Hurt H Jr. *Engineer Who Studied Motorcycle Accidents*, Dies at 81. New York Times 2009 Dec 4; B13.
11. Liptai L, Cecil J. *Forensic Engineering Application of the Scientific Method*. Orlando, FL: National Academy of Forensics Engineers; 2010.
12. Campbell JP. Speaker Recognition: A Tutorial. *Proceedings of the IEEE* 1997; 85(9):1437–1462.
13. Merriam TVN, Matthews RAJ. Neural computation in stylometry: An Application to the Works of Shakespeare and Marlowe. *Literary and Linguistic Computing* 1994; 9(1):1–6.
14. *United States v. Starzecpyzel*. Southern District New York (1995 Mar 31).
15. *United States v. Byron Mitchell*. US District Court for the Eastern District of Pennsylvania (1999 July 13).
16. Goodwin A. *Vehicle Automation Technologies of Today and the Future*. www.cnet.com; 2010 Mar 17. 2011 Sept 7. http://reviews.cnet.com/2300-10863_7-10002833.html (accessed September 20, 2011).
17. Joseph J. *Motorcycle Airbags. NTSB Motorcyle Safety Public Forum*. Washington, DC; 2006 Sept.
18. Science and Technology Committee. *Seventh Report, The Forensic Science Service*, 2011. http://www.publications.parliament.uk/pa/cm201012/cmselect/cmsctech/855/85502.htm (accessed October 14, 2011).
19. Silverman B. *Research and Development in Forensic Science: A Review*, 2011. http://www.homeoffice.gov.uk/publications/agencies-public-bodies/fsr/forensic-science-review/ (accessed October 20, 2011).
20. *MPs Criticize Government Plans for Forensive Science Service Closure*, 2011. http://www.parliament.uk/business/committees/committees-a-z/commons-select/science-and-technology-committee/news/110701-fss-report-published/ (accessed July 5, 2011).

12

Jurisprudence

ARW Forrest[1] and RT Kennedy[2]

[1]*Department of Chemistry, University of Sheffield, Sheffield, UK*
[2]*New Mexico Court of Appeals, Albuquerque, New Mexico, USA*

12.1 Introduction

> ***Jurisprudence:*** *1. a) Knowledge of or skill in law. b) The science which treats of human laws (written or unwritten) in general; the philosophy of law. 2. A system or body of law (1, p. 636).*

This chapter, concerning the 'science' underlying the Jurisprudence Section of the Academy of Forensic Sciences is, in important respects, no less than a discussion of what puts the 'forensic' in forensic science. In each of the other discipline-specific chapters in this book, the nature of applying a particular scientific or technical methodology to resolve some question of legal relevance is explained. A recent book on the forensic laboratory describes forensic science as 'the testing of evidence' [2]. The word 'evidence' holds the key, both to forensic science and the 'discipline' of jurisprudence, the latter requiring evidence to support the proof of legal conclusions.

Physicist David Goodstein points out in his chapter in the new Federal Courts' *Reference Manual of Scientific Evidence* that 'evidence' in the law is something subject to precise rules governing what may or may not be considered, or what constitutes proof of a fact necessary to a decision. To a scientist, 'evidence' is something that loosely supports a proposition but is something less than proof, as a scientific paper will talk about '[e]vidence for (or against) . . . ' (3, p. 51). Similarly, in jurisprudence, a 'law' is a product of some human authority, not a law of nature that requires discovery but which is immutable against human authority or intervention (although observations may disprove a scientific 'law' and suggest a better description of natural phenomena [3]).

Such disparate meanings for basic words between the legal and scientific worlds [3] are but a small part of the myriad of problems encountered when science goes to court. The jurisprudence of forensic science encompasses thousands of legal decisions, scholarly articles in law reviews, and books. It also includes 'skill in

Forensic Science: Current Issues, Future Directions, First Edition. Edited by Douglas H. Ubelaker.

law', including trial law. Lawyers are introduced to scientific and expert evidence in law school. From that introduction, they quickly learn that technical evidence and expert witnesses are an integral part of the resolution of legal cases.

The American Academy of Forensic Sciences states that its mission is to provide 'leadership to advance science and its application to the legal system' (4, p. 1). The adjective 'forensic' is simply defined as 'pertaining to or belonging to the courts' (5, p. 488). Without the needs of a legal system to understand evidence in its cases through ways that scientific technique or thinking can assist, there is no use for forensic science. Science is thus 'applied' to the legal system. Within the American Academy of Forensic Sciences, the forensic scientists can hone their abilities to identify and explain phenomena. However, it is the lawyers who provide the knowledge and information necessary to:

- use science within the legal system;
- put evidence before a trier of fact to be legal proof;
- employ the legal rules for the acceptance of expert opinion to explain the evidence; and
- argue for the admission of the evidence into the trial record.

The legal system, however, is a crucible of human problems in need of ordered solutions. A legal investigation, unlike a scientific investigation, is finite in time and scope and reflects as much truth as the limited amount of time is allowed to make a final decision. Dr. Goodstein points out that the scientific search for truth has 'no time limits, and no point at which a final decision must be made'. (3, p. 52)

As human knowledge progresses, that knowledge makes its inexorable way into the courts as both the root of, and solutions to, the disputes that are brought before them. The courts and legal practitioners are faced with much having to do with new ideas about 'science', and are manifestly in need of crafting adequate responses to scientific and technical problems that are both complicated and pervasive. One writer on the subject of personal injury law talks about a 'revolution of rising expectations' among lawyers that new scientific discoveries in medicine and genetics will expand the law's reach into new areas of liability (6. p. 7). Significant tension is resulting from the collision between a landslide of new scientific knowledge and a legal system that needs it but might not be prepared to fully understand it. US Supreme Court Associate Justice Stephen Breyer, in his Introduction to the new Third Edition of the Federal Courts' *Reference Manual on Scientific Evidence*[1] [7] summed up the aspiration of the law with regard to scientific evidence:

> *'In this age of science, science should expect to find a warm welcome, perhaps a permanent home, in our courtrooms. The reason is a simple one. The legal*

[1] The Federal Reference Manuals, First through Third Editions, are available to be downloaded for free from the Federal Judicial Center, www.fjc.gov.

> *disputes before us increasingly involve the principles and tools of science. Proper resolution of those disputes matters not just to the litigants, but also to the general public – those who live in our technologically complex society and whom the law must serve. Our decisions should reflect a proper scientific and technical understanding so that the law can respond to the needs of the public.'* (7, p. 2)

However, the other side is succinctly stated in the 2009 Report of the National Research Council of the National Academies of Science, *Strengthening Forensic Science: A Path Forward*[2], regarding the tension that develops when courts strive properly to resolve questions relating to scientific and technical matters:

> *'Lawyers and judges often have insufficient training and background in scientific methods, and they often fail to fully comprehend the approaches employed by different forensic science disciplines and the strengths and vulnerabilities of forensic science evidence offered during trials.'* (8, p. 238)

The roots of the tension are the divergent natures of the legal and scientific disciplines and their differing ways of investigating and of resolving investigations. With regard to the forensic sciences, their provenance and continued existence between law and science provides a virtual triumvirate of challenges.

Truly, forensic science belongs to the courts. The need of a legal advocate to supplement or explain first-hand testimony or the relation of primary evidence to the particular case regularly brings persons with expertise in various scientific and technical fields before legal tribunals of all sorts. The 'principles and tools of science' are brought to bear on very real legal problems that have profound effect on individuals and on our society at large. Traditionally, the 'forensic sciences' are employed to answer questions about evidence. Overwhelmingly, the clients of forensic scientists are law enforcement agencies, and most forensic scientists are employed by some government agency either involved with law enforcement or charged with a duty toward the criminal justice system (8, p. 86).

As scientific knowledge has exploded, its application to criminal justice has similarly exploded with the development of crime laboratories in the 1920s (9, p. 6). It is this role to which this chapter will pay primary attention – forensic science applied in the criminal law. Many of the principles and the issues discussed, however, are equally applicable to forensic science in civil matters.

A definition of jurisprudence evokes both the philosophy and 'science' of the law, as well as the institutions of the legal system itself, including its decisions. Jurisprudence also encompasses the craft of the lawyers themselves [10]. As a discipline, the law is not the sort of 'science' that allows for much comparison with the forensic science disciplines that make up the American Academy of Forensic Sciences. Lawyers are seldom scientists; indeed, the law's relationship with science

[2] Hereinafter referred to as the "NRC Report". This report is also frequently referred to in the literature as the "NAS Report". Both designations, NRC/NAS refer to the same publication.

as a way to explain phenomena that have legal significance has been fraught with much fear, loathing, pain and inefficiency. However, as the law has used scientific evidence, it has discovered and created problems with the use of forensic science – for instance, both analytical deficiencies in employing the scientific process and manipulation of scientific evidence by forensic scientists on the stand.

Much has been written by lawyers on this subject, Dagan and Kreiter's recent essay [10] being but one example. As a result, changes have been wrought in both the structure and the practice of the law to respond to the effects those problems might have on the process of doing justice. The jurisprudence regarding scientific evidence changes as the courts set and revise normative standards for the conduct of trials and the admission of evidence of a high quality. It requires certain conduct on the part of both the forensic scientists who testify and the legal proponents of their evidence.

To some extent, as with the *Melendez-Diaz* case [11], the courts apply existing statutory and constitutional rights, such as the right to examine the actual scientist who did the work whose result is being offered as evidence, to ensure fairness in the presentation of forensic evidence. In England and Wales, the Law Commission has recommended the introduction of a statutory 'Reliability Test' to be applied to the introduction of criminal evidence in criminal cases. This would allow the opinion evidence of an expert to be admissible only if it is sufficiently reliable to be admitted. The Law Commission has recommended this a part of statutory codification of the common law of criminal expert evidence [12].

In other cases, such as the *Daubert* case [13] mentioned below, courts employ procedural rules to ensure the quality of evidence that is proffered. Thus, the law alters the way in which forensic science is practiced in the laboratory and is presented from the witness box. The comprehensive effect of Jurisprudence – the discipline – on the forensic sciences is alpha and omega; forensic science grew out of an application of basic scientific knowledge to questions of criminal activity presented in court [14,15], and the law makes the rules by which that knowledge may, or may not, be received to promote the finding of truth in court to serve the goal of a just result.

This chapter, then, will have three parts. The first will introduce the reader to the Jurisprudence Section in the context of the Academy and its history. The second will briefly discuss the way in which scientific and technical expertise have been used by the law and admitted in court, including increasing realization by the courts of problems both with forensic science and with the way it has been used. The third will attempt to provide some view to the future, with an eye toward how the legal system (and political system) might promote better understanding of forensic science within the law. This will include the development of what legal requirements for standards of practice and proficiency for the forensic sciences may be needed to ensure the continued admissibility of forensic science evidence in court.

The authors specifically commend to the reader the sources of information noted in this chapter, but particularly the NRC Report *Strengthening Forensic Science in the United States* and the joint NRC/Federal Judicial Center *Reference Manual on Scientific Evidence*, Third Edition, which for the first time undertakes to inform the Federal Judiciary about forensic identification science.

12.2 The Jurisprudence Section of the Academy

As a section of the Academy, jurisprudence is both small and mighty; it could reasonably be described as the salt in the Academy's melting pot [16]. While itself not possessing the expertise collected in the other ten sections, the Jurisprudence Section represents the sum of what they have to offer, in both figurative and literal senses. Within the Academy, the Jurisprudence Section provides a human face for the conduit by which forensic science enters the legal system. Questions of admissibility of scientific and technical expertise in court rest solely in judicial hands and are controlled by the law. The section also has provided internal leadership and advice to the Academy in legal matters as they have arisen, and its members have frequently represented the Academy and its policies outside of the Academy in court and in public. Jurisprudence Section members also represent attorneys and judges to the Academy as the creators, consumers and judges of the forensic sciences.

12.2.1 History and leadership

The Academy's Jurisprudence Section, through its membership and representation, can demonstrate a pervasive presence and influence in many aspects that are central to the relationship between forensic science and the law in the United States. This has proven to be so, both within the Academy, as the section representing the law and lawyers to the other sections, and as member-lawyers take their knowledge and experience from the AAFS to the community at large and help to forge that relationship in the courts, in politics and in academia. In the last three decades, the forensic sciences have faced a sea-change in scientific practice driven by the introduction of DNA evidence and its underlying scientific methodology. Similarly, developments in the law and its view of science have had a profound impact on the use of science in court, and are continuing to do so today. Fellows and members of the Jurisprudence Section have been present at virtually every turn of this process.

The history of the American Academy of Forensic Sciences, from its origins at a medicolegal conference in the late 1940s to the international organization it is today, owes much to the influences of the lawyers who were present at its inception, and to the continuing resources and leadership provided by the Jurisprudence Section. The first Constitution for an 'Institute of Law-Science Relationships' was drafted by Orville Richardson, a St. Louis, Missouri Attorney and was presented at the First American Medicolegal Congress in 1948. Richardson subsequently served as secretary of the AAFS Jurisprudence Section and was a member throughout the 1960s [17]. At subsequent planning meetings, the nascent American Academy of Forensic Sciences proposed as its goals (17, p. 7):

1. Raise the standards of investigative techniques and the quality of testimony in court.

2. Engender the confidence and respect of the judiciary.
3. Promote the betterment of medicolegal testimony
4. Create and foster confidence by the courts in scientific-legal proof.
5. Raise the standards of reliability in investigations and testimony of men [sic] who do this work.
6. Exchange information and encourage research and dissemination of knowledge.
7. Encourage enlightened legislation.
8. See the ends of justice attained.
9. Improve the participation of scientists in the attainment of justice.

In the years that have followed, the Jurisprudence Section has been active in pursuing these goals. As the Academy adopted a definition of forensic science as 'the study and application of all the sciences to the purposes of the law', as proposed by then-president and jurisprudence fellow Oliver Schroeder (1963–4), the Academy was to become a body promoting a broad and inclusive view of scientific expertise that would serve the legal system. As the nature of jurisprudence concerning the presentation of expert and scientific evidence in court has changed, the section has been a resource to provide a legal perspective to forensic practitioners to assist in their understanding of the law and its processes.

Controversy has swirled around the forensic sciences and the Academy itself in the past 60-plus years. As the 1990s dawned, the perceived scientific certainty underlying DNA identifications was held up as a yardstick for forensic practice in supporting convictions, as well as exonerating the innocent both prior to trial and, unfortunately, after trial, by exposing many failures of forensic evidence and expert malfeasance. The United States Supreme Court issued its *Daubert* decision, purporting to make judges 'gatekeepers' of scientific evidence through viewing scientific evidence through the employment of the scientific method itself [13]. As scrutiny of the use of science in court increased, the assumptions underlying many of the forensic disciplines themselves were called into question (18, p. 61).

The Academy's Jurisprudence Section has provided assistance, explanation, education and evidence to its fellow Academy members at every turn, as well as providing resources both to the Academy leadership and to the public to aid understanding of the issues and arguments in those continuing debates. Jurisprudence fellows both presented evidence to the committee and served as reviewers to the 2009 National Academies of Science report *Strengthening Forensic Science In the United States: A Path Forward* [8], and fellows also serve on the Presidential Subcommittee on Forensic Science that was formed as a result of the report. Two fellows, both law professors and authors of a pre-eminent textbook on scientific evidence [19], contributed the chapter on 'Forensic Identification Evidence' to the Reference Manual on Scientific Evidence for the US Federal courts (18, p. 56–127).

As the United States Congress has undertaken to legislate with regard to the forensic sciences, Jurisprudence fellows and members have also provided assistance by attending meetings, giving testimony concerning the use of forensic science in court, and the needs of the legal community with regard to ensuring validity and reliability of the forensic sciences for adjudicative purposes. Jurisprudence Section members are directly involved with educating the legal profession and other public groups about the forensic sciences. Many Jurisprudence members serve as members of forensic advisory boards in many states, or consult with legislative bodies and the courts on aspects of the interplay of science and the law. Many are instructors to legal groups; some teach at the National Judicial College concerning scientific and legal issues.

As experts have come to testify more often, and their contact with the legal system increases, the adversary system of justice has produced much friction between forensic practitioners and attorneys. The Jurisprudence Section is very concerned with minimizing such friction within the Academy by assisting forensic practitioners to understand better the law governing expert testimony and scientific evidence. Within the Academy, Jurisprudence Section members continually present information about the law and trial practice in educational programs, such as deposition and case preparation workshops, within the Academy to assist scientists with the process of preparing for, presenting evidence in and surviving the court experience. Beyond that, Jurisprudence Section members are frequently sought out to participate in other sections' scientific sessions at the AAFS annual meetings to present a legal perspective in most, if not all, of the other ten sections in the Academy of Forensic Sciences. Other sections may seek to be involved in presentations and workshops organized by the Jurisprudence Section.

The legal and scientific arenas which underlie 'forensic science' have different characteristics, governing principles and interests, which often results in friction between the law and science. The reliance on precedent – the doctrine of '*stare decisis*' – is central to Anglo-American legal practice, but has resulted in a far less fluid approach to new and developing information than science itself.

As long ago as 1934, Professor Robinson wrote an article in the *Yale Law Journal*, urging lawyers to be more 'scientific' and noting a drag that the law placed upon 'the adventurous (scientific) spirit of the times' [20]. More recently, in his address to the Conference on Forensic Science for the 21st Century, US Circuit Judge Harry T. Edwards, a co-chair of the NRC committee that issued *Strengthening Forensic Science*, pointed to the appellate opinion in *U.S. v. Havvard*, 260 F. 3d 597 (7th Cir., 2001), in which an appellate court reported that an FBI fingerprint examiner had reported that the 'error rate for fingerprint comparison is essentially zero'. Two years later, another appellate court cited that opinion approvingly noting that 'essentially zero' error rate, *U.S. v. Crisp*, 324 F.3d 261, 269 (4th Cir. 2003). Thus, judges, in reviewing trials, create evidence by precedent and perpetuate other courts' foibles.

Only in more recent years, in the US, has the idea that any expertise brought to bear on a legal matter carry with it an imprimatur of 'general acceptance' in a relevant scientific field – the *Frye* standard – given way to an attempt to view expertise based in 'science' according to the precepts of the scientific method. This latter doctrine, announced in 1993's *Daubert* decision (13, p. 579), has since been applied in the US federal courts to non-scientific expertise, or expertise based in other more artful or experiential disciplines [21]. Further development of the jurisprudence of science has come to emphasize the utilization of objective standards of practice in generating expert opinions and findings that reflect the standards employed by scientific disciplines themselves [22]. Outside the courtroom, however, legal practitioners continually seek novel ways of explaining the circumstances of the case at hand using expertise of a scientific, and sometimes not-so-scientific, nature.

Within the Academy, the Jurisprudence Section serves two communities:

- First, the community of lawyers and judges, through its intimate contact with the other sections and their expertise at meetings, helps structure legal knowledge of the inquiries necessary for making rulings on objections to expert evidence where it might be challenged (23, p. 43).
- Second, the section assists the forensic science community of the Academy to understand the necessity of presenting qualified, methodologically sound and reliable evidence as is understood by the consumers of its services – lawyers and the courts.

In addition to being the practitioners of the law, and the end-users of forensic science information, the profession of law deeply informs the structure of the AAFS itself. Lawyers practice under ethical constraints that compel faithfulness to the fairness of the tribunal, avoidance of conflicts of interest and the exercise of independent professional judgment. Transgressions of legal ethical norms can result in the loss of professional licensure.

Jurisprudence Section members significantly contributed similar concepts to their role in drafting the Academy's Ethical Standards, which, among forensic organizations, stand at the forefront for their comprehensive scope, gravity and thoroughness of process. As more attention is given to strengthen forensic science in the United States, Jurisprudence Section members advise forensic research groups on the subject of professional ethics in the forensic sciences. For example, Jurisprudence Section member Christine Funk serves on the Education, Ethics and Terminology Working Group of the Presidential Subcommittee on Forensic Science of the National Science and Technology Council (NSTC) [24]. Partly as a result, the AAFS enjoys high professional regard as an organization committed to enforcing high ethical and professional standards.

The Jurisprudence Section is about 170 members strong, but this size belies the influence the section has had both in the Academy and in the general field of forensic science. Nine past presidents of the AAFS have come from the Jurisprudence Section (and another two or three presidents from other sections have also had degrees in law). Five distinguished fellows of the Academy come from Jurisprudence, as does the AAFS's 2012 Gradwohl Laureate, Professor James Starrs.

Section fellows include well-known professors and authors in forensics, medicine, evidence law, and criminal law. Prominent prosecuting attorneys mingle with past presidents of the National Association of Criminal Defense Lawyers (US), who in turn share their interest in the jurisprudence of science with equally prominent civil law attorneys. Both trial and appellate judges are fellows within the section. Jurisprudence Section members have been influential in helping to develop other sections within the Academy, such as Odontology and Engineering.

Given that the AAFS was originally conceived to include practitioners of the forensic sciences throughout North America, and has grown to include members from an additional 62 countries besides, it is harder to talk about a single jurisprudence of forensic science evidence. In a major respect, the jurisprudence – the development of the common law by judicial decisions – concerning the admission in court of scientific and expert testimony has shaped the conversation concerning what constitutes the various sorts of expertise that come to court labeled as 'forensic science'.

The inclusion of members from legal systems that depart from the Anglo-American adversarial model, who are used to a more inquisitorial system such as is more common in civil law-based jurisdictions on the European continent and elsewhere, is rapidly expanding the view of how best to handle the use of scientific and technical expertise in the courts. As Kessler puts it: '[t]*he adversarial and inquisitorial models are distinguished primarily by whether the parties or the court control three key aspects of the litigation: initiating the action, gathering the evidence, and determining the sequence and nature of the proceedings.*' [25]

The common denominators, however, are the use of expertise in the service of justice and the need of courts to have access to the most reliable, up-to-date and objective expertise to assist in making critical decisions. The debate over what the parameters of 'good' forensic science might be is central to the legal use of this evidence. As the representative in the AAFS of the legal system itself, the Jurisprudence Section is critical to the mission of the Academy.

The section's involvement in being the go-between between lawyers and courts, on one hand, and forensic practitioners on the other, make it a resource to both sides concerning the standards and methods by which expert testimony is admitted by a court for consideration by the fact-finder in a case. The section is in a unique position to recognize and teach that the divide between the various methodologies and techniques of science with regard to questions such as validation, standards for both scientific and personal practice, and the ultimate acceptance of their results for use in the courtroom, are all considerations that are profoundly rooted in the legal process represented by the Jurisprudence Section.

As the law's view of the forensic sciences goes forward, the Jurisprudence Section has been required to deal with a number of issues which will be discussed below. As authors, we were directed to discuss 'key scientific issues' in our discipline. As the lawyers of the Academy, our discussion is about how the law and the courts are to deal with the myriad of issues that arise in *all* forensic science disciplines.

All controversies regarding forensic science relate to the law. Similarly, the nature of the legal response to problems that arise consists of trying to find ever better ways of understanding science, of ensuring the quality of forensic science evidence and of promoting the quality of legal practice when dealing with this evidence, so that justice may ultimately be served. Our discussion follows.

12.3 Forensic science in court: current issues, current problems

12.3.1 History, epistemology, conflict

There is a joke that goes as follows: a surgeon, a physicist and a lawyer are arguing over what is the oldest profession. The surgeon says that surgery is the oldest profession because God took a rib from Adam to fashion Eve. The physicist then claims that physics is the oldest profession, because it created order from the primordial chaos before Adam and Eve. The lawyer shrugs and simply asks, 'Who do you think created all the chaos?' [26]

Commonly, legal scholars looking at the issues and problems concerning the use of various forms of expertise in court speak of a long-standing uneasiness between scientific experts and lawyers [27,28]. One distinguished writer has described it as 'a kind of shotgun marriage between the two disciplines [in which] both are obliged to some extent to yield to the central imperatives of the other's way of doing business, and it is likely that neither will be shown in its best light.' (7, p. 52–53)

Courts in the Middle Ages used tradesmen as experts, empanelled as an expert jury, to help judge matters within their specialized knowledge (27, p. 42). This kept the tradition of commonly accepted knowledge within a trade or profession as a hallmark of acceptability of the evidence its members gave. Later, experts would be called to assist when a court was unable to reach a conclusion, as in a mayhem case where the opinion of surgeons was sought [29]. As scientific and technical knowledge expanded, lawyers found new ways of presenting evidence of this knowledge to explain phenomena relevant to matters before the courts. Then, as now, the forensic expert would be called to be a witness to provide an explanation of, or perspective to, physical or percipient evidence that had been provided by primary witnesses (it is a rare thing that the subject matter of any given lawsuit occurs within the ability of the expert to observe the occurrence firsthand).

Rather than dwell on controversies between disciplines supporting the evidence given in court, the role of the expert in court is to give testimony concerning various methodologies that are beyond the knowledge or understanding of the normal fact-finder (judge, jury or administrative officer), and evidence is presented by persons recognized by rules of the tribunal to be 'experts' in the field [30]. The expert's role in court is primarily to interpret and give an opinion about the significance or meaning of evidence in the context of the case. In British and American courts, in criminal matters, a person can be called as an expert by one side of the case or another but, in European courts, the experts are frequently employed by the court itself.

Because of the expert's superior knowledge, the tribunal, whether in an adversarial or inquisitorial system, must have reason to trust the reliability of the expert's application of his knowledge to the case. Since the knowledge the expert possesses is, by definition, beyond that of the trier of fact, and frequently beyond that of the judge or lawyers, it is inevitable that mischief has occurred.

In 1885, an editorial in *Nature* reflected:

> *'A well-known lawyer, now a judge, once grouped witnesses into three classes; simple liars, damned liars, and experts. He did not mean that the expert uttered things which he knew to be untrue, but that by the emphasis which he laid on certain statements, and by what has been defined as a highly cultivated faculty of evasion, the effect was actually worse than if he had.'* [31]

Throughout history, trust in an expert's integrity and objectivity has been slow to build in the legal community and the society at large. 'Because an expert can be found to support almost any position, most commentators agree on the need for judicial review and control, but there is no consensus on how to achieve these objectives' [32]. This problem has two sides, however. The first is what responsibility an adversary legal system bears that encourages the calling the expert witness as a part of a partisan enterprise, while the second is the failure of the expert to fully and fairly give an exposition of all of the relevant considerations to his or her testimony. We examine each in turn, as each has generated various legal problems and responses.

12.3.2 'Science and forensic science': tension within 'science'

The field of forensic science disciplines is divided into two large groups. One applies the methodology and techniques of traditional academic scientific disciplines (e.g. biology, medicine, chemistry) to solve questions that arise in the criminal law. These disciplines are regarded as employing established research, experimentation and validation of their techniques using the scientific method.

The second group represents the 'forensic' disciplines that have grown more or less directly from what has been the *ad hoc* need of the law to prove the connection between evidence and/or between persons by interpreting observed patterns for use in criminal prosecutions [14]. These disciplines are variously viewed as dealing with forensic identification or individuation – matching an unknown person or thing to an item of evidence in the case.

Although having a demonstrable relationship with traditional academic science, these disciplines frequently are more training- and experience-based, with roots in law enforcement and criminology. The sum of forensic science reflects such variation in 'scientific underpinning, level of development, and ability to provide evidence to address the major types of questions raised in criminal prosecutions and civil litigation' (8, p. 7) that further distinction would not be productive. Our discussion, therefore, will be concerned with the general reaction of the law to its growing awareness of the scientific process, not intramural arguments between scientists.

A recent editorial in *Nature* entitled 'Science In Court' stated, 'Forensics has developed largely in isolation from academic science, and has been shaped more by the practical needs of the criminal-justice system than by the canons of peer-reviewed research.' [33] Furthermore, standards for practice in the discipline have been mostly fragmented between agencies, jurisdictions and laboratories. Additionally, the adversary nature of the forensic sciences, and their roots in police laboratories, has contributed to a culture that lacks transparency for evaluating practices and methods, both between the forensic sciences and other 'academic' science [33]. This should not be surprising:

> '[F]*orensic scientists' major purpose is to provide a service to a client by answering specific questions about evidence. The primary clients of the vast majority of forensic scientists are law enforcement agencies. Most forensic scientists are employed directly by law enforcement agencies. Their role in litigation is typically, and often exclusively, to provide evidence in support of criminal prosecutions.'* (15, p. 113–114).

Some writers have commented on a perceived divide between 'forensic science' and 'real' science [34]. A purist may hold that 'science is prediction before, not explanation afterwards.' [35]. Forensic science, by its very nature, is 'explanation afterwards'. It is not surprising that the term 'police science' is included in an internet listing of oxymorons [36]. The NRC Report itself commented: 'The law's greatest dilemma in its heavy reliance on forensic evidence . . . concerns the question of whether – and to what extent – there is science in any given forensic science discipline' (8, p. 9).

This tension, and its interjection into the use of forensic science evidence in court, has its roots in the way the law understands, or fails to understand, science. It is a product of the greater scientific and technical capabilities being made available to the law enforcement community, the increased scrutiny of the law being applied to all things 'scientific', as described below, and 'the revolution of rising expectations' that such advances can be suitably applied.

For lawyers and the courts, the question is not what is 'real science', but whether:

1. any forensic discipline producing a result that is offered as evidence in court is based on reliable scientific methodology that is capable of accurately analyzing evidence and reporting findings; and

2. the extent to which practitioners in a particular forensic discipline rely on human interpretation that could be tainted by error, the threat of bias or the absence of sound operational procedures and robust performance standards (8, p. 9).

As the capability to produce a result may outstrip the ability to account reliably for the process that produced it, a tension of a different sort arises – where the forensic technique is inevitably compared with more traditional science. Lawyers are instructed by the courts to regard evidence relating to science through a scientific lens. The US Supreme Court's *Daubert* decision told us that evidentiary reliability will depend on scientific validity and that we are to focus 'solely' on the expert's principles and methodology, not on their conclusions (12, p. 590–591, 595). Lawyers, being sensitive to any controversy that affects the ability of an opponent's evidence to be open to question, predictably have seized on the debate between the 'academic' and 'forensic' sciences as a rhetorical tool.

12.3.3 'Legal science VS. scientific science': the law muddles along

Almost universal agreement exists on the existence of 'fundamental differences between the legal and scientific processes' [28]. We recommend Dr. Goodstein's chapter in the *Reference Manual* as an example of a scientist explaining, comparing and contrasting the science of law with the science of science (7, p. 38–55). Justice Blackmun, in his majority opinion in *Daubert v. Merrill-Dow*, intoned:

> '[T]*here are important differences between the quest for truth in the courtroom and the quest for truth in the laboratory. Scientific conclusions are subject to perpetual revision. Law, on the other hand must resolve disputes finally and quickly.'* [13].

Yet the NRC report, *Strengthening Forensic Science*, points out that the adversarial process relating to admitting and excluding scientific evidence is not 'suited to the task of finding "scientific truth"' (8, p. 12). It also states bluntly:

> *'Lawyers and judges often have insufficient training and background in scientific methods, and they often fail to fully comprehend the approaches employed by different forensic science disciplines and the strengths and vulnerabilities of forensic science evidence offered during trials.'* (8, p. 238)

Since before Galileo went on trial in 1633 – in a legal proceeding based on his knowledge of science – jurisprudence and science have struggled with their ways of finding truth. How did the law proceed from Galileo to *Daubert*? The need of the law to reach the best answer to a case in a finite amount of time has led to many legal devices for using persons possessing expertise to enhance the fact-finding process. However, as a popular legal textbook informs law students, 'Data disclosed in the laboratory or testing facility has no real meaning in law until presented to the trier of fact.' (37, p. xii) A brief look at how the law has come to analyze forensic science in the context of its general approach to science is interesting.

The use of experts by the law has moved from a consensus of tradesmen in the Middle Ages, judging matters unique to their trades, to an *ad hoc* approach where experts were only allowed to apply their methods to the exact facts of a case by way of answering hypothetical questions and being forbidden to opine on any ultimate issue in the case (an 'ultimate issue' might be, 'In my expert opinion, the matching tire prints between the defendant's car and the victim's body indicate the defendant's presence at the crime scene.').

In the 1920s, a test for the admissibility of the result of a novel scientific technique based on its 'general acceptance' in a 'relevant scientific community' developed. By the 1980s, that test finally gained its own acceptance by courts. Throughout, lawyers have tried to develop better ways of ensuring reliable testimony based on valid use of expertise. As scientific and technical knowledge expanded exponentially, new ways of assessing opinions from science required new tools. Much of the impetus for change in the law came from the explosion of cases seeking to impose liability for injuries whose causation could only be explained through the use of advanced expertise in the sciences. Toxicology, epidemiology, engineering, chemistry: the explosion of litigation begat an explosion of experts appearing in court.

The traditional suspicion of experts' motivations and biases in testifying from an adversary position in a case exerted pressure on both the public and the legal community to take a close look at expert witnesses. One immensely popular response was Peter W. Huber's book, *Galileo's Revenge* [38], which coined the phrase 'junk science', excoriating the effect of expert shopping on the quality of science used in court and asserting that: '[t]he legal establishment has adjusted rules of evidence . . . so that almost any self-styled scientist, no matter how strange or iconoclastic his views, will be welcome to testify in court' (38, p. 3).

Pointing to the progression from court-appointed experts (the tradesmen) to experts hired by the parties, Huber pointed out the moderating influence of *Frye's* 'general acceptance' test for scientific admissibility on the law, and its demise to the enacting of the Federal Rule of Evidence that allowed an expert to testify on no greater basis than whether his or her 'scientific, technical or other specialized knowledge will assist the trier of fact to understand the evidence or to determine a fact' (38, p. 15). Huber's book was widely read, and has been cited by many federal and state courts in the US. One observer wrote:

> *'Expert scientific information is relevant to, even decisively important in, a rapidly growing percentage of decisions throughout civil and criminal law. Most judges and juries, however, are not sufficiently familiar with relevant scientific fields to be able independently and reliably to bring scientific information to bear on their decisions. Instead, they must solicit and defer to the judgments of expert scientific witnesses.'* [39]

Academic calls for a cohesive legal framework for the admission of expert testimony about scientific and technical matters, such as Bert Black's in 1988, began to concentrate on the relevancy of expert testimony to the case, as distinguished by the validity of the reasoning leading to an expert conclusion and the reliability of the conclusion (32, p. 599). Most of this discussion concerned large-scale civil litigation; very little made its way to criminal cases.

Then, in 1993, a case involving the epidemiology and toxicology of the drug Bendectin® was decided by the United States Supreme Court: *Daubert v. Merrill-Dow Pharmaceuticals*. In *Daubert*, the Supreme Court decided, first, that *ad hoc* tests for admissibility of scientific evidence had fallen to the enactment of the Federal Rules Of Evidence (particularly Rule 702 that governs expert opinions) and, second, that where those opinions concerned scientific matters, judges should act as gatekeepers and assess the reliability of the testimony with reference to science itself. The reaction to this case was significant. The Federal Judiciary responded with its first *Reference Manual*, which was directed mostly to complex civil litigation.

The AAFS convened, and has continued to convene, hundreds of workshops and presentations on the application of *Daubert* to the forensic sciences. Law reviews discussed the implication of *Daubert* in the courts in thousands of articles. Legal textbooks on evidence law and scientific evidence were re-written to reflect the new understanding of how science would be presented in courts.

Although *Daubert* directed judges to scrutinize scientific evidence through its reliable employment of scientific methods, it was not routinely employed to judge the quality of forensic science evidence, nor can it be shown to have contributed much to improving the use of forensic science evidence in criminal cases (12, p. 106). Part of this has to do with the relative resources available to civil litigants and criminal defendants to challenge forensic science evidence (8, p. 107). Another has to do with many courts simply applying former methods of admissibility analysis, mainly general acceptance (8, p. 107–109).

In both cases, the promise that *Daubert* might extend its protection to the innocent person accused of a crime, in the form of insisting on valid and reliable scientific evidence supporting any conviction, has been largely chimeral (40, p. 896). The fact that unreliable forensic science has continued to be routinely admitted, while reliable expert evidence offered by the defense has been routinely excluded [41], might not have been demonstrated, and the forensic sciences not critically examined, save for a single type of scientific evidence that came to prominence in the early 1990s – the use of DNA analysis. Cases using DNA analysis have presented a model of scientific reliability that has led to almost universal acceptance by courts; it has held both the forensic sciences and justice system up for scrutiny and has provided a catalyst for future developments in both science and law, designed to improve both.

12.3.4 DNA and the US courts

DNA profiling, touted in the 1980s as the ultimate forensic identification tool, came to court having been extensively researched, tested and validated by scores of scientists as to its fundamental proposition:

> '[T]*his evidence is so powerful because all persons other than identical twins have a unique genetic pattern and the enormous variability of DNA makes it possible to distinguish among individuals. It is the nature of nuclear DNA that gives DNA evidence its enormous probative value.*' (41, p. 1129)

Berger goes on to point out that:

> *'Although courts are now so convinced that DNA evidence is admissible that they generally treat all challenges as going to the weight of the evidence, there may, nevertheless, be instances when DNA evidence should be found inadmissible or insufficient to convict, as when fraud has infected a particular laboratory.'* [41].

The amount of research and testing proving the underlying truth of DNA profiling had developed amid a culture of scientific procedure, quality assurance protocols and proficiency testing that accompanied the evidence and the witnesses who testified about it into the courtroom. Early cases, such as *People v. Castro* [42], which excluded DNA evidence owing to concerns related to failures within the testing laboratory to utilize generally acceptable methods to obtain reliable results after a 12-week hearing (8, p. 99), were followed by *United States v. Yee* [43]. *Yee* involved a six-week hearing, where eminent scientists were called to provide expertise to ensure that the DNA science would be valid and reliable [8].

Interestingly, the experience of one expert in *Castro* was such that he declined ever again to testify in court. However, he agreed to assist Magistrate James Carr as the court's expert to help the court understand the scientific parameters of the evidence in *Yee*.

In 1992 and 1996, the US National Academy of Sciences issued two recommendations on the use of DNA in forensic science [8]. The end result of this process was an unprecedented amount of attention and expertise being brought to bear on a forensic science to clarify the nature of matching profiles, the methods of analysis and the requirements for quality assurance and laboratory performance protocols [8]. Matching a suspect to the crime, if DNA said it was so, became almost universally admitted by the courts.

Because of this exceptional scientific foundation for the admission of evidence, DNA scientific evidence has created a 'gold standard', whose characteristics have been held up to other forensic sciences. With DNA, matching was accomplished by employing population genetics – also validated by much research – to establish the likelihood of a random match (18, p. 60). This was a concept not generally employed by other forensic sciences. The issue of blind proficiency testing began to bleed into questions posed to other forensic science disciplines, as did the broad regulation of DNA science by the scientific community, in marked contrast to the lack of universal standards for analysis employed by other forensic disciplines (18, p. 81).

The format in which DNA evidence is presented to the court in England and Wales has been codified to avoid the 'prosecutor's fallacy' following a number of Appeal Court cases [44,45].

The other side of this certainty, however, is what has led to the present groundswell of minute scrutiny of forensic science and calls for forensic science reform in the United States and other countries. If we are to assume that the forensic sciences have at their root the goal of helping to convict the guilty, and the justice system exists to protect the innocent, then the forensic sciences also exist to exonerate the innocent when possible. It is in this role – exonerating the innocent –

in which DNA has played a perhaps its most important role with regard to the jurisprudence of forensic science. In fact, it has sparked a 'reappraisal of the trustworthiness of the testimony of forensic identification experts' [18].

By 1987, criminal defendants were requesting DNA testing to exonerate themselves at trial, while others were seeking post-conviction relief based on the new science [46]. As of this writing, more than 250 persons convicted of crimes have been exonerated and released.

A 2008 article analyzing post-conviction exonerations reports that in 57% of the DNA exonerations, faulty forensic science played a significant role [47], including false hair or fiber comparisons (48, p. 73), misleading testimony, fabricated test results and faulty analysis [49]. The NRC Report *Strengthening Forensic Science* similarly concluded that there is no doubt that faulty forensic science has contributed to wrongful convictions (8, p. 42).

The parade of DNA exonerations that began in 1989 thus began to focus attention on the other forensic sciences and their ability to deliver the expert conclusions of near-certainty that had been accepted for years by the criminal justice system and the courts. Two other consequences quickly developed: a revelation that forensic science and forensic laboratories did not always stand on the firmest of theoretical or methodological foundations, and the unmasking of a number of cases of outright forensic fraud. *Strengthening Forensic Science* states:

> *'The insistence of some forensic practitioners that their disciplines employ methodologies that have perfect accuracy and produce no errors has hampered efforts to evaluate the usefulness of the forensic science disciplines.'* (8, p. 47)

The questioning of forensic science disciplines was compounded by closer scrutiny of forensic practitioners; some were revealed to have testified not only in a misleading fashion, but had falsified evidence (8, p. 45). The integrity of crime laboratories themselves has been called into question, and investigations of the laboratories have shown lax standards, inadequate quality control and breakdowns in handling evidence, documenting tests conducted, as well as failure to use correct calculations for computing statistical frequencies (8, p. 44–45). The NRC Report commented, 'although cases of fraud appear to be rare, perhaps of more concern is the lack of good data on the accuracy of the analyses conducted in forensic science disciplines and the significant potential for bias that is present in some cases' (8, p. 45).

The three factors that a prominent author termed 'DNA, *Daubert*, and the lab scandals' [50] have spawned a resolve to reform the use of forensic sciences in the courts. The NRC explained the problem, and it does not solely rest on the forensic scientists:

> *'The bottom line is simple: in a number of forensic disciplines, forensic science professionals have yet to establish either the validity of their approach or the accuracy of their conclusions, and the courts have been utterly ineffective in addressing this problem.'* (8, p. 53)

If the justice system itself is unable to discern problems and avoid injustice in light of what it believes are rigorous evidentiary standards, the legal system itself must be examined for how it, too, might improve.

With the explosion of DNA profiling, increased attention to the validity of forensic science increased. In 2004, while promoting an expansion of DNA profiling, the US Congress enacted the Justice For All Act [51], also setting in motion a required mechanism by which allegations of negligence, misconduct, false testimony or inadequate forensic analysis on the part of agencies receiving federal funds for forensic science pursuits would be investigated. The American Bar Association made recommendations for improving crime laboratories and legal expertise with regard to handling forensic evidence in 2004 [52], and it has adopted standards for use of DNA evidence [53]. In 2005, the US Congress, specifically noting that much information existed 'concerning the requirements in the discipline of DNA', declared that there was a need for further study of the larger forensic science community and commissioned the National Academy of Science to conduct a study (8, p. 1). That initiative resulted in the NAS Report *Strengthening Forensic Science*.

12.3.5 The jurisprudential problem: law must come to grips with science and itself

As DNA has helped to show past problems with the use of forensic science at trial, a universal observation concerning the use of forensic science in the courts is that legal education and the practice of law are scant preparation for lawyers and judges to assess the quality of scientific and expert evidence fairly. Some of this is undoubtedly accounted for by the adversarial imperative of trial, contributing to both attorneys and witnesses neglecting a full and fair exposition of the evidence as one side of a case might present it. As the courts might set standards for judges' *Daubert*-ordered gatekeeping that require the trial judge to evaluate the scientific validity and reliability of scientific testimony to the task presented by the case, it remains a fact that *Daubert* is not an instruction book, and that judges are ill-equipped to do so. The NAS report comments, early on, that post-trial exonerations show the 'danger of giving undue weight to evidence and testimony derived from imperfect testing and analysis' or 'imprecise or exaggerated expert testimony' (8, p. 4).

12.4 Likely directions to improve the jurisprudence of science

The shame of the failure of forensic science, and its use by lawyers, to advance justice is made manifest by the exoneration cases. The indictment of current scientific and judicial practice is particularly acute in those cases where faulty, misleading or outright fraudulent scientific evidence was uncovered as a contributing factor to a wrongful conviction. Disagreements exist on the extent of injustice owing to faulty forensic testimony (8, p. 42), but no one denies that injustices have occurred. The most important questions for jurisprudence in coming years will be 'in which direction we want to skew the risk of error?' (54, p. 895), and in what substantive reforms can the legal profession engage on its own account? The latter subject will involve not only reforms for the use of forensic science in court, but the initiation of legal changes to the structure and practice of forensic science itself.

In the US, the accused is presumed to be innocent until proven otherwise by evidence establishing his or her guilt beyond a reasonable doubt, and it is the prosecution's (government's) duty both to collect and to present that evidence. As a result, that high burden of proof also assigns the risk of erroneous evidence be borne by the government – if the evidence is faulty, the accused is freed. Constitutional, statutory and procedural protections exist to protect the innocent. These include the right to present a defense, confronting and cross-examining witnesses, compulsory process to secure witnesses for trial, rules of evidence and case law such as Daubert, designed to facilitate the rules' use, procedures for practice by the state from investigation through appeal.

Against these safeguards exists institutional bias ('tunnel vision' [54, p. 898–899] – resistance to change), the competitive pressure of 'fighting crime' and systemic choices in everything from police procedure through judicial rulings that belie those protections. Reforms have been suggested, some of the most salient of which will be discussed below.

In an inquisitorial system, where an investigating judicial officer directs the police investigation and collection of evidence, rather than a prosecuting advocate, some biases might be mitigated; this is an idea foreign to those of us steeped in Anglo-American adversary jurisprudence. In the third series of the long running French TV police series 'Engrenages' ('Spiral') a *Juge d'Instruction* makes some interesting comments about the advantages of the inquisitorial system over the adversarial system, saying that in the inquisitorial system all are equal before the law and that justice cannot be bought, as sometimes appears to be the case in an adversarial system [55].

Either way, there is overwhelming evidence that the forensic sciences are in need of closer scrutiny as to their underlying validity and practices, the procedures by which their work is done and the qualification and conclusions of its practitioners. The NRC report *Strengthening Forensic Sciences* was a confirmation of the problem, but also a prescription for sweeping reform.

The NRC report made a number of recommendations for reform of the forensic sciences and the jurisprudence with which the courts work with forensic evidence. These recommendations were endorsed in whole by the Academy [56]. The Academy committed itself to 'particularly emphasize, endorse and promote the following principles:

1. All forensic science disciplines must have a strong scientific foundation.
2. All forensic science laboratories should be accredited.

3. All forensic scientists should be certified.
4. Forensic science terminology should be standardized.
5. Forensic scientists should be assiduously held to Codes of Ethics.
6. Existing forensic science professional entities should participate in governmental oversight of the field.
7. Attorneys and judges who work with forensic scientists and forensic science evidence should have a strong awareness and knowledge of the scientific method and forensic science disciplines' (56, p. 2).

To further these goals, the Academy proposed taking the following courses of action:

1. To form a committee at the request of the Board of Directors to review the scientific basis of a forensic science technique in collaboration with relevant sections, professional organizations, certifying bodies and other interested persons to assess the question, and make a report to the board.
2. Assessing whether the Academy might contribute to the standardization of terminology in the forensic science disciplines and in making definitions of the terminology readily available.
3. Assist in improving the knowledge of attorneys and judges with respect to forensic science and scientific evidence by developing a curriculum covering all forensic science disciplines, with entries dependent upon proficiency, through the cooperation of various legal and forensic professional organizations (56, p. 2).

These goals reflect an observer's comment that:

> *'It would also undoubtedly be helpful to recommend methods for efficiently educating willing scientists in the ways of the courts, just as it would be helpful to develop training that might better equip judges to understand the ways of science and the ethical, as well as practical and legal aspects of scientific testimony.'* [57]

The NRC report suggested reform in three primary areas: improving education, methods, practice and performance in the forensic sciences; strengthening independent oversight of forensic science practice, including certification of practitioners; and promoting education among lawyers and judges. All of these goals are implicated by jurisprudence, as the legislating process of making new law and regulation is as much a part of the law as is the application of law to evidence.

12.4.1 Forensic science goals

A prominent author on the subject of scientific evidence (58, p. 148) reminds us that the two important questions to be answered when considering the admission of forensic science evidence are:

1. is the technique scientific?
2. do practitioners of the technique avoid interpretations that become tainted by error, bias, or the lack of proper procedures?

Unfortunately, forensic science stands indicted of 'lacking methodological rigor, openness and cautious interpretation of data' and, in some of the disciplines, a 'significant potential for bias' (8, p. 45). Bias is not limited to the science and scientist: in the legal system, unreliable evidence of this sort is routinely admitted when offered by the prosecution, while some types of reliable defense evidence are routinely excluded [58].

The implications of the NRC report for the forensic sciences are to promote collaboration between forensic and academic sciences to attend to the validation of techniques employed by those disciplines that have not done so. This process requires a degree of transparency and exposure by forensic science of some matters that it has previously considered closely guarded. *Strengthening Forensic Science* rejects self-evaluation and development of standards within the forensic community. Judges and lawyers, thanks to documents such as the *Reference Manuals* and *Strengthening Forensic Science*, now have in their hands documents (and the underlying documentation in the footnotes) demonstrating the problems with individual forensic methods, as well as the language and conceptual scientific foundation provided as a reference by DNA profiling.

The very language of forensic science must change. Where there has been inconsistency in the use of words like 'match' and 'consistent with', some objective qualification will have to unite the terms across disciplines. The assurances, common in the past, that a correlation of evidence was a 'near certainty' or 'only a 1% possibility of error' are likely to be rejected – indeed, a 2007 court decision in Maryland rejected fingerprint evidence on the basis that it was 'untested, unverifiable, and purports to be infallible' [59]. Although the holding was reversed on appeal in 2009, such arguments to trial courts are to be expected more frequently thanks to the comments on friction ridge identification found in *Strengthening Forensic Science*.

The 'infallibility' statement that is frequently uttered by experts speaks to another forensic problem with which lawyers are increasingly familiar, that of 'contextual bias'. Blind testing of exemplars to 'match' questioned samples without knowing their source is a result of the fingerprint debacle of the Mayfield case, where an Oregon attorney's fingerprints were mistakenly 'matched' to prints found on a bombed train in Madrid, Spain (58, p. 153).

Proficiency testing of laboratories and scientists to ensure quality of analyses is a solution in which the law may also become involved; in 1991, Prof. Jonakait wrote an article calling for the regulation of forensic laboratories [60]. At that time, it was asserted in rebuttal that his information was dated and that the forensic sciences were engaging in self-regulation that would be sufficient to avoid the problems he noted [61]. Now, almost two decades later, an authoritative government committee has identified many of the same problems and has called for many of the same reform. It has elicited similar tepid responses [62]. This time, however, the United States Congress is paying attention, and jurisprudence may participate in the formation of laws regulating forensics. Accreditation of laboratories and certification of forensic scientists would also thus be a product of lawmaking.

12.4.2 Law and the courts

Where the quality of judicial determinations of the admissibility of evidence may falter and fail, the quality of representation given to criminal defendants to fight the injustice that results has not been any better. In his recent book, *Convicting the Innocent: Where Criminal Prosecutions Go Wrong* [63], Brandon Garrett points out that, in most of the DNA exonerations studied, the defendant was represented by a publicly-funded attorney and no scientific defense was undertaken (alibi being predominant) [64]. Given that fact, the education of defense counsel in forensic science and the jurisprudence of expert testimony could produce a two-fold reward. First would be an immediate benefit to the accused in the way cases would be investigated, prepared and litigated. Second, given the volume of cases involved, increased scrutiny of the forensic sciences and pressure on the judiciary to live up to the promise of *Daubert* and *Kumho Tire* legal evidentiary standards might incite better judicial education on these subjects and closer attention to rulings from the bench.

Another frequently mentioned problem is the opacity with which many forensic programs present to the world around them. The failure, or refusal, to provide data on the quality of the work performed in the laboratory and the quality of the analysts who perform it, is widely observed. This attitude is rooted in both an adversary view of the world, born of the relationship between most forensic programs and the government, and also in a resistance to being held up to comparison with 'academic' science. The NRC Report *Strengthening Forensic Science* strongly suggests making forensic laboratories and programs independent of law enforcement agencies, and enforcing standards for accrediting their practices and the proficiency of their analysts. Further suggestions include compelling forensic services to be provided equally to both sides of criminal cases, as has been done in the UK Forensic Science Service, and as is practiced by the University of North Texas's Institute for Human Identification.

The legal profession's broadening view of science should also, if good will can prevail, encourage openness to conducting validating research of the underlying assumptions of various forensic disciplines – especially those involved with individuation, or matching of evidence to cases. A final aspect of trial practice is that of the 'discovery process'. This aspect of criminal procedure is seeing some movement to reform as the importance of complete information concerning forensic analyses to the preparation of cases becomes better known. The occasional resistance of forensic laboratories (sometimes prosecutor-assisted) to providing their information as part of the prosecution's duty to allow discovery of evidence in the case to the defense is increasingly ill-regarded. Attorneys are becoming more conversant with the methods used to produce forensic evidence. Forensic scientists who have interfered with this process, by turning over less information than they have, have been personally fined, and evidence has been lost to the prosecution [65,66].

The last impact of jurisprudence that bears mentioning as a likely avenue of change within the law is the ongoing debate about forensics and the Confrontation Clause of the US Constitution. As mentioned above, a criminal defendant has the right to confront and cross-examine the witnesses brought to court by the prosecution. Together with the realization that some forensic laboratories do not follow proper standards, methods, or employ adequate quality control procedures, there exists the knowledge that analysts make mistakes. As a result, the common practice of having a surrogate witness testify for the actual analyst who performs the work in the laboratory has resulted in courts finding that the criminal defendant in such a case has been deprived of the right to adequately confront his accuser – namely, the analyst working for the state.

In the *Melendez-Diaz* [67] and *Bullcoming* [68] opinions, issued in the past three years, the US Supreme Court has taken a tight line to protect the right of the accused to adequately address the evidence against him or her. Jurisprudence Section members have already participated in workshops and panel discussions within the Academy on the subject. The subjects of conversation have included how to comply with the requirements of the law, and possible solutions on the laboratory's part, perhaps to front-load information given to the defense prior to trial and perhaps to forestall the need for trial testimony about the forensics. It is with these sorts of dialogue that the Jurisprudence Section can excel in its contributions to the Academy.

Finally, many comments about the education of judges and lawyers in the ways of science in general and forensic science in particular are contained in this chapter. The Jurisprudence Section is made up of lawyers and judges who have already made a commitment to learning about these disciplines. At the same time, Jurisprudence Section members are demonstrably committed to educating forensic scientists how better to deal with the legal system and to communicate in language that both can understand.

12.5 Conclusion

Jurisprudence is the essence of the law but it is also, in a fundamental way, the essence of the forensic sciences as well. The law continues to change to reflect the continuing changes of society's needs. The issues of concern for the quality of trial evidence are universal among all justice systems that seek to administer a rule of law that is accepted by its people as capable of producing reliable, predictable and legitimate resolutions of its cases. It requires quality forensic evidence in service of this goal and process. Its partnership with the Academy of Forensic Sciences, in service to both the legal and scientific communities, serves the membership of both as well as the overall goals of the Academy. We are proud and honored to be able to walk these most interesting roads together.

References

1. *Oxford English, Dictionary*. Third American Printing. Oxford, UK: Oxford University Press; 1973.
2. Mozayani A, Noziglia C. *The Forensic Laboratory Handbook Procedures and Practice*. Totowa, NJ: Humana Press; 2011.
3. Goodstein D. *How Science Works, Reference Manual on Scientific Evidence*. 3rd ed. Federal Judicial Center and National Research Council. Washington, DC: National Academies Press; 2011.
4. American Academy of Forensic, Sciences. *Policy & Procedure Manual*. http://aafs.org/sites/default/files/pdf/PPMCompleteAugust11Rev8-29-11.pdf (accessed February 2, 2012).
5. Anderson WS, editor. *Ballantine's Law Dictionary*. 3rd ed. Rochester NY: Lawyer's Cooperative Publishing Co.; 1969.
6. Rabin RL. Tort Law in Transition: Tracing the Patterns of Sociolegal Change. *Valparaiso University Law Review* 1988; 23:1–32.
7. National Academy of Sciences and the Federal Judicial, Center. *Reference Manual on Scientific Evidence*. 3rd ed. Washington, DC: National Academies Press; 2011. www.fjc.gov/library/fjc_catalog.nsf (accessed February 2, 2012).
8. National Research, Council. *Strengthening Forensic Science in the United States: A Path Forward*. Washington, DC: The National Academies Press; 2009.
9. Saferstein R. *Criminalistics: An Introduction to Forensic Science*. 6th ed. Upper Saddle River, NJ: Prentice Hall; 1998.
10. Dagan H, Kreitner R. The character of legal theory. *Cornell Law Review* 2011; 96(4):671–692.
11. *Commonwealth v. Melendez-Diaz*, 557 U.S. 1256, 129 S. Ct. 2527 at 2537 and 2536, 174 L.Ed. 314 (US Supreme Court) (2009).
12. The Law Commission. *Expert Evidence in Criminal Proceedings in England & Wales*. London: The Stationery Office, 2011. http://www.official-documents.gov.uk/document/hc1011/hc08/0829/0829.pdf (accessed November 30, 2011).
13. *Daubert v. Merrell Dow Pharmaceuticals, Inc.*, 509 US 579, 113 S. Ct. 2786, 125 L. Ed. 469 (US Supreme Court) (1993).
14. Quarino L, Houck MM. Saving Us From Ourselves: Re-creating Forensic Science. *Forensic Magazine*, 2008 Jan 2. http://www.forensicmag.com/article/saving-us-ourselves-re-creating-forensic-science (accessed January 19, 2012).
15. Thompson WC. A sociological perspective on the science of forensic DNA testing. *UC Davis Law Review* 1997; 30(4):1113–1136.

16. Matthew5:13. *The Holy Bible. Authorized King James Version.* New York: Oxford University Press; 1972.
17. Joling R. *History of the AAFS Jurisprudence Section.* Unpublished manuscript, 2008.
18. Gianelli P, Imwinkelried EM. Reference Guide on Forensic Identification Evidence. In: *Reference Manual on Scientific Evidence.* 3rd ed. Washington, DC: National Academies Press, 2011; 55–128.
19. Gianelli P, Imwinkelried E. *Scientific Evidence.* 4th ed. Newark, NJ: Lexis-Nexis; 2007.
20. Robinson ES. Law – An Unscientific Science. *Yale Law Journal* 1934; 44:235–267.
21. *Kumho Tire Co. v. Carmichael*, 526 U.S. 137 (US Supreme Court) (1999).
22. Risinger DM. Defining the 'Task at Hand': Non-Science Forensic Science after Kumho Tire V. Carmichael. *Washington and Lee Law Review* 2000; 57(3):767–800.
23. Berger MA. Evidentiary Framework. In: *Reference Manual on Scientific Evidence.* 1st ed. Washington DC: Federal Judicial Center, 1994; 39–117.
24. National Science and Technology Council, Committee on Science, Subcommittee on Forensic Science. *Education, Ethics& Terminology Inter-Agency Working Group.* http://www.forensicscience.gov/iwg_ethics.html (accessed December 12, 2011).
25. Kessler AD. Our Inquisitorial Tradition: Equity Procedure, Due Process, and the Search for an Alternative to the Adversarial. *Cornell Law Review* 2005; 5(90).
26. Holmgren B. The Expert Witness. *New England Law Review* 2003; 36:593–600.
27. Hand L. Historical and practical considerations regarding expert testimony. *Harvard Law Review* 1901; 15:40–58.
28. Developments in the law – Confronting the New Challenges of Scientific Evidence. *Harvard Law Review* 1995 May; 108(1481).
29. (1353) Lib.Ass 145. Cited in the later case of *Buckley v. Rice Thomas* (1554) 1 Plowd. 118.
30. Rule 702, Testimony by Experts Federal Rules of Evidence, US Government Printing Office, 2009.
31. The whole duty of a chemist [editorial]. *Nature* 1885; 33(839):74.
32. Black B. A Unified Theory of Scientific Evidence. *Fordham Law Review* 1988; 56:595–1233.
33. Science in court [editorial]. *Nature* 2010; 464(7287):325.
34. Jonakait RN. Real and Forensic Science. *Shephard's Expert & Scientific Evidence Quarterly* 1993; 1:435–445.
35. Hoyle F. *The Black Cloud.* London: Penguin Classics Edition; 1998.
36. Skiles A, Shore J. *The Internet's Best List of Oxymorons.* http://www.atlantamortgagegroup.com/oxymoronlist.htm (accessed November 30, 2011).
37. Moenssens AA, Henderson CE, Portwood SG. *Scientific evidence in civil and criminal cases.* 5th ed. Westbury, NY: Foundation Press; 2007.
38. Huber PW. *Galileo's revenge: junk science in the courtroom.* New York, NY: Basic Books; 1991.
39. Brewer S. Scientific Expert Testimony and Intellectual Due Process. *Yale Law Journal* 1998 April; 107:1535–1681.

40. Findley KA. Innocents at Risk: Adversary Imbalance, Forensic Science, and the Search for Truth. *Seton Hall Law Review* 2008; 38(3):894–973.
41. Berger MA. Expert Testimony in Criminal Proceedings: Questions *Daubert* Does Not Answer. *Seton Hall Law Review* 2003; 33(4):1125–1140.
42. *People v. Castro*. 545 N.Y.S 2d 985 (N.Y. Sup. Ct.) (1989).
43. *U.S. v. Yee*, 134 F.R.D. 161 (N.D. Ohio) (1991).
44. *R v Doheny and Adams*, Cr App R 369 (1997).
45. *R v Deen* [1994], *The Times*. 10 January 1994.
46. Innocence Project. *Browse the Profiles*. http://www.innocenceproject.org/know/Browse-Profiles.php (accessed December 16, 2011).
47. Garrett BL. *Judging Innocence. Columbia Law Review* 2008; 108(1):101–190.
48. Connors EF, Lundregan T, Miller N, McEwen T. *Convicted by juries, exonerated by science: case studies in the use of DNA evidence to establish innocence after trial*. Washington, DC: US Department of Justice, National Institute of Justice; 1996.
49. Garrett BL, Neufield PJ. Invalid Forensic Science Testimony and Wrongful Convictions. *Virginia Law Review* 2009; 95(1):1–97.
50. Gianelli P. Forensic Science: Under the Microscope. *Ohio Northern University Law Review* 2008; 34(315, 316).
51. Title I of the Omnibus Safe Streets and Crime Control Act of 1968, Part BB, codified at 42 U.S.C. Ï 3797k(4).
52. American Bar Association. *Index to ABA Criminal Justice Policies, By Meeting*. http://www.americanbar.org/groups/criminal_justice/policy/index_aba_criminal_justice_policies_by_meeting.html (accessed November 30, 2011).
53. American Bar, Association., Criminal Justice Standards, Committee. *ABA standards for criminal justice: DNA evidence*. 3rd ed. Washington, DC: American Bar Association; 2007.
54. Findley KA. Guilt vs. Guiltiness: Innocents at Risk: Adversary Imbalance, Forensic Science, and the Search for Truth. *Seton Hall Law Review* 2009; 39(3):893–973.
55. IMDb. Engrenages. http://www.imdb.com/title/tt0477507/episodes#season-3 (accessed February 2, 2012).
56. American Academy of Forensic Sciences. *The American Academy of Forensic Sciences Approves Position Statement in Response to the National Academy of Sciences' 'Forensic Needs' Report* [Press Release]. Colorado Springs, CO; 2009 September 4.
57. Omenn GS. Enhancing the Role of the Scientific Expert Witness. *Environmental Health Perspectives* 1994; 102(8):674–675.
58. Berger MA. Evolving trends in forensic science. *Tennessee Journal of Law & Policy* 2010; 6:147–160.
59. *State v. Rose*, No. K06-0545 (Md., Baltimore Co.) (Oct 2007).
60. Jonakait RN. Forensic Science: the need for regulation. *Harvard Journal of Law & Technology* 1991; 4:109–191.
61. Letters Forum: Forensic, Labs., *Harvard Journal of Law & Technology* 1991; 5:1–17.
62. Bono JP. The blame has run its course. Strengthening forensic science investigation. Texas Bar Journal 2011; 74(7):592–596.
63. Garrett B. *Convicting the innocent: where criminal prosecutions go wrong*. Cambridge, MA: Harvard University Press; 2011.

64. Garrett BL. Appendix: The Defense Case in DNA Exonerees' Cases. In: Garrett BL.*Convicting the Innocent: Where Criminal Prosecutions Go Wrong*. Cambridge, MA: Harvard University Press; 2011. www.law.virginia.edu/pdf/faculty/garrett/convicting_the_innocent/garrett_defense_case_appendix.pdf (accessed November 30, 2011).
65. *State v.Blenden*, 748 So.2d 77 (Miss.) (2000).
66. *State v. Meza*, 50 P. 3d 407 (Ariz. App) (2002).
67. *Melendez-Diaz v. Massachusetts*, 129 S.Ct. 2527 (2009).
68. *Bullcoming v. New Mexico*, 131 S. Ct. 2705 (2011)

13

Global forensic science and the search for the dead and missing from armed conflict: the perspective of the International Committee of the Red Cross

Morris Tidball-Binz

International Committee of the Red Cross, Geneva, Switzerland

13.1 Introduction

In August 2011, members of a remote community in the mountainous north-western region of Libya approached the International Committee of the Red Cross (ICRC) for forensic assistance in recovering their dead relatives. They had been reportedly killed two months earlier by members of the opposing warring party and buried in a mass grave in a minefield. The bereaved explained that the recovery and proper burial of their loved ones was for them far more important than any other humanitarian assistance they were receiving, including medical care for those wounded in armed conflict. They insisted on the urgent need for forensic expertise, which was unavailable in the region as a result of the ongoing conflict, in order to fulfill their right to know the fate of their loved ones and to mourn and bury them according to their traditions and religion.

At the end of that month, two forensic specialists from the ICRC, together with an ICRC mine clearance team, carried out the emergency humanitarian forensic operation as requested by the community, which was thus able to mourn and bury their dead (Figure 13.1).

Why are the dead so important for those affected by armed conflict? And what is the background for growing global awareness about the role of forensic sciences for

Forensic Science: Current Issues, Future Directions, First Edition. Edited by Douglas H. Ubelaker.

Figure 13.1 Libya, August 2011: The ICRC carrying out an emergency forensic operation for the recovery and identification of people killed and reported missing during the armed conflict. The area around the burial site was cleared from mines and unexploded ordnance by the ICRC's weapon contamination specialists.

fulfilling humanitarian goals? This chapter provides some clues for answering these questions. It outlines the purpose, framework and some recommendations for carrying out forensic investigations into the dead and missing from armed conflicts and it also summarizes some of the trends and opportunities for global forensic sciences observed by the ICRC's forensic services with regard to humanitarian forensic sciences.

13.2 The International Committee of the Red Cross

The International Committee of the Red Cross (ICRC) is an impartial, neutral and independent organization whose exclusively humanitarian mission is to protect the lives and dignity of victims of armed conflict the world over [1]. The ICRC was established in 1863 as a result of the campaign launched by Henry Dunant, a Swiss citizen from Geneva, after witnessing the battle of Solferino (1859) in present-day northern Italy, where thousands of wounded French, Austrian and Italian soldiers died for lack of adequate medical care. Dunant's account of that battle and his proposals for protecting and assisting the war wounded led to the founding of the ICRC and to the creation of national relief societies in each country (e.g. the American Red Cross, the Turkish Red Crescent, etc.).

The ICRC, the national societies and the societies' International Federation of the Red Cross and Red Crescent Societies form the International Red Cross and Red Crescent Movement, with more than 800 million members. In situations of armed conflict, the ICRC directs and coordinates the international relief activities of its Movement partners.

The ICRC is also at the origin of international humanitarian law, a body of international public law applicable at times of armed conflict, which seeks to reduce the suffering caused by war and impose limits upon the way it is waged.

A major part of international humanitarian law (IHL) is contained in the four Geneva Conventions (GC) of 1949 (see below). Nearly every state in the world has agreed to be bound by them. Since their adoption, the Conventions have been supplemented by three Additional Protocols (AP), which address emerging challenges in the protection of victims of conflict. International humanitarian law is different, albeit complementary, to international human rights law, another branch of international public law that was developed for protecting individuals and groups against abuses of the State [2].

The ICRC's permanent international mandate for its humanitarian activities derives from the 1949 Geneva Conventions, their Additional Protocols and from the Statutes of the Movement.

The ICRC is presently active in more than 80 countries in all regions, it employs more than 13,000 staff around the world and its work is coordinated from its Geneva headquarters in Switzerland. The ICRC is a non-profit organization, funded principally by governments, regional organizations, national Red Cross and Red Crescent societies, municipal authorities, the private sector and members of the public. It remains a private organization governed by Swiss law and is strictly independent in its governance and operational decisions.

Since its creation, the ICRC has received numerous awards for its work in favor of all victims of war, including three Nobel Peace Prizes (in 1917, 1944 and 1963). Also, the first Nobel Peace Prize was awarded in 1901 to the founder of the ICRC, Henry Dunant, who shared it with Frédéric Passy, a French pacifist.

In 2003, the ICRC acquired forensic capacity to promote and help implement forensic best practices applied to humanitarian operations, principally for the location, recovery, identification and management of the remains of those killed or missing as a result of armed conflict.

At the time of writing, the ICRC's Forensic Unit is composed of 12 professionals from the fields of medicine, anthropology, archaeology, dentistry and criminalistics. These forensic services focus primarily on building local and regional forensic capacity and fostering regional and international cooperation among forensic practitioners and institutions, especially among countries affected by armed conflict and catastrophes, to help prevent and resolve the humanitarian consequences of these events. In addition, the ICRC carries out a range of forensic training activities, often in collaboration with forensic institutes and universities. It helps to cultivate and promote standards of best practice, and it also develops specialized tools, including disaster-victim identification (DVI) equipment and software.

The ICRC has built up a global network of forensic practitioners and institutions offering advice on, and often participating in, ICRC forensic humanitarian activities. In 2010, a group of leading forensic scientists from around the world, and representing several disciplines in forensic sciences, were invited by the organization to stand as the ICRC's Forensic Advisory Group.

In 2011, the International Association of Forensic Sciences (IAFS) presented its first Human Rights Award to the ICRC 'for the outstanding contribution given to the use of forensic sciences in the protection and promotion of human rights' [3].

13.3 The dead and missing from armed conflict

The four Geneva Conventions (GC I to IV) [4], their Additional Protocols I and II (AP I and AP II) [5] and customary international law [6] contain various provisions dealing with mortal remains, gravesites and the missing from armed conflict.

The treatment of the dead from armed conflict is often a very sensitive issue. For example, pictures of desecrated dead US servicemen, dragged through the streets of Mogadishu, Somalia, in 1993, provoked global outrage and condemnation [7]. More recently, the removal of a dozen graves and a memorial for Soviet soldiers from WWII in Tallinn, Estonia, led to massive protests which left one person dead and eroded the relationship between Estonia and Russia [8].

International humanitarian law specifically protects the dignity of the dead and provides for their identification. It sets out detailed obligations to governments and parties to armed conflicts with regards to the treatment of the dead and their burial sites, the collection of information on the deceased, their identification, the registration and protection of grave sites and the repatriation of remains [9]. In addition, the right of human beings not to lose their identity after death is nowadays universally recognized [10].

On a more fundamental level, and in terms of humanity, a universal mark of civilization is that the dead are identified, that something is known about how they died, and that funerary rituals are carried out.

Some countries, notably the United States of America, have created institutions exclusively dedicated to recovering, identifying and properly managing the remains of their soldiers killed in action [11]. It is hoped that more countries around the world develop similar commitments towards their fallen servicemen and servicewomen. In parallel, more attention needs to be paid to the remains of non-combatants who die as a result of armed conflicts.

In addition to the loss of human life, one of the tragedies resulting from armed conflict is that large numbers of people go missing and cannot be accounted for.

The missing:
The need for information

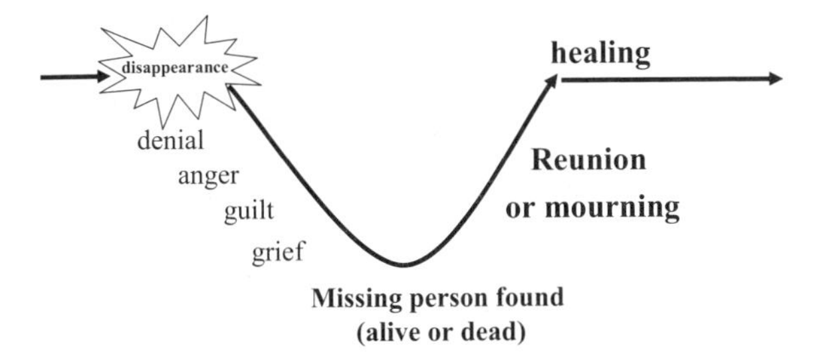

Figure 13.2 Graphic summary of the principal events and phases affecting families of the missing. The angst caused by uncertainty about the whereabouts and fate of their loved ones may continue for as long as the bereaved receive no credible information about their missing relatives. Finding the person, alive or dead, is often indispensable for the healing process and for overcoming the suffering caused by the disappearance.

For the families of the missing, the combination of lack of news and uncertainty about the fate of their loved ones amounts to unbearable suffering. The unresolved grief often remains active as long as 'closure' is not possible (Figure 13.2).

For example, in Spain, thousands of grieving families of the missing from the civil war which took place between 1936 and 1939 are still searching for the remains of their relatives, to mourn them according to their cultural traditions [12]. Such closure requires credible information and, if the person is dead, the recovery of the remains.

The core element of international humanitarian law that applies to address this problem is the families' right to know the fate of a relative who is missing (Article 32 of AP I). This means that families have a right to know whether the missing relative is dead or alive – and, if the relative is dead, they have a right to know something about how the death came about and the whereabouts of the deceased.

13.4 A brief history: from Katyn to the Balkans

Forensic investigations have not always been a part of the aftermath of conflict. Only in the last 30 years have forensic specialists been called to work in such contexts with any regularity. There are, however, some landmarks in this development.

The exhumation and examination of the victims of the Katyn Forest Massacre in World War II represents the first comprehensive use of forensic pathological and scientific techniques to evaluate a mass killing in war [13]. Poland had been divided between Nazi Germany and the Soviet Union during the war, and 15,000 Polish officers captured by the Soviets were kept in three special camps at Kozielsk, Starobielsk and Ostaszków in eastern Poland. Between 3rd April and 2nd May

1940, the officers were reportedly taken from these camps by train and trucks to a forest in Katyn near Smolensk, in present-day Russia. Their hands were bound behind their backs and all were killed by pistol shots in their backs. They were buried in unmarked ditches. Only 448 were known to have survived.

In June 1941, Germany invaded the Soviet Union and the eastern part of Poland. In April 1943, the German Minister of Propaganda, Joseph Goebbels, announced the discovery by a Special Medical Judiciary Commission of the bodies of 4,443 Polish officers in graves in the Katyn Forest. The findings were supported by an International Medical Commission made up of specialists from 12 countries not at war with Germany. Goebbels decided to use the discovery to embarrass the Anglo-American-Soviet coalition. The Soviet Union responded by alleging that the Polish officers had been killed by German troops after the German invasion of June 1941. Thus, the forensic issue at stake – which was not resolved until after the war – was the timing of the deaths.

Subsequently, in 1951, a United States Congressional Select Committee undertook an investigation. It 'unanimously agree(d) that the evidence . . . proves conclusively and irrevocably the Soviet NKVD committed the massacre of Polish Army officers in the Katyn Forest near Smolensk, Russia, not later than the spring of 1940.' [14] In September 1989, Mikhail Gorbachev admitted Soviet responsibility for the mass killings at Katyn.

As regards identification of the victims, a technical commission of the Polish Red Cross was involved in the exhumation and identification of many of the bodies of the Polish officers in 1943. This was overseen by the German authorities. Many of the remains were identified by documents or other items and were re-buried in a dignified manner in marked mass graves.

The complex history of this atrocity and its subsequent investigation shows that the issue of people missing in conflict can easily be politicized with little thought for the affected families. This stark truth underscores the need for professionalism of forensic scientists involved in this kind of work.

The first recorded systematic use of forensic sciences for the recovery and identification of large numbers of victims of conflict took place in the 1980s in Argentina, where thousands of people were abducted, killed and buried in unmarked graves during the military regime that ruled the country between 1976 and 1983. Following the return to democracy in late 1983, it became apparent that the challenge of locating the burials and recovering and identifying the remains was beyond the capacity of existing local forensic institutions.

This prompted the families of the missing and the newly established National Commission on Disappeared Persons to seek international forensic assistance, for which they contacted the American Association for the Advancement of Science (AAAS). In response, the AAAS organized a delegation of forensic specialists, including members of the American Academy of Forensic Sciences (AAFS), who traveled to Argentina in June 1984. They observed and assisted in the scientific investigation of the missing and the identification of disappeared children who had been abducted by the military, and they offered valuable advice to the Commission and to relatives of the missing. The findings of the delegation were published in the December 1984 edition of the *American Journal of Forensic Medicine and Pathology*, dedicated to human rights and the forensic scientist [15].

The delegation also carried out some forensic casework and helped to kick-start and train a group of Argentinean scientists who would go on to form the Argentine Forensic Anthropology Team, (*Equipo Argentino de Antropología Forense*, EAAF) [16]. That mission of American forensic scientists to Argentina pioneered international forensic assistance in the field of humanitarian forensic sciences, and it served as a model for international cooperation in this novel field.

The EAAF became the first ever non-governmental organization dedicated exclusively to forensic investigation of the missing. In the following years, often in cooperation with the AAAS and Physicians for Human Rights – USA, and with the participation of leading forensic scientists (notably Dr. Clyde C. Snow, an American forensic anthropologist), the EAAF helped train other similar teams in Latin America and other regions, and has since carried out hundreds of missions the world over. The experience gained in Argentina and elsewhere was later applied to large-scale investigations for the search for the missing in the Balkans, Rwanda and other contexts. It also helped to shape applicable international standards, including the 1995 *Guidelines for the conduct of United Nation inquiries into allegations of massacres* [17].

The outcome of this kind of forensic work in the 1980s and early 1990s went beyond resolving cases in specific contexts. Because of its visibility and potential media interest, it raised the expectations of families and communities for justice and for identification of remains. This became particularly apparent during the violent disintegration of Yugoslavia in the early 1990s, a war that provided another major milestone in the history of forensic sciences and armed conflict. The International Criminal Tribunal for the Former Yugoslavia (ICTY) was established pursuant to Security Council Resolution 827 on 25 May, 1993. It was the first international criminal tribunal to try war crimes and crimes against humanity since the Nuremberg and Tokyo Tribunals at the end of World War II.

The Deputy Prosecutor of the ICTY explained the purpose behind exhuming the many mass graves: 'Following the exhumations . . . all the bodies underwent autopsies . . . to determine the cause and manner of death and the demographic profile of the victims' [18]. The reference to the 'demographic profile of the victims' implies that identification of the victims was not seen as a prerequisite for concluding whether an accused person was guilty of murder. For example, the exhumation of a person's skeletal remains with arms tied and a gunshot wound to the back of the head may provide sufficient evidence for a successful prosecution, even if the name of the dead person is not known. The international criminal justice system can fulfill its responsibilities without needing to identify many of those who have died. Furthermore, the task of identification was not part of the mandate of the ICTY. The forensic specialists who examined the exhumed remains for the ICTY had neither the time nor the resources to initiate a process of identification, and certainly not to deal with the families.

In 1995, as part of the Dayton Peace Accords, which brought peace formally to Bosnia and Herzegovina, there was an international commitment to deal with the issue of nearly 30,000 people reported missing in that particular context. The International Commission on Missing Persons (ICMP) was created in 1999 and given a mandate to search for the missing. Soon after, the ICMP acquired forensic capacity and it adopted a DNA-led screening and identification process, which has since assisted in the identification of more than 15,000 persons reported missing from the Balkans wars [19]. The ICMP is also active in other regions, where it helps the authorities to develop their capacity to investigate the missing, including from disasters.

The pioneering role of the ICMP, in applying high throughput DNA analysis for the identification of victims of armed violence, has proved as relevant and valuable for fulfilling humanitarian purposes as the innovative role of the EAAF in applying forensic anthropology to the search for the missing. Both organizations, which presently advocate for and use a combination of methods, including forensic genetics and anthropology for their investigations, should be credited for their outstanding contributions to the promotion of global forensic sciences applied to humanitarian objectives.

13.5 A global agreement on best practices: the 2003 International Conference on The Missing and Their Families

The recommendations from the 2003 International Conference on The Missing and Their Families consolidated lessons learned from the past, helped to launch humanitarian forensic sciences and, it is hoped, added a new landmark in the development of global forensic sciences applied to addressing the consequences of armed conflict.

The right of families to know the fate of their missing relatives and the need for the proper management of the dead from armed conflict and other forms of violence were highlighted in a process launched by the ICRC between 2002 and 2003 [20]. This aimed at bringing together the lessons learned worldwide for the prevention and resolution of the missing and at raising awareness of the issues among the community of states and other key actors – including forensic scientists – and to provide political and practical solutions to the tragedy of the missing. The recommendations were adopted by the international community during the International Conference of Governmental and Non-Governmental Experts on The Missing and Their Families, which was held in Geneva, Switzerland in February 2003 [21].

The ultimate objective of the Conference was to ensure that families of missing people are assisted in finding their loved ones, that all the dead from armed conflict are managed in a proper and dignified way, and to prevent people becoming unaccounted for in the future.

The Conference recognized that forensic sciences play a pivotal, and often essential, role for the recovery and identification of the dead and the documentation of facts, which may lead to accountability of perpetrators and reparations for the victims and – some argue – may also help prevent their repetition. It also recognized that forensic sciences can help to uphold the families' right to know the fate of their missing relatives.

For many, this was a new concept, as forensic sciences have been traditionally regarded as focusing mostly on domestic criminal investigations, with little or no application to humanitarian activities.

In the contexts of armed conflict or other situations of widespread armed violence, the responsibility to investigate deaths, identify the victims and inform the families, in theory, lies squarely with the authorities. However, all too often (either through lack of capacity or lack of will), this responsibility is not fulfilled. As a result, international forensic cooperation is called upon increasingly by the concerned communities, the authorities, and also by international agencies, to help monitor and safeguard provisions of international humanitarian law.

In carrying out such operations, however, forensic practitioners often face unique challenges, including ethical and legal dilemmas and technical constraints seldom experienced in their domestic settings. The recommendations on forensic best practices from the Conference, and the experience gained since then by the ICRC's forensic services, can provide useful and practical guidance for those involved or willing to participate in humanitarian forensic work [22].

The recommendations from the Conference have proved to be particularly valuable in contexts with ill-defined or inexistent regulatory frameworks, but they have also shown to be valuable in contexts with well-defined and active normative, legal and institutional settings, such as in the case of some countries in Latin America, which have incorporated the recommendations into their legislation. Colombia is one such example: its *Plan Nacional de Búsqueda de Personas Desaparecidas* [23] (National Plan for the Search of Missing Persons*)*, adopted in 2007) is based on the ICRC recommendations.

The recommendations also provide elements for designing, implementing and assessing strategies for the investigation of the missing. In addition to the need for a normative framework, proper coordination and particular health and safety considerations, this underlines the importance of ensuring the legitimacy of these investigations (i.e. end-users' acceptance) (Figure 13.3). Legitimacy will often depend on adequate communication and confidence-building between investigators and the bereaved and their affected communities.

The recommendations include advice and guidelines on multidisciplinary team efforts for the recovery, management and identification of the dead in war-torn and challenging contexts, including:

- the roles, duties, responsibilities and applicable ethical standards for forensic practitioners and teams;

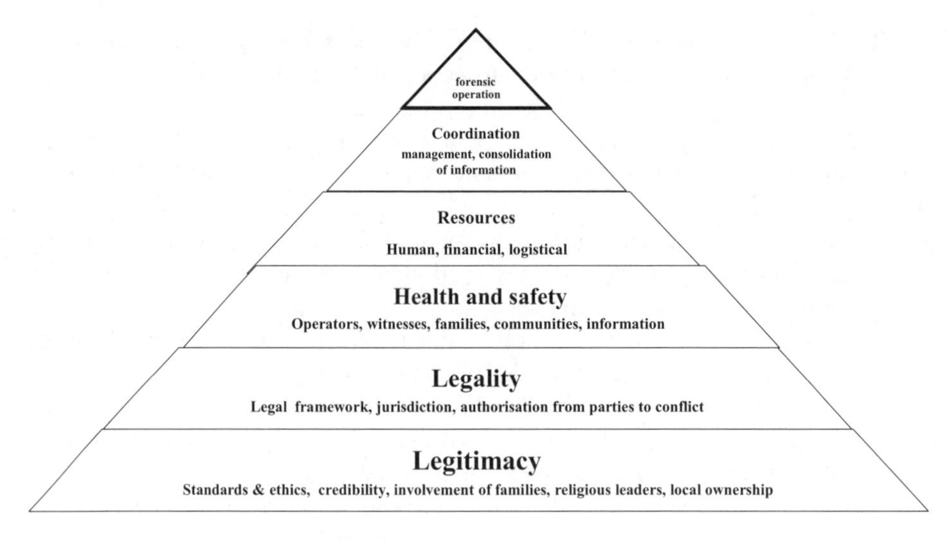

Figure 13.3 The successful planning and implementation of forensic investigations into the missing from armed conflicts and catastrophes requires due consideration of a number of factors relevant to this particular kind of investigations, as summarized in the ICRC's strategic and planning pyramid, which also illustrates the relative importance of the different factors. For example, investigations which are well understood and supported by the families, their communities and authorities will enjoy strong legitimacy (i.e. a wide-based pyramid) which, in turn, helps improve the sustainability of the process.

- guidelines for documentation and storage of human remains and associated evidence;
- the use of different forensic disciplines, methods and criteria for forensic human identification;
- principles for ethical, effective and efficient information management, including the collection and comparison of ante-mortem and post-mortem data;
- advice on the relationship between forensic practitioners and bereaved families and communities.

In the years which followed the Conference, it became growingly apparent that many of the recommendations related to forensic sciences are also valuable for the investigation of the missing in catastrophes (i.e. disasters beyond expectations). The ICRC has helped to develop simple guidelines for their application in catastrophes, especially when ideal disaster victim identification procedures, as prescribed by Interpol [24], are delayed or simply not possible [25]. The ICRC guidelines and corresponding forms were developed jointly with the Pan-American Health

Organization and the International Federation of Red Cross and Red Crescent Societies, and in consultation with Interpol [26]. They are compatible and also complementary with Interpol's DVI Guidelines, which are regarded as the international 'gold standards' for disaster victim identification.

13.6 Roles and responsibilities of forensic practitioners participating in humanitarian operations

As recognized during the 2003 International Conference on The Missing, and confirmed by the ICRC's experience thereafter, the role that forensic specialists play in their domestic context cannot be automatically extrapolated to an international context involving investigations into the dead and missing from armed conflict [22,27,28]. These usually require special considerations, including with regards to:

- the professional profile of forensic practitioners;
- health and safety needs of the forensic staff;
- the use of specialized guidelines;
- capacity and preparedness for carrying out swift forensic investigations in challenging contexts, including recovery and analysis of human remains and familiarity with human forensic identification procedures.

13.6.1 General recommendations

Practitioners involved in investigations into the dead and missing from armed conflict and catastrophes require clarity on four principal questions:

1. What is the purpose of the investigation (e.g. humanitarian, under provisions of IHL)?
2. Under whose authority and/or jurisdiction does the investigation take place (e.g. national or international agency)?
3. Which institution/s is he/she responsible to?
4. What are the operational implications and particular needs for the investigation (including those with regards to logistics)?

A clear framework or, in its defect, agreed standards and recommendations, will help the practitioner respond to key questions related to any investigation into the dead and missing, including:

- Who has legal possession of a body and unidentified human remains?
- Who is in charge of death investigations in the different operating environments (e.g. military or civilian court, truth commission, international agency, etc.)?
- Which is the rule of evidence required for identification? (i.e. on the balance of probabilities or beyond reasonable doubt?)
- Who signs death certificates and hands over the body?

Forensic practitioners working in contexts involving missing persons must demonstrate a level of professionalism that goes beyond simply assuring adequate standards of practice at their domestic level.

In addition to the immediate forensic questions at hand, a highly professional approach to the sensitive issue of human remains can even have the effect of providing the necessary basis for a much-needed dialogue between two parties locked in conflict. Also, professionalism among forensic specialists involved in such investigations can be a major factor in promoting international humanitarian law, international human rights law, accountability and a process of reconciliation.

In particular, when applied to investigations into the dead and missing from armed conflict, professionalism implies a degree of respect, impartiality and neutrality that transcends conflict. For this purpose, adequate guidelines can serve to empower forensic specialists working in new, difficult or highly political circumstances.

Forensic practitioners involved in humanitarian operations in foreign countries should also endeavor, whenever possible, to consult their local colleagues and forensic institutions in those countries. Where necessary, they should promote local forensic capacity-building and foster international forensic cooperation for empowering the sustainability and local ownership of the investigations.

Governments and parties to armed conflict have well-defined and often clear duties (both domestic and international) towards the dead and the missing from armed conflict. Forensic practitioners interested in working in such cases should familiarize themselves with those obligations (e.g. as enshrined by international humanitarian law – see above), among other things, to prevent any confusion of roles and the misuse or manipulation of the results of their work, which might affect or undermine their own professional standing.

Forensic practitioners should also advocate for and recommend that, once identified, human remains are swiftly returned to their families, either by the authorities or by some other relevant stakeholder who is competent to do so. Thought must be given to the whole process before the forensic practitioner becomes involved; it cannot be assumed that the entire chain of responsibility that usually exists in a domestic context is in place and will be effectively operational in war-torn contexts.

In short, forensic practitioners involved in investigations into the dead and missing from armed conflict must prove to be qualified, competent and experienced, and to have a good understanding of the frameworks and challenges related to the particular humanitarian forensic operations.

Practitioners have an ethical duty to observe and record all information potentially relevant to the identification of human remains under their responsibility, and to advocate actively for an identification process in the investigation. They should also familiarize themselves with, and be respectful of, the provisions of local culture and religion regarding the treatment of the dead and funeral rituals, and should consider the families' rights and needs before, during and after the investigation. Special consideration should be given to the management and disposal of unidentified remains in a way appropriate to the context.

Forensic physicians also have a duty to abide by the ethics of their medical profession as they apply to the investigations into missing persons [29].

The ICRC's own recommendations and operational best practices [20,21] complement and expand existing references and guidelines for investigations into the missing, including the 1991 *UN Manual on the effective prevention and investigation of extra-legal, arbitrary and summary executions* (*Minnesota Protocol*) [30], the 1995 *Guidelines for the conduct of United Nation inquiries into allegations of massacres* [17] and *Interpol Disaster Victim Identification (DVI) Guide* [24].

One of the shared objectives of these guidelines is, whenever possible, to subject human remains to one examination only. Examinations should not have to be repeated because previous examinations were incomplete. The same examination should serve to establish the cause and probable manner of death and should record the information needed to make a positive identification.

The ICRC has also developed guidelines for specific fields of humanitarian forensic sciences, such as DNA analysis [31], as well as for particular end-users, including members of armed forces [32] and first responders [25] and on special topics, such as wound ballistics [33]. At the time of writing, it is finalizing a set of guidelines on the investigation and prevention of deaths in custody, prepared jointly with the Universities of Geneva and Berne, Switzerland and with the International Centre for Prison Studies, King's College, London. This will be published by ICRC press in 2013.

Existing best practice guidelines, including those described in this chapter, have universal application. They were designed to help forensic practitioners involved in humanitarian operations to act within the ethical and professional boundaries applicable to investigations into the dead and missing, especially where the legal and institutional frameworks for forensic investigations are collapsed or eroded as a result of armed conflict. Best practice guidelines should, wherever possible, accommodate local skills and expertise, and they should also contemplate the need for local capacity-building, including the training of non-specialists and partially qualified personnel. Such training should be planned in the early stages of any mission or project.

13.6.2 Recommendations for forensic teams, contracts, employers

The national or local authorities, including forensic institutions, have ultimate responsibility for the management, recovery and identification of human remains.

However, in contexts involving missing persons as a consequence of armed conflict, others may have to undertake this role and bring forensic specialists to the area, albeit following and respecting the general legal and institutional frameworks of the country where the investigation is taking place. This often requires that the investigation be placed under the responsibility of, and even the direction of, a local authority, who may or may not be familiar with required forensic procedures. A clear definition of roles and responsibilities will help to avoid misunderstandings during the operations.

When forensic practitioners move from their everyday domestic context to work in an international context and foreign jurisdiction, often for the first time, a forensic professional with the necessary qualifications, skills and experience should be assigned to be in charge of the team responsible for the examination of the remains.

The findings, conclusions and reports signed by the forensic specialist in charge should be both formally valid and accepted as credible by local officials, the families and tribunals (national or international).

Before embarking on a humanitarian mission, forensic practitioners should consider the following questions for their employer (whether a government service, international organization or other agency) and for the relevant authorities:

1. Which domestic and international legal frameworks apply?
2. Which part of the proposed work may be inappropriate if performed by a foreigner? For example, it may be lawful for a foreign specialist to watch a post-mortem examination being performed by a local pathologist, but not actually to carry out an autopsy.
3. Does the forensic practitioner have any kind of legal immunity (e.g. from testifying in court about confidential findings)?
4. Do the authorities recognize the forensic specialists' qualifications?
5. Is the contract with the employer recognized by the authorities?
6. Is the work in fact being done in a context that is unlawful, or might be deemed to be unlawful, in domestic law?
7. What is the mandate and legal standing of the employer (if not the authorities) in the context?
8. Can it be assumed that such work performed under the mandate of an international organization (e.g. the ICRC, United Nations, etc.) automatically preempts domestic law?
9. Has the forensic work been incorporated into any political process which the forensic practitioner may need to know about?

The concerned forensic practitioner must fully understand the employer's mission, including the broad legal and political framework and implications of its operations and whether the employer is recognized as competent and credible and is willing to work with others. The employer must also provide adequate background information about the context, including political, cultural and security information.

Clarity should also be required about the ultimate purpose of the forensic investigation. For international agencies, this may include:

- the protection of promotion of human rights and the investigation of violations (e.g. organizations involved in human rights advocacy, such as Amnesty International, Physicians for Human Rights and Human Rights Watch);
- the promotion of international humanitarian law while preserving neutrality and impartiality (for example, the ICRC);
- the question of criminal accountability (e.g. the International Criminal Court).

A contract by which a forensic practitioner is engaged to work in a context involving dead and missing persons from armed conflict should include the following points:

1. An affirmation of professional qualifications.
2. A commitment to work by standard guidelines and standard operating procedures relating to recovery, management, analysis and identification.
3. A clear definition of who bears ultimate responsibility for the recovery and identification of human remains and for issuing death certificates, if the authorities are unable or unwilling to do this.
4. An assurance that the employer has obtained or will obtain full authorization and security clearance from the authorities and other relevant stakeholders.
5. A reference to the handling and preservation of all evidence, including chain of custody procedures.
6. A clear indication of whether or not the forensic specialist is expected to present findings in court.
7. A commitment that health and safety procedures will be followed.
8. A commitment that adequate insurance coverage – such as malpractice insurance – is provided for all eventualities, as the coverage pertaining to the specialist's domestic work may not apply
9. In keeping with standard forensic practice, an agreement that the practitioner has the right to copy documents and photos for which he/she was responsible, subject to an undertaking of confidentiality and acknowledgement that copyright lies with the employer.

Figure 13.4 Libya August 2011: a stockpile of unexploded ordnance and mines recovered from battlefields. Weapons contamination posed serious health and safety risks to the forensic recovery and identification of those killed and missing in armed conflict. Forensic operations often required assistance from explosive ordnance disposal specialists.

In short, the terms of reference or contract must ensure that the employer's mandate is compatible with the ethical and professional practice of forensic specialists. For humanitarian forensic investigations, the mandate must be underpinned by the principles of humanitarian action, including independence, neutrality and impartiality. The employer must recognize the roles and responsibilities outlined above and the need to adhere to best practice guidelines.

Consideration of the points outlined above will help forensic practitioners to work within a safe, professional and ethical framework, while promoting the application of international humanitarian law and international human rights law and, at the same time, minimizing distress to the bereaved.

13.7 Health and safety issues

Security issues, including contamination from weapons and chemical, biological, radioactive and nuclear hazards (CBRN), are a frequent source of concern during investigations into the dead and missing from armed conflict (Figure 13.4). These serious threats require appropriate management for ensuring satisfactory health and safety conditions for the investigators. Expertise in CBRN threats and weapon contamination should be consulted for advice in the preparation of missions to the field and, if required, for assistance during operations [33].

Forensic practitioners involved in humanitarian operations for the ICRC or other components of the International Red Cross and Red Crescent Movement (see above) during armed conflict must use the emblem (e.g. Red Cross, Red Crescent) when required. This provides protection, under international humanitarian law, to humanitarian actors; the targeting of emblem-carriers is a war crime.

Other risks for forensic operators may arise from the findings of the investigations and might lead to threats and attacks against the investigators by those under investigation. Adhering to strict principles of independence, neutrality and impartiality will reduce (but not eliminate) such risks.

Working in contexts of armed conflict is often stressful. Special guidelines have been developed to help minimize both the danger and stress faced by humanitarian staff in armed conflict [34,35].

Forensic specialists and their support staff, as well as staff working with the families of missing people, might find their health affected by the work they are doing. Because of the emotionally charged nature of such work, especially the interaction with the bereaved, it is essential that staff who have to inform families of the death of a relative, or who have to return personal effects or remains, are duly trained and briefed for this particular job [36].

13.8 Responding to constraints

In contexts of active armed conflict, forensic practitioners involved in humanitarian operations may sometimes find themselves unable to perform a full and detailed examination (e.g. a complete post-mortem) due to extreme constraints, including threats to their security. It may be necessary, under such difficult conditions, to carry out succinct examinations with little time and without access to mortuary and laboratory facilities.

In such pressing circumstances, the forensic specialist may find that 'the best may be an enemy of the possible'. However, performing an abbreviated forensic analysis or examination (e.g. an external examination of a cadaver versus its full autopsy) may be fully compatible with professional conduct and existing guidelines (such as those developed by the ICRC), given the constraints, provided the supporting reasons are duly noted and recorded.

The objective of such an abbreviated examination is to collect and preserve as much information and evidence as possible within the constraints of the case, with a view to maximizing the chances of later analysis and identification and helping to fulfill the humanitarian goal of the operation.

A forensic pathologist is best qualified to perform such an abbreviated examination of recently dead bodies, while forensic anthropologists might be best equipped to so examine skeletal remains.

13.9 The identification of human remains

In normal circumstances, the identification of a body is an integral part of any investigation into a suspicious death, and it goes hand in hand with ascertaining the cause and manner of death.

However, in the case of investigations into the dead and missing from armed conflict, especially when these involve victims of atrocities, establishing their cause and manner of death may be far easier to ascertain than their identity. In effect, in many such cases, identification of the victims is often the most difficult and resource-intensive forensic task. As a result, simple recovery procedures may readily satisfy the need for evidence of the cause and manner of death of the deceased, even if these are not identified. This can lead to the unfortunate situation whereby remains are recovered and their cause and manner of death established but, because the process of identification is much more resource- and time-consuming, the bodies remain unidentified.

This situation is unacceptable from a humanitarian point of view, as it prevents the bereaved from knowing the fate of missing relatives or from recovering their remains [37].

In humanitarian investigations, human remains are usually examined and identified by means of a team effort comprising several specialties, principally archaeology, anthropology, pathology, dentistry and genetics, and working under the overall responsibility of a professional with the necessary qualifications, skills and experience required for overseeing the identification procedures. This role is usually assigned to a forensic pathologist or physician, as this reflects the legal requirements in most parts of the world, including for the signature of death certificates.

It is the responsibility of the head of the forensic team to decide which methods of identification are most appropriate in a given context and applicable legal requirements. It is also his/her responsibility to ensure that relatives, the community and the authorities are well informed throughout the investigation and are given realistic expectations of the process, including on the methods used and the timeframes for the recovery and identification of remains. The need for relatives to view the bodies of their loved ones as part of the grieving process should be respected and adapted to the context [38,39] (Figure 13.5).

Figure 13.5 Libya, August 2011: an ICRC forensic operation in Libya. Relatives of the missing and members of the community participate, on their request, in the visual recognition of the exhumed remains of their relatives and acquaintances, whose identity was confirmed on the basis of unique identifiers, including dental traits.

In order to help to ensure quality work and the respect for professional standards in challenging contexts, processes for the forensic identification of human remains must be implemented according to a framework agreed upon by all actors involved.

Such a framework should contemplate all topics to be dealt with by the concerned actors, ideally before starting any forensic recovery and identification process. It should include clear regulations and procedures, duties and responsibilities, including for:

- the chain of custody of evidence;
- the possession of remains;
- the issuing death certificates;
- the release of identified bodies;
- the storage of unidentified remains.

It should also define the required protocols for recovery, ante-mortem and post-mortem data and sample collection, autopsies and identification.

The collection of ante-mortem data required for forensic identification of victims should include a standard questionnaire for collecting data, training of the staff collecting data and considerations for the collection of biological samples for DNA profiling. The management and analysis of information on human remains and missing people, especially in cases involving large numbers of individuals, requires appropriate databanks and software (including those developed by Interpol and the ICRC) for disaster victim identification and for the analysis and identification of victims of armed conflict respectively.

Forensic human identification, including the methods and criteria selected, must be adapted to each context, including the type of case (e.g. open, closed or mixed group of victims), the logistical and security constraints and the resources available for the investigators, and they will be framed by applicable law and practice. However, the selection of methods used should always be based on sound scientific and practical considerations within the identification process strategy defined for the corresponding context. Whenever possible, the procedures used should be validated for quality assurance and control.

When forensic DNA analysis is deemed necessary for identification, this must be carried out, whenever possible, by laboratories working under accredited standards (e.g. ISO/IEC 17025:2005 standard for testing and calibration laboratories) [40]. The handling of human remains, samples and data must be done in accordance with the rules regarding informed consent and the protection of personal data, including medical and genetic information [31].

In order to minimize the chances of misidentification, methods commonly used for the forensic identification of large numbers of victims of armed conflict have evolved over the last two decades with the growing use of scientific means of identification (i.e. primary methods), to help ascertain identification beyond reasonable doubt.

It is accepted best practice today, and recommended by the ICRC, to use whenever possible at least one primary method of identification and to corroborate the findings with another primary method or with secondary traits. The use and interpretation of different methods and results towards a reliable identification process requires a multidisciplinary effort which should produce integrated reports of their findings.

13.10 Community and family involvement in a recovery and/or identification process

Whenever possible, forensic investigations into the dead and missing from armed conflict should be carried out in constant interaction with the concerned communities and with the families or their representatives, who have the right to be kept informed throughout. This right is protected under international humanitarian law [41].

The notion of 'family', including its composition and the roles of those included in it, may vary considerably in different cultural contexts. This needs to be borne in mind in order to help ensure the best possible interaction with the families of the missing persons, including the optimal collection of ante-mortem data and reference samples. The participation and information of the bereaved is usually essential for fulfilling the humanitarian goal of the investigation, including for ensuring the understanding and acceptance of results, regardless of whether these satisfy expectations or not.

Communities and families often become deeply traumatized when their members are killed or go missing in the context of armed conflict, and the process of identifying remains may simply add to the trauma if not managed properly. The social and psychological impact of investigations on communities and families should, therefore, be assessed as part of their planning. Investigators also must be aware that any undignified or unskillful handling of remains and information (real or perceived, according to local customs and sensitivities) may further traumatize the families of the missing and imperil the investigation.

The extent to which the communities or families are involved in the investigation and its different stages should be decided on a case-by-case basis, following consultation with the bereaved, community and religious leaders and other concerned stakeholders. The decision should always consider the possible benefit for the families, the avoidance of any interference with the investigation and security issues.

In addition, the concerned communities or families may:

- provide valuable information on the location of human remains or burial;
- aid in the identification of some remains after their recovery, including through the recognition of unique traits and belongings;
- offer valuable advice on particular health and safety issues.

Informing the communities and families about the exhumation requires a communication strategy. Realistic appraisals must be given of the outcome and information should be up-dated regularly. Investigators should be aware that:

- family observation of some stages of the forensic investigation (e.g. exhumation) may lead to easier acceptance of the results;
- religious or community leaders should be contacted and consulted in some contexts, including for the approval and authorization of forensic procedures (e.g. handling of human remains).

In addition, building trust and establishing a good relationship with the families of those missing is essential for the adequate collection of ante-mortem data and reference samples. Such collection may not require the direct involvement of forensic specialists, in which case those responsible for their collection need to be properly trained and monitored by forensic practitioners for quality control and assurance.

In principle, the aim should be to carry out only one interview per family of a missing person (even if conducted in several phases). Multiple interviews and subsequent requests for reference samples or further information on the missing individual might further traumatize the family.

Psychosocial support for the bereaved and concerned communities should be systematically planned and made available if required as an integral part of any forensic investigation into the dead and missing from armed conflict [42].

13.11 Emerging trends and opportunities in humanitarian forensic sciences

Since the 2003 Conference, the ICRC has observed growing recognition worldwide about the importance of the proper management and identification of the dead and missing from armed conflict and catastrophes, for fulfilling the right of families to know the fate of their missing loved ones and to mourn their dead. This has prompted a shift of paradigm in some contexts affected by armed conflict, such as Colombia and Iraq, concerning the main objective of large-scale forensic investigations into the dead and missing. The focus has changed from being primarily on establishing the cause and manner of death of victims to ensuring their proper management and identification.

The growing use of multidisciplinary forensic teams for carrying out these investigations the world over is promoting the development of several fields of forensic sciences applied to humanitarian investigations, including archaeology, criminalistics, anthropology and genetics.

The increase in demand worldwide for proper investigations into the missing has led to a growth in the number of forensic practitioners and institutions dedicated to these investigations, including specialized governmental institutions in dozens of countries around the world, intergovernmental and international organizations, such as the United Nations and the ICRC respectively, as well as national and international non-governmental organizations. It is expected that the demands for forensic practitioners available to work in this field will continue to grow in the following years.

The need for standards of forensic best practice, ethics and quality assurance and control, such as those from the ICRC, require that these be disseminated, promoted and supported. National and international forensic institutions and bodies, such as the AAFS and the International Association of Forensic Sciences, can play a leading role in this. Some national standards and recommendations for forensic best practice, such as the 2009 report *Strengthening Forensic Science in the United States* [43], may also offer valuable guidance for practitioners and institutions involved in international humanitarian investigations, and these should be disseminated accordingly. Similarly, national guidelines on forensic best practices, such as those developed in the USA by scientific working groups on several forensic disciplines, may be of interest to practitioners and institutions in other countries and regions.

The development in global forensic sciences applied to investigations into the dead and missing from armed conflict and catastrophes has highlighted the need for ensuring local (i.e. national) implementation of forensic best practices, including for quality assurance and control, that will satisfy the public, including the families of the missing, who increasingly expect accurate, reliable and timely results.

For example, in 2006, the Chilean government launched reforms of its national medicolegal services to help ensure reliable identifications of the missing from the military régime which ruled the country between 1973 and 1990. To help in this process, the Chilean government called for advice from national and international experts, including members of the AAFS, as well as the ICRC. It also ensured transparency for the bereaved families, who were consulted and kept informed throughout the process of reforms. As a result of this process, the Chilean medicolegal services (*Servicio Médico Legal de Chile*) is now regarded as a leading institution in the region for its capacity to investigate the missing [44].

However, in many countries around the world, forensic practitioners are unavailable or unable to cover all needs, especially those arising from extreme situations such as catastrophes and armed conflicts.

In those situations, first responders, including Red Cross or Red Crescent volunteers, can play an important role in supporting concerned forensic practitioners and institutions. This, of course, requires ensuring their training, guidance and material support towards acceptable standards and results. Forensic practitioners and institutions play an essential role in this process, including for the development of necessary guidelines, such the those on crime scene management developed by the United Nations Office on Drugs and Crime [45] and the manual for the management of the dead in disasters developed by the ICRC, PAHO/WHO and the IFRC.

For example, in the aftermath of Cyclone *Nargis*, which battered Myanmar (Burma) in May 2008, it was the first responders there who helped to ensure the proper and dignified management of the dead in that country. They received emergency training and equipment from ICRC forensic experts, as well as the ICRC manual mentioned above (translated into Burmese), to accomplish this task. Their timely action helped many of the bereaved to recover and mourn their missing loved ones, and it also helped to convince the authorities into supporting further development of local forensic capacity to better respond to future catastrophes.

After the devastating earthquake which struck Haiti on January 12, 2010 first responders, including civil defense officials, Red Cross volunteers and mortuary staff, played an important role in managing the dead in a country which was lacking medicolegal services. They received advice, training and equipment from the ICRC to carry out this task, which helped to ensure that many of the dead, including the unclaimed – such as inmates who died in the prison of Port Au Prince – were managed properly and in a dignified way (Figure 13.6).

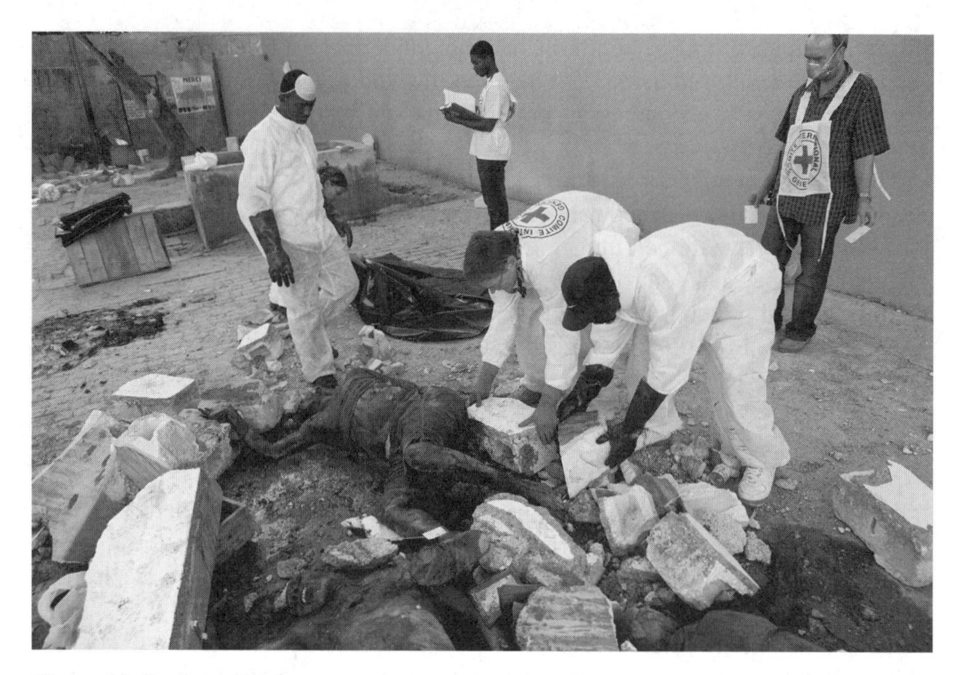

Figure 13.6 Haiti: the ICRC carrying out an emergency forensic operation to recover and help identify the remains of inmates who died in the central prison of Port Au Prince during the earthquake of 12 January 2010.

Worldwide recognition of the role and need for forensic sciences in the investigation of the dead and missing from armed conflicts and catastrophes has been accompanied over the last decade by a noticeable increase in international cooperation, capacity-building and opportunities for research and development in this field.

International cooperation in forensic capacity building for the management and identification of the dead from armed conflict and catastrophes (as pioneered by the AAAS in Argentina in 1984) is now observed in all regions of the world, with the United States of America still leading the way – including, for example, through the International Criminal Investigative Training Assistance Program of the US Department of Justice [46].

A growing number of countries affected by humanitarian tragedies and which have developed their forensic capacity accordingly, such as the cases of Chile and Colombia in Latin America, are growingly involved in regional and international forensic cooperation, offering their forensic experience and resources to advise, train and support other countries in the region. There is also growing understanding on the part of those offering forensic assistance that local forensic capacity-building requires shifting from models of substitution towards effective transfer and exchange of knowledge, technology and tools, especially for those working in under-resourced contexts. This approach offers true opportunities for local forensic practitioners in affected countries to become future sources of advice, training and research, thus benefiting the wider forensic community.

A good example of local forensic capacity-building and the implementation of best practices in the Mediterranean region is the work carried out by the Committee on Missing Persons (CMP) in Cyprus [47]. Since re-launching its activities in 2004, the CMP has framed its work in line with the recommendations from the 2003 Conference, for which it received advice and support from the ICRC and a group of international forensic experts, including from Argentina, Canada, Colombia, Ireland, the United Kingdom and the United States. This aimed to help develop a local team of forensic scientists, under the CMP's mandate, to investigate the missing in the island, dating from the 1960s and 1970s, for humanitarian purposes.

The CMP and its all-Cypriot forensic team has since become a model of best practice for its integrated approach applied to the recovery and identification of the missing, in full consideration of family needs. In addition, the CMP is the only official bi-communal project which is fully operational in Cyprus, and it may therefore also stand as a model for peace-building in the war-divided island.

The forensic investigation into the dead and missing from armed conflict and catastrophes offers opportunities for context-specific research and development of forensic methods best suited to particular contexts and needs, including in areas such as:

- new exploratory and excavation techniques and methods;
- the management of large forensic operations;
- population-specific standards and frequencies;
- particular taphonomic dynamics and dating of human remains;
- war-wound ballistics;
- the integration of different identification methods in resource-challenged contexts, etc.

Finally, the need for improved communication, coordination and cooperation between forensic practitioners investigating the missing calls for empowering a truly global forensic community. The fifth meeting of the ICRC-sponsored *Ibero-American Network of Medico-Legal Institutes*, held in September 2011 in Madeira, Portugal, during the 19th triennial meeting of IAFS, is an example of this. This network of medicolegal and forensic institutes from Latin American countries, Portugal and Spain is helping to improve bilateral forensic cooperation between those countries, including for disaster response. The network, which was launched in 2007, has set an example for similar initiatives in the Asia/Pacific region and in Africa, which are helping bring together forensic practitioners and institutions from those regions, often for the first time and with the shared goal of building a truly international forensic community.

13.12 Conclusion

The dignified and professional management of the dead and of human remains, and the clarification of the whereabouts and fate of persons missing from armed conflict and catastrophes, is a humanitarian priority and an obligation under international law.

Much has been achieved over the past two decades; the forensic recovery and identification of tens of thousands of victims worldwide has given those victims' families a possibility of closure. The global forensic community's contribution to this remarkable achievement has been immense but much remains to be done, with hundreds of thousands of victims of war and disasters still waiting for recovery and identification the world over.

Developments in the field of forensic sciences applied to humanitarian goals over the last decades have been impressive. They now offer novel opportunities for the forensic profession for building global forensic sciences and for furthering their contribution for addressing the consequences of war and catastrophes all over the world.

References

1. *The mission of the International Committee of the Red Cross.* http://www.icrc.org/HOME.NSF/060a34982cae624ec12566fe00326312/125ffe2d4c7f68acc1256ae300394f6e?OpenDocument (accessed October 14, 2011).
2. *International humanitarian law and International human rights law: Similarities and differences.* ICRC Advisory Services, Geneva, 2003. http://www.ehl.icrc.org/images/resources/pdf/ihl_and_ihrl.pdf (accessed October 2, 2011).
3. *ICRC receives award for contributions to humanitarian forensic sciences.* http://www.icrc.org/eng/resources/documents/news-release/2011/forensics-news-2011-09-14.htm (accessed October 6, 2011)
4. *The Geneva Conventions of August 12 1949.* Geneva: ICRC publications; 2002.
5. *Protocols Additional to the Geneva Conventions of 12 August 1949.* Geneva: ICRC Publications; 1977.
6. Henckaerts J-M, Doswald-Beck L. *Customary International Law, Vol. 1: Rules.* Geneva: Cambridge University Press; 2006.
7. Rebels drag soldiers' bodies through Mogadishu streets. *The Guardian* 2009 March 21. http://www.guardian.co.uk/world/2007/mar/21/1 (accessed October 2, 2011).
8. Estonia to remove Soviet memorial. *BBC News* 2007 January12. http://news.bbc.co.uk/2/hi/europe/6255051.stm (accessed October 5, 2011).
9. Petrig A. The war dead and their gravesites. *International Review of the Red Cross* 2009; 874:341–369.
10. Interpol, ICPO-Interpol General Assembly, 65th session, resolution AGN/65/RES/13. 1996. https://www.interpol.int/Public/ICPO/GeneralAssembly/Agn65/Resolutions/AGN65RES13.asp (accessed October 10, 2011).

11. (US) Joint POW/MIA Accounting Command (JPAC). http://www.jpac.pacom.mil/ (accessed October 5, 2011).
12. Renshaw L. *Exhuming Loss, Memory and Mass Graves of the Spanish Civil War.* Walnut Creek, California: Left Coast Press; 2011.
13. Debons D,Fleury A,Pitteloud J-F, editors. *Katyn and Switzerland: Forensic Investigators and Investigations in Humanitarian Crises*, 1920–2007. Geneva: Georg; 2009.
14. *US Congress Hearings on Katyn War Crime.* http://www.kanada.net/war/us_congress_doc.html (accessed October 5, 2011).
15. Snow C, Levine L, Lukash L, Tedeschi L, Orrego C, Stover E. The investigation of the human remains of the "disappeared" in Argentina. *American Journal of Forensic Medicine and Pathology* 1984; 4: 297–303.
16. The Argentine Forensic Anthropology Team (Equipo Argentino de Antropología Forense, EAAF). http://www.eaaf.org/ (accessed October 3, 2011).
17. Office of Legal Affairs, United Nations. *Guidelines for the conduct of United Nations inquiries into allegations of massacres.* New York: United Nations Publications; 1995.
18. Blewitt G. The role of forensic investigations in genocide prosecutions before an international tribunal. *Medicine, Science and The Law* 1997; 37(4): 288.
19. ICMP – International Commission on Missing Persons. http://www.ic-mp.org/ (accessed October 4, 2011).
20. ICRC Report: *The Missing and Their Families. Summary Conclusions arising from Events held prior to the International Conference of Governmental and Non-Governmental Experts* (19–21 February 2003). http://www.icrc.org/eng/assets/files/other/icrc_themissing_012003_en_10.pdf (accessed October 6, 2011).
21. ICRC Report: *The Missing and their Families.* ICRC publications, Geneva; 2003 http://www.icrc.org/eng/resources/documents/misc/5jahr8.htm (accessed October 6, 2011).
22. Tidball-Binz M. Forensic Investigations into the Missing: Recommendations and Operational Best Practices. In: Schmitt A, Cunha E, Pinheiro J, editors. *Forensic Anthropology and Medicine, Complementary Sciences From Recovery to Cause of Death.* Totowa, New Jersey: Humana Press; 2006; 383–407.
23. Plan Nacional de Búsqueda, Comisión de Búsqueda de Personas Desaparecidas, Bogotá, 2007. http://www.comisiondebusqueda.com/Documentos/PLAN%20NACIONAL.pdf (accessed October 5, 2011)
24. *The DVI Guide, Interpol* (available in electronic version only). http://www.interpol.int/INTERPOL-expertise/Forensics/DVI-Pages/DVI-guide (accessed October 2, 2011).
25. Morgan O, Tidball-Binz M, VanAlphen D. *Management of dead bodies after disasters: A field manual for first responders.* Washington, DC: PAHO/WHO, ICRC, IFRC; 2006.
26. Tidball-Binz M. Managing the dead in catastrophes: guiding principles and practical recommendations for first responders. *International Review of the Red Cross* 2007; 866: 421–442.
27. *The Missing: Progress Report*, ICRC publications, Geneva; 2006. http://www.icrc.org/eng/resources/documents/publication/p0897.htm (accessed October 1, 2011).
28. *Missing Persons: A hidden tragedy*, ICRC publications, Geneva;2007. http://www.icrc.org/eng/assets/files/other/icrc_002_0929.pdf (accessed October 1, 2011).

29. *World Medical Association statement on forensic investigations of the missing, adopted by the WMA General Assembly, Helsinki, 2003*. Main page of the WMA. http://www.wma.net/e/policy/m34.htm (accessed October 12, 2011).
30. *United Nations Manual on the Effective Prevention and Investigation of Extra-Legal, Arbitrary and Summary Executions*, U.N. Doc. E/ST/CSDHA/.12; 1991. http://www1.umn.edu/humanrts/instree/executioninvestigation-91.html (accessed October 5, 2011).
31. *Missing people, DNA analysis and identification of human remains: a guide to best practice in armed conflicts and other situations of armed violence*. Geneva: ICRC publications; 2009.
32. *Operational best practices regarding the management of human remains and information on the dead by non-specialists, for all armed forces, for all humanitarian organizations*. Geneva: ICRC publications; 2004.
33. *Wound ballistics: An introduction for health, legal, forensic, military and law enforcement professionals*. Geneva: ICRC publications; 2008.
34. Bierens de Haan B. *Humanitarian action and armed conflict: coping with stress*. Geneva: ICRC; 2001.
35. Roberts DL. *Staying alive: Safety and security guidelines for humanitarian volunteers in conflict*. Geneva: ICRC; 1999.
36. *Dealing with the Stress of Recovering Human Dead Bodies*. US Army Center for Health Promotion and Preventive Medicine. http://nvmassfatality.com/images/Dealing-with-the-stress-of-recovering-human-dead-bodies.pdf (accessed October 3, 2011).
37. Stover E, Shigekane R. The missing in the aftermath of war: When do the needs of victims' families and international war crimes tribunals clash? *International Review of the Red Cross* 2002; 84 (848):845–847.
38. Chapple A, Ziebland S. Viewing the body after bereavement due to traumatic death: qualitative study in the UK. *BMJ* 2010 April 30; 340:c2032.
39. Tarassenko S. Benefits of viewing the body. *BMJ* 2010 May 29; 340: c2757.
40. ISO/IEC 17025: *2005 – General requirements for the competence of calibration and testing laboratories*. http://www.iso.org/iso/catalogue_detail.htm?csnumber=39883 (accessed October 10, 2011).
41. Tidball-Binz M. *An interview: 'Missing persons: providing support to the grieving'*. http://www.icrc.org/eng/resources/documents/interview/missing-interview-100423.htm (accessed October 3, 2011).
42. *Accompanying families of missing persons, a field guide*. Geneva: ICRC publications; 2011. In press.
43. National Research Council. *Strengthening Forensic Science in the United States: A Path Forward*. Washington, DC: The National Academies Press; 2009.
44. Entrevista a Andrés Aylwin. www.sml.cl/portal/pdfs/Revista_Octubre2008_sml.pdf (accessed October 2, 2011).
45. United Nations Office on Drugs and Crime. *Crime scene and physical evidence awareness for non-forensic personnel*. New York: United Nations; 2009.
46. US Department of Justice. *International Criminal Investigative Training Assistance Program (ICITAP)*. http://www.justice.gov/criminal/icitap/ (accessed October 13, 2011).
47. Committee on Missing Persons in Cyprus – CMP. http://www.cmp-cyprus.org/nqcontent.cfm?a_id=1 (accessed October 9, 2011).

Further reading

Joyce C, Stover E. *Witnesses from the grave*. Boston: Little, Brown and Company; 1991.

Ferllini R, editor. *Forensic Archaeology and Human Rights Violations*. Springfield: Charles C. Thomas; 2007.

Steadman DW, editor. *Hard Evidence, Case Studies in Forensic Anthropology*. Upper Saddle River: Prentice Hall; 2003.

Blau S, Ubelaker DH, editors. *Handbook of Forensic Anthropology and Archaeology*. Walnut Creek: Left Coast Press; 2008.

Payne-James J, Busuttil A, Smock W, editors. *Forensic Medicine, Clinical and Pathological Aspects*. London: GMM; 2003.

Kimmerle EH, Baraybar JP, editors. *Skeletal Trauma: Identification of Injuries Resulting from Human Rights Abuse and Armed Conflict*. Boca Raton: CRC; 2008.

Adams B, Byrd JE, editors. *Recovery, Analysis and Identification of Commingled Human Remains*. Totowa: Humana Press; 2008.

Argentine Forensic Anthropology Team (Equipo Argentino de Antropología Forense, EAAF). *Argentine Forensic Anthropology Team 2007–2009 Triannual Report*. New York: EAAF publication; 2010. http://www.eaaf.org/ (accessed October 5, 2011).

Forsythe PD, Rieffer-Flanagan BA. *The International Committee of the Red Cross, A Neutral Humanitarian Actor*. London: Routledge; 2007.

14

Forensic systems and forensic research: an international perspective

DN Vieira

University of Coimbra, Coimbra, Portugal

14.1 Introduction

The world is a mosaic of different realities – geographic, social, cultural, economic, political, legal, and so on. These differences occur even in parts of the globe where geographic, socio-cultural and political proximity might lead us to expect greater homogeneity. Europe (and particularly the 27 countries of the European Union) is a clear example of this. In fact, despite many attempts at harmonization, the countries that make up the European Union continue to form a mosaic of different realities on various levels.

These discrepancies are inevitably reflected in the organization and functioning of the forensic systems established in different countries [1]. In fact, among the 193 countries recognized by the United Nations, there are enormous differences in the way the forensic services are structured and organized. There are even some countries that have no forensic service providers at all, either because they are too poor and underdeveloped, or because their small size does not justify the creation of departments of their own, leading them to depend instead upon those of neighboring countries (Luxembourg is a case in point).

What is more, a country's level of economic and socio-cultural development is not always directly proportional to the development and structuring of its forensic services. Many of the so-called 'first world' countries reveal marked gaps and deficiencies in the organization and functioning of their forensic services, while other countries that are considered less developed – or even underdeveloped – may

Forensic Science: Current Issues, Future Directions, First Edition. Edited by Douglas H. Ubelaker.

have remarkably sophisticated forensic systems, at least as regards the quality and training of the specialists involved.

These national differences are evident on a number of different levels. Firstly, there are differences in the very concepts and terminology used. For example, there are many countries in the world where the term 'forensic science' does not even exist. These are usually countries of French influence, where the term 'medicolegal sciences' is traditionally used to refer to what many other countries (particularly those of Anglo-Saxon influence) call the 'forensic sciences'. These differences may become more pronounced in specific areas, such as forensic medicine. The term 'legal medicine' is used in some places to refer to forensic medicine in general while, in others, it is restricted to medical or health law.

Other countries have attempted to overcome these ambiguities by referring to the broad area as 'forensic and legal medicine', a solution that has in fact been chosen by one of the international journals in the field, the *Journal of Legal and Forensic Medicine*. However, the discussion remains heated about the correct terms to be used in many areas [1].

These national differences are also reflected in forensic procedures and practices, and in the training and preparation of the professionals involved. Taking again the example of forensic medicine, in some countries, the specialists that work in this field (called 'forensic pathologists') only perform forensic autopsies and issue expert opinions relating to cadavers and death. In other countries, forensic pathology may also involve examinations of living persons within the sphere of criminal law. In still others, the same activity extends even further, covering all medical examinations that have forensic implications, such as personal injury evaluations for insurance compensation. In these countries, practitioners are more often referred to as 'forensic medical specialists' rather than 'forensic pathologists'.

There is also a great deal of variety between countries as regards the training of such professionals, even when the skills attributed to them are similar. For example, in countries where forensic medicine is not very highly regarded as a medical specialty, any doctor – even one without any specialist training – may be called to intervene in the forensic sphere. Then there are others where it brings a social status equivalent to, or even superior to, the other medical specialties, and where specialist training is highly structured, sometimes lasting as long as five years.

Similarly, there is a great variety of approaches in the field of forensic practice, even when the concept and content of the area is similar. For example, the situations and rules that determine when a forensic autopsy takes place vary considerably from country to country. In some, all cases of violent death are automatically referred for autopsy. In others, the decision will be taken by a coroner (who may or may not have adequate training) or by a magistrate, or it may depend upon the whim of an individual doctor.

In some countries, the forensic medicine specialist or forensic pathologist (depending on the term used) is involved in the investigation from the outset, examining the scene where the body was found as part of the autopsy procedure. In others, however, that never happens, or the examination of the scene and clothing, like the complementary histological analyses, is the responsibility of different professionals or institutions who do not liaise, and who therefore produce different reports that are never compared. In some other countries, the medical specialist has the autonomy to request any complementary tests that s/he considers scientifically justifiable for the autopsy. In others, however, this can only be done with authorization from the organization or individual responsible for the process (often a magistrate). Hence, the medical expert may have to conclude his/her report without having had the opportunity to undertake specific examinations that might have been fundamental for the decision.

There are also great differences in the way these systems are organized, and the more developed countries do not always offer the best examples. Depending upon the country, forensic services may be overseen by the Ministry of Justice, Ministry of Home Affairs (or its equivalent), Ministry of Education (i.e. the universities), Ministry of Health, Ministry of Sciences, etc. – or they may be split between various ministries.

There are even countries where forensic science has gradually shifted into the private sector. In some countries, all forensic services are centralized in a single structure – a National Institute of Forensic Sciences or National Institute of Legal Medicine (which, despite being so named, may nevertheless involve many areas of forensic science beyond the strictly medical). In others, there may be multiple institutes operating simultaneously on a national, regional or even municipal level, while others may have centralized laboratories belonging to the scientific police (which frequently bring together all areas of forensic science except the medical), or diverse forensic laboratories (which may or may not belong to the various police forces), operating in areas that may overlap with the services provided by medical legal institutes and universities. In many cases, this lack of interconnection has led to an enormous duplication of human and technological resources, resulting also in a substantial loss of efficiency.

Thus, there is an incredible variety in the way the forensic services are structured and, indeed, in their respective missions. In some countries, their activity is totally restricted to the provision of forensic evaluations, while in others it also involves teaching (at undergraduate and postgraduate level) and scientific research (often enshrined in law).

14.2 What is the ideal system?

We might ask, then, if there exists an ideal medical legal and forensic system – a perfectly balanced approach that could function as a model for everyone? The answer is obviously 'no'. There are no perfect systems. All have their advantages and their drawbacks. Given the great differences in the geographic dimension, economic power, socio-cultural traditions, organizational structures and legal systems of different countries, there can be no single forensic system of global reach. Indeed, any forensic system can be a good system, provided that it guarantees competence, objectivity, independence, impartiality, quality, speed and prudence.

And in situations where the forensic services are divided between different departments and institutions, close collaboration and coordination are also required to avoid duplicating resources and dispersing expertise.

Despite this, experience has clearly shown that the best results are achieved by countries that have created unified forensic science institutes, where all forensic professionals work together, maximizing the effectiveness of facilities, resources and expertise. Indeed, these countries stand out from the others in the international panorama. All areas of forensic science today are highly interdisciplinary; no one can have access to all the knowledge and perspectives that are necessary for the success of a forensic investigation. For this reason, there needs to be joint coordinated work involving total collaboration between various professionals. Competition and disputes between different forensic agencies and institutions only jeopardize the investigation and generate individual and collective frustrations.

But a quality forensic system cannot limit itself to the mere provision of expert evaluations, however well that is done. Excellence can only be achieved if this activity is also linked to teaching and scientific research [2]. Indeed, these three areas need to be developed simultaneously, as they complement each other and generate new potential. There is nothing better than teaching to ensure the constant updating of knowledge, as the expert tries to respond to the anxieties, doubts and demands of the learner; and there is nothing better than scientific research to develop expertise and stimulate progress in the forensic sciences, as well as ensuring an awareness of the potential and limitations of forensic activity.

In truth, the forensic sciences offer enormous potential for scientific research, particularly when they are oriented towards the general community [3–6]. What better approach is there for understanding the crime and accident rates in our societies? And who is better qualified than the forensic scientist to study phenomena such as drug abuse, suicide and domestic violence, or to detect risk factors and causes of sudden infant or adult death? Understanding does not imply acceptance; rather, it is an important step in the promotion of more effective prevention and control measures.

14.3 Why research is often absent

Research in the forensic sciences can also have significant impact on other areas of knowledge. As mentioned above, the forensic services in many countries do not always have a research component and, as a result, these countries continue to show low or negligent levels of scientific productivity. Reasons include:

1. *Staff shortages and excess workload.* Many forensic departments have too many demands made on them for the number of specialists available. Overloaded with expert evaluations, specialists have no time to devote to scientific research.

2. *Lack of professional and economic incentives.* In many departments, research counts for very little on the curriculum of forensic experts, contributing much less to career advancement and salary than expert evaluations.
3. *Isolation from academic communities.* Many forensic departments have no connections with the academic world (in some cases, even an aversion to it), leading to a level of isolation that is not propitious to scientific research. Experience shows that forensic departments with university connections generally have higher levels of research productivity and quality.
4. *Lack of training in scientific research.* Although many countries now have very well-structured training programs in the various forensic careers, most do not offer any kind of training in scientific research. This is something that could be included in the initial training of any forensic specialist, as well as a strong accent in ethical behavior. In fact, there is nothing worse than a fraudulent forensic scientist [7].
5. *The fact that scientific research is not an institutional priority.* Most forensic organizations and institutions are interested above all in the volume of expert evaluations provided. Unfortunately, scientific productivity is not properly valued by the heads of forensic departments or their supervisory institutions in many countries.
6. *Shortage of funds.* The funding of scientific research is always a complex and difficult matter, and funding bodies do not always give priority to projects in the forensic sphere. Indeed, forensic science is one area of knowledge that tends to be overlooked by international and national funding institutions. As department budgets are also usually quite limited, it is difficult to make funds available for scientific research, which is not generally considered to be a priority.
7. *Language problems.* Today, most research is conducted and disseminated in English, to the disadvantage of those that do not have sufficient mastery of the language.
8. *Legal restrictions upon the development of scientific research based on forensic investigations.* In many countries, there are legal obstacles preventing the use of samples collected during forensic investigations for the purpose of research and also blocking access to expert reports for retrospective studies.

14.4 The importance of research and cooperation

Research should be a strong commitment for departments. There are many new areas open to research in the forensic field and still many researches to be done in classical forensic themes [8–14]. Even case reviews still may have importance [15].

The promotion and development of research should today be a concern, even a priority, for any forensic department, but particularly for those that wish to achieve standards of excellence. This is an area where international cooperation is absolutely essential. Cooperation brings all kinds of benefits [16,17]:

- maximizing the cost-effectiveness of facilities and equipment;
- enabling the pooling of different kinds of knowledge, experience and perspectives;
- facilitating access to forensic samples and situations;
- creating synergies through the integration of different realities.

However, there is still a long way to go, and a substantial change in mindset is required before it can be fully realized. This will affect all the parties involved:

- The directors of forensic departments, who must begin to see research as one of their main objectives, rather than as secondary to the provision of expert evaluations.
- The forensic professionals themselves, who must realize that their everyday activity can become stagnant if it is not informed by scientific research, which can only improve their knowledge and performance.
- Scientific funding bodies, which must understand the extraordinary potential offered by the forensic sciences for the various branches of knowledge.
- And, finally, the institutions that oversee forensic departments, which have to stop viewing these merely as forensic service providers and begin perceiving them also as important centers for teaching and scientific research.

It is also true that changes are necessary in the way we evaluate the quality of research and promote it. The value of impact factors and citation indexes needs an urgent discussion and review, and research should not only focus essentially in the areas that potentially give more visibility and citations [18–21]. In the main, this will have a negative impact in the progress of forensic sciences.

14.5 Conclusions

International cooperation today has an important role to play in the field of forensic research. However, this cooperation is not easy to achieve, and this is particularly true for those that operate in more isolated parts of the world, who may be limited by a lack of financial resources and the physical, technological and human structures necessary for research. In this sense, national and international scientific societies can play an important role by facilitating contacts and the exchange of knowledge, and also by stimulating joint projects.

The recent creation of the International Network for Forensic Research is a move in the right direction. It links forensic researchers throughout the world, enabling contacts to be made and joint research to be carried out, thus taking maximum advantage of potential and synergies.

References

1. Madea B, Sauko P, editors. *Forensic Medicine in Europe*. 1st ed. Lübeck: Schmidt-Römhild; 2008.
2. Madea B, Saukko P. Forensic medicine as an academic discipline – Focusing on research. *Forensic Science International* 2007; 165:87–91.
3. Madea B, Saukko P, Musshoff F. Tasks of research in forensic medicine – different study types in clinical research and forensic medicine. *Forensic Science International* 2007; 165:92–97.
4. Koller M, Lorenz W. Study types and study issues in clinical forensic medicine. *Forensic Science International* 2007; 165:98–107.
5. Maurer HH. Demands on scientific studies in clinical toxicology. *Forensic Science International* 2007; 165:194–198.
6. Schemeling A, Geserick G, Reisinger W, Olze A. Age estimation. *Forensic Science International* 2007; 165:178–181.
7. Disney RHJ. Fraudulent forensic scientists. *Journal de Médecine Légale Droit Médical* 2002; 45:225–230.
8. Madea B Musshoff F. Postmortem biochemistry. *Forensic Science International* 2007; 165:182–184.
9. Cattaneo C. Forensic anthropology: developments of a classical discipline in the new millennium. *Forensic Science International* 2007; 165:185–193.
10. Drummer OH. Post-mortem toxicology. *Forensic Science International* 2007; 165: 199–203.
11. Schneider PM. Scientific standards for studies in forensic genetics. *Forensic Science International* 2007; 165:238–243.
12. Peters FT, Drummer OH, Musshoff F. Validation of new methods. *Forensic Science International* 2007; 165:216–224.
13. Grellner W, Madea B. Demands on scientific studies: Vitality of wounds and wound age estimation. *Forensic Science International* 2007; 165:150–154.
14. Henssge C, Madea B. Estimation of the time since death. *Forensic Science International* 2007; 165:165–171.
15. Madea B. Case histories in forensic medicine. *Forensic Science International* 2007; 165:111–114.
16. Tambuscio S, Boghossian EE, Sauvageau A. From abstract to publication: the fate of research presented at an annual forensic meeting. *Journal of Forensic Sciences* 2010; 55(6):1494–1948.
17. Pao ML. Global and local collaborators: a study of scientific collaboration. *Information Processing and Management* 1992; 28(1):99–109.
18. Jones AW. Which articles and which topics in the forensic sciences are most highly cited. *Science and Justice* 2005; 45:174–180.

19. Jones AW. Crème de la crème in forensic science and legal medicine – the most highly cited articles, authors and journals 1981–2003. *International Journal of Legal Medicine* 2005; 119:59–65.
20. Jones AW. The distribution of forensic journals, reflection on authorship practices, peer-review and role of the impact factor. *Forensic Science International* 2007; 165:115–128.
21. Coleman R. Impact factors: use and abuse in biomedical research. *The Anatomical Record* 1999; 257:54–57.

15

Summary and conclusions

Douglas H. Ubelaker

Department of Anthropology, National Museum of Natural History, Smithsonian Institution, Washington, D.C.

15.1 Introduction

As discussed in the introductory chapter, contributing authors within this volume were provided only very general guidelines regarding the content of their contributions. Their responses present an insightful, detailed and authoritative account of key professional issues within the forensic sciences. The chapters reveal the diversity of activity within the forensic sciences, while providing international context and perspective. They also discuss the deep academic roots of the forensic disciplines, relevant technological developments, current issues and future directions.

This volume is the first comprehensive presentation of the forensic sciences written by an international group of practicing professionals representing the American Academy of Forensic Sciences. Perspective is provided from all 11 sections of the American Academy of Forensic Sciences and from two of our international plenary speakers at the 2012 AAFS meeting in Atlanta, Georgia. The authors have collectively sought to present a full range of issues and perspectives from the diverse field of forensic science. All have succeeded.

15.2 General forensics

The General Section of AAFS formed in 1967; by 2011, it had 711 members.

General forensics is placed first in the chapter line-up for two major reasons. First, within the AAFS, it represents specialists in 18 accepted sub-disciplines of forensic science that are not represented clearly in the other ten sections. As noted by the authors, the General Section is considered to be the 'gate-keeper' of the Academy and bears the responsibility of determining whether new areas of forensic

Forensic Science: Current Issues, Future Directions, First Edition. Edited by Douglas H. Ubelaker.

activity are sufficiently robust and scientific to merit AAFS recognition. This vetting process has always been valuable in Academy history and in the forensic sciences more generally, but it is especially important today. With forensic science under increasing scrutiny from the legal community and beyond, it is crucial that we maintain high standards and recognize forensic applications that reflect solid, legitimate science. Colleagues in our General Section make sure that our forensic gate is well maintained.

The second reason for presenting general forensics early in the volume is the section's crime scene investigation, medicolegal death investigation and forensic nursing components. As succinctly stated by the authors, no one starts until they finish. Through these activities, forensic science is introduced into the process of recovery, analysis and interpretation. Other scientific disciplines cannot attain the best results unless the evidence has been properly recovered and documented.

The importance of their work cannot be overstated, since these investigators are responsible not only for the recovery of primary evidence but also for maintaining its integrity. The context of crime scene and medicolegal data collection is vital in most areas of the forensic sciences, so final interpretation frequently depends upon the quality of the initial documentation and observations. Attention to detail, with an open mind and innovative thinking, is also important in this practice. A thoughtful investigator may notice a potential clue that all others have overlooked because it is not the usual sort of expected evidence.

The authors also note appropriately that activity in their area of the forensic sciences has been highlighted by the popular media. While television portrayals of crime scene investigation have generated significant public exposure and have likely attracted bright young minds to the field, they also have created public perceptions of forensic science activity that are fictitious. The 'CSI effect' involves unrealistic expectations among members of juries and beginning students. Reality is, of course, more mundane and less dramatic than the stories presented on television. Paying attention to the popular understanding of our field, however, allows us to negotiate public relations challenges more intelligently.

A special focus on forensic nursing is included in this chapter, as the field represents an area of major scholarly interest and growing international recognition. Objectivity is an issue throughout the forensic sciences, but the thoughtful discussion of potential sources of bias in forensic nursing is particularly praiseworthy. The discussion shows that, to maintain objectivity, practitioners must be aware of the way in which employment context and case history can affect professional judgment.

15.3 Criminalistics

The Criminalistics (formerly Police Science) Section of the AAFS formed in 1950. In 2011, the Criminalistics Section was the largest section of the Academy, with 2,585 members.

The field of criminalistics includes laboratory management and the analysis of illicit drugs, DNA, fire debris and trace evidence. The discussion of management in this section shows the importance of organizational and quality control issues in modern forensic science. Effective management of the forensic process or forensic laboratory is paramount in meeting demands for quality, timely analysis. This discussion also reveals the growing business component of forensic science work. Budgets, personnel issues, equipment maintenance and improvement, deadlines, procedure definitions, backlogs and politics all need to be addressed in effective management.

The discussion on illicit drugs reveals the complexity of work in this area and how it has evolved over time. Technological advances clearly have been important factors in this evolution, as well as political, legal and social concerns regarding illicit drugs. Readers should note the need for scientific certainty in this field. The authors write that 'there must be no reasonable doubt as to the identity of the drug(s) present.' Such accuracy demands high laboratory standards and effective technology, as well as excellent management as discussed above.

This section presents a thorough and informative discussion of DNA analysis. The forensic introduction of molecular techniques in the late 1980s revolutionized identification procedures and impacted related areas of forensic science. The authors show how this field enjoys robust research efforts aimed at automation, improvement in quality control, faster analysis and the ability to work on more limited evidence.

The authors emphasize that fire scene investigation and the laboratory analysis of fire debris require thorough documentation firmly anchoring interpretation in science and solid reasoning. Knowledge is key here, not only of the science related to the analysis of recovered materials but also of fire dynamics and related complexities. Research continues to play a major role, as does the development of accepted international standards.

The thoughtful discussion of the analysis of trace evidence reveals how minute evidence can contribute to forensic interpretations with proper analysis. Small fibers of clothing or carpet, soil, paint, hair, glass, pollen and many other materials may prove to be fundamental to case interpretation. This area of forensic science is challenging, not only due to the small size of the evidence, but also to the variety of potential materials involved. Interpretation requires knowledge of a vast range of comparative materials, sophisticated technology and strict laboratory procedures. Since the evidence is small, the threat of contamination is large and, thus, the laboratory environment and procedures must be controlled and effective. Analysis of trace evidence can produce very useful information, as illustrated by the case studies provided.

15.4 Forensic pathology

The Pathology/Biology Section of AAFS was formed in 1950. In 2011, this section had 889 members.

Within AAFS, forensic pathologists are concentrated in our Pathology/Biology Section, along with specialists in clinical pathology, entomology, botany, forensic medicine, veterinary forensics and molecular analysis. The nature of this volume precludes a detailed presentation of each of these specializations, but all of them make valuable contributions to the field.

The chapter's presentation of forensic pathology highlights the powerful role of technology in advancing the practice of autopsy. Regarding molecular analysis within forensic science, we primarily think of individual identification, as discussed in detail within the chapter on criminalistics (Chapter 3). However, discussion within the forensic pathology contribution reveals how molecular testing augments the forensic autopsy in other ways. If resources are available, molecular analysis can detect cardiac disease and diseases of genetic origin, as well as the presence of certain infectious agents. Such analysis can positively impact the quality of the autopsy, providing beneficial information to the living relatives of the decedent.

Technological advances in radiologic imaging have also contributed to the practice of forensic pathology. This chapter surveys the advantages and disadvantages of the three primary imaging modalities: fluoroscopy, radiography and computed tomography (CT). These tools can be employed to determine if dissection/autopsy is needed, to assist autopsy, or even to substitute for conventional autopsy – a use described as 'virtual autopsy'. While these imaging systems are extremely important advances, their use clearly calls for the interpretation and judgment of a properly trained forensic pathologist. The authors also note the potential impact of the technology of magnetic resonance imaging (MRI) in advancing approaches in autopsy.

This chapter demonstrates the profound impact of new research and technology on forensic pathology. It also reinforces the message in Chapter 3 of the need for sound management and judgment in the use of resources. As attractive as these technological advances are, those in authority have to examine their financial resources and make difficult decisions regarding whether the cost of this technology is commensurate with the gains potentially offered. Such decisions are especially challenging for smaller facilities with limited resources.

15.5 Forensic anthropology

The Physical Anthropology Section of the AAFS formed in 1972. In 2011, this section had 427 members.

Although forensic anthropology is represented by its own section within AAFS (physical anthropology) and is usually thought of as a distinct discipline, it actually involves diverse applications and methodologies. Anthropologists enhance recovery operations thought to involve human remains. In the laboratory, they use scientific methodologies to determine if materials recovered are of human origin and to establish the biological profile of sex, age at death, ancestry and living stature. Anthropologists also can make positive identification, primarily from radiographic evaluations of skeletal structures, and can detect evidence of foul play. Trauma analysis calls for thorough training and understanding of post-mortem taphonomic alterations, developmental anatomical variation and the healing process as it relates to the detection and interpretation of injuries sustained earlier in life. Since methodologies vary in all of these areas of analysis and application, anthropologists need broad training and experience to be effective.

As noted in this chapter, the applications of forensic anthropology within the forensic sciences have broadened considerably in recent years. While anthropologists continue to do their work from university and museum environments, increasingly they find employment in medical examiner offices, government programs and organizations that pursue investigations of mass disasters and international violations of human rights. Anthropologists have proven useful in these endeavors, not only for their knowledge of archeological techniques and human anatomy, but also for their skills in statistical analysis, broad population perspective and strong research orientation.

15.6 Forensic toxicology

The Toxicology Section of the AAFS formed in 1950. In 2011, Toxicology had 487 members.

This chapter provides detailed perspective on an area of forensic science that enjoys broad recognition of its research vitality and academic linkages. One challenge faced by all in forensic science, but especially in highly technical disciplines such as toxicology, is the need for clarity in report writing and the avoidance – or at least clarification – of jargon. The authors contend that forensic reports should be understandable to non-specialists, especially given the importance of these documents in different areas of the legal process.

Reports in forensic science tend to be highly variable in structure and terminology. While some variation is normal and acceptable, efforts to improve standardization and communication effectiveness are needed. The authors of the toxicology chapter have articulated this need, but it can also be echoed across the forensic sciences.

The authors also note the importance of adequate training to ensure the quality of future practitioners. While some of this need can be met by in-house laboratory training and accreditation and certification procedures, linkages with academia are important as well. Effective academic linkage not only helps to address training and education issues, but also paves the way for research initiatives and related activities that could go a long way to address current concerns with forensic science.

15.7 Odontology

The Odontology Section of AAFS formed in 1970. In 2011, its membership totaled 429.

For many years, forensic odontologists have provided rapid, cost-effective and accurate personal identifications of deceased individuals through careful comparisons of the dental attributes of recovered remains with ante-mortem dental records. While forensic odontologists continue to prove their worth in individual cases needing identification, they are increasingly involved in identification efforts following mass disasters or events leading to multiple deaths. Sophisticated techniques have evolved to make both ante-mortem and post-mortem information available and suitable for comparison. As noted by the authors, odontologists also are needed for age assessment of the living, especially among the young, when age is questioned and important to establish. While several issues and research needs remain to be addressed in this field, the existing procedures are based on solid science and produce results that can be trusted.

This chapter also provides open and insightful discussion of bitemark analysis, an area of professional activity that has attracted some concern both within and outside the field of forensic odontology. The thoughts expressed pave the way for advancement as the key issues are isolated and addressed.

Odontologists also work closely with professionals in other fields of forensic science, especially forensic pathology and anthropology. The overlap and collaboration with anthropology is especially rich in the area of age estimation. As stated by the authors, research advances in dental age estimation will likely be fueled by activity in both of these fields.

15.8 Forensic psychiatry and forensic psychology

The Psychiatry and Behavioral Science Section (formerly Psychiatry) of AAFS formed in 1950. In 2011, this section had 135 members.

Professionals in this field apply their science to a variety of problems relating to mental conditions, ranging from criminal responsibility issues to false confessions. The authors of this chapter take us through the complexities of professional work in this area of forensic science, including a rich discussion of relevant historical developments. Clearly, attempts to understand both mental illness and more generalized aspects of human behavior have very deep roots. The development of this field has been augmented through activity in professional organizations, including the AAFS.

Prospective students of forensic psychiatry and forensic psychology will find this chapter especially illuminating, due to the way it presents current and potential forensic applications. Activity in this field remains firmly anchored in traditional procedures, but the authors discuss important developments in neuroscience and psychopathy studies that are likely to attract future research attention.

15.9 Forensic document examination

The Questioned Documents Section of AAFS formed in 1950. In 2011, its membership was 197.

Authors of this chapter provide a succinct, comprehensive summary of the history, current practice, issues and future directions of forensic document examination. Faced with diverse evidence, forensic document examiners utilize technology and apprenticeship training in case applications. This field clearly has international participation, since the basic problems addressed are shared globally, and materials such as business machines, pens and ink are frequently produced and used in many different countries.

Considerable perspective is provided on the impact of legal developments and other challenges to the practice of forensic document examination. While some of these challenges have created tension within the field, discussion indicates positive attributes as well. The points raised have stimulated research, including ongoing projects aimed at clarification of accuracy and reliability issues. In addition, a certification program dating back to the 1970s has aided the field's advancement, as well as an active scientific working group and international efforts to standardize methods and incorporate new technology. While the field evolves to address new technology, the forensic document examiner continues to be a much-needed and valued contributor to the practice of modern forensic science.

15.10 Digital and multimedia sciences

The Digital and Multimedia Sciences Section of AAFS formed in 2008. Its membership total was 90 in 2011.

The AAFS section for digital and multimedia sciences was formed in recognition of this rapidly growing field. Technology is central to this area of the forensic sciences. New technology is part of the forensic solution as novel approaches emerge to resolve the complex and varied issues addressed by practitioners in this field. However, new technology can also be problematic, as experts strive to keep pace with rapidly evolving developments in audio, video and image representations. By definition, this field is international in nature, since development of computers, software, communication equipment and related materials is global in scope. This chapter provides an overview of forensic topics in this area, emphasizing the nature of technological development and the associated legal issues. Practitioners in this field also address cybercrime, an area of growing international concern.

Although most think of digital and multimedia science issues as being very recent phenomena, the authors of this chapter trace the history of general digital approaches back to the ancient Mayans, Chinese and others in centuries past. The chapter also summarizes key technological developments that have culminated in the current diversity of digital media. Although a scientific working group has proved to be useful in communication and advancing the field, challenges remain regarding training, certification, accreditation, international communication and keeping up with rapidly changing technology.

15.11 Engineering sciences

The Engineering Sciences Section (previously Engineering) formed in 1980. In 2011, its membership was 156.

The chapter on forensic engineering sciences emphasizes the science-based problem-solving approach, diversity of applications and global variation in practice. In a general sense, industrial development and incidents involving motorized vehicles have led to many diverse applications of forensic engineering. The common denominator in these applications appears to be public safety, with engineering specialists making important contributions to litigation focusing on product failure and/or misuse.

Global perspective in this chapter is provided from North America, the United Kingdom, the Philippines, Bangladesh and Poland. These discussions document international variation in training, certification, case applications, compensation and methodological approaches. While global variation in practice occurs, common ground is apparent in the underlying science and concern for public safety.

15.12 Jurisprudence

The Jurisprudence Section of AAFS formed in 1950. In 2011, its membership was 190.

A book on forensic science would be incomplete without some reference to the legal context of the application of our methodology. It is this legal context that separates our work from the more general academic scientific interests. For new recruits in the forensic science arena, this context requires some training and adjustment from the academic classroom or the research environments to which they may be more accustomed. While it is important that scientists inform lawyers and judges about both the capabilities and limitations of our science, it is equally essential that scientists be informed about the legal parameters of forensic science.

This chapter on jurisprudence presents a balanced review of the legal/science interface that regularly takes place within AAFS and within the general field of forensic science. The Jurisprudence Section represents a vital and valued component of the AAFS, providing legal perspective on a variety of issues and developments.

Other chapters within this volume have discussed modern developments within the legal community affecting the practice of forensic science. These range from the positive accounts of applications of DNA technology to personal identification, to more critical and challenging interactions with other scientific areas. The chapter on jurisprudence provides necessary perspective to these issues. This discussion documents historical development in the legal arena and the dynamics of science/legal interaction. This dynamic continues and, while it generates tension at times, ultimately it strengthens forensic science.

15.13 International humanitarian applications

Recent years have witnessed a surge of global interest in forensic science applied in a humanitarian context to problems related to human rights violations. Few applications of our science are more compelling or more complex. While forensic scientists in a variety of specialized areas are needed to participate in these international endeavors, those interested should proceed with caution, mindful of the nature of the commitment and the personal, professional and political factors of involvement.

This chapter provides a useful and timely summary of the nature of this work with a perspective based in a deep history of involvement, as well as leadership in team development and organization. An overview is provided about the history and nature of forensic involvement in these issues. Special attention is appropriately given to the Forensic Unit of the International Committee of the Red Cross, an emerging leader in international humanitarian applications of forensic science. The chapter also mentions other key organizations such as the Equipo Argentino de Antropología Forense (EAAF) and the International Commission on Missing Persons (ICMP), which have provided leadership on similar global issues. Any colleague who is considering involvement in international human rights initiatives should read this chapter carefully.

Guidelines useful to anyone considering participation in such a project include considerations of issues relating to health and safety, community involvement and procedures of identification. The increase in worldwide demand for forensic activity, as outlined in this chapter, is very real. Those who respond to this demand need to be aware of the potential time commitment and the complexities of involvement outlined in this chapter.

15.14 International forensic systems and forensic research

While the world's forensic scientists share methodology, problems in need of attention and research databases, some regional heterogeneity persists. For those working in distant places and different cultures, it is important to recognize this variation and to attempt to obtain an understanding of the society in which the work is performed. Practicing forensic science without a thorough understanding of the cultural context of the work invites errors and potentially can compromise competence. The team approach to international humanitarian forensic projects can be useful, especially if at least one team member understands the local culture.

This chapter presents an overview of the variation in the international forensic landscape. Globally we share methodology, the same kinds of evidence, problems and general goals but, in reality, variety persists in resource availability, training, research, and organization. While some variation is inevitable in consideration of the major existing differences in history, language, culture and economic situations, this chapter documents its magnitude and how it impacts the practice of forensic science. Timely discussion is also focused on the varied attitudes toward the value of research and linkages with academia.

15.15 History

Many chapters of this book speak to the very deep roots of the forensic sciences. Indeed, evidence of interest in forensic applications and this type of science relate back to the initial periods of recorded history and early law proceedings.

The American Academy of Forensic Sciences was founded in 1950, following prior meetings of the First American Medicolegal Congress in January 1948 and a Steering Committee gathering in New York City in October 1948 [1]. Initial membership was approximately 90 persons, organized into the seven initial sections. In 2011, membership in our now 11 sections has grown to 6,296. Throughout this 61-year period, the AAFS has provided leadership in the development of the forensic sciences through its annual meeting, publications, grants and other activities and programs.

15.16 Training

Discussion within this volume relates considerable variation in the training and preparation required for professional activity in the forensic sciences. Such diversity is particularly high within the chapter covering the General Section (Chapter 2). Minimum requirements for crime scene investigators within the United States range from a high school diploma to a master's degree. For coroners, this variation is also noteworthy. While some coroners hold medical degrees, the requirement in some states is only adequate age and state residency. Clearly, variation in training is more pronounced within a global context.

The chapter on anthropology (Chapter 5) addresses another key issue in training within the forensic sciences, namely whether preparation should be general or specific to the methodology employed. This chapter notes that students entering the field encounter a variety of academic training options within both forensic science and anthropology. However, the question remains: should a student entering this field begin by acquiring a broad academic background in fundamental anthropology, or seek specialized training in forensic applications early in their career?

The existence of these varying tracks produces graduates with very different training and perspectives. Broad training in general anthropology positions the student for teaching opportunities in academic anthropology departments and provides broad interpretive capabilities with regards to casework. Such broad anthropological training, however, might preclude extensive exposure to practices in other areas of forensic science, including legal procedures. In contrast, the forensic anthropologist trained specifically in the applied area of anthropology might forego the broad anthropological background that has, historically, proved to be so useful in the development of professionalism within that field.

While these issues are articulated most directly within the anthropology chapter, they relate to other areas of forensic science as well. In all areas of forensic science, we need practitioners who understand case applications, the relevant methodology and the related quality controls. It can also be argued that, in addition, we need practitioners who fully understand the area of science they are working within, as well having as a working knowledge of other relevant scientific pursuits. Such broad training and perspective maintains linkages with mainstream scientific academia and encourages thoughtful research initiatives. The argument for such broad training also relates to recently debated concerns about the scientific basis of some methodology within current forensic science. Clearly, the road to minimize such concern will require practitioners to obtain a clear understanding of the general scientific areas involved.

Evidence of this variation in training can also be found within the requirements of admission to the entry level Associate Member status of the various sections of the AAFS. Minimal degree requirements vary from the baccalaureate or equivalent for criminalistics, digital and multimedia sciences, engineering sciences, general, questioned documents and toxicology. Physical anthropology requires at least a master's degree. Odontology requires a dental degree (DDS, DMD or equivalent). The Pathology/Biology Section requires a medical degree (MD, DO, DVM) or PhD. Psychiatry and behavioral science requires the PhD or PsyD for clinical psychologists, the MD or DO degree for psychiatrists and the PhD for research psychologists. Associate membership degree requirement for the Jurisprudence Section consists of a law degree, a license to practice law or the equivalent. Most sections have additional requirements of specific training, related work experience, AAFS meeting participation and others.

15.17 Evidence issues in forensic science

As noted in the chapters from the General Section, Criminalistics and, to some extent, Anthropology, proper recovery of evidence and its documentation at the scene are essential to successful forensic analysis and interpretation. Evidentiary quality controls begin with recovery and are maintained through proper laboratory management. While all agree that strict quality control is necessary, in reality decisions have to be made regarding the effective use of resources. Such decisions are especially important internationally where, in some regions, ideal technology and resources may not be available. These issues are addressed in the ICRC humanitarian applications chapter (Chapter 13), which documents situations in which solutions are found without compromising effectiveness.

The sequence of evidence analysis also can be critical, especially when destructive or evidence-altering methodology is employed. One specialist may desire to remove the soil from evidence to ease the examination, but that soil might represent evidence to another specialist.

Prior to specimen removal for DNA analysis, the necessary measurements and observations of the intact specimen need to be recorded and preserved. Does the cut on a human bone discovered by the anthropologist represent perimortem sharp-force trauma or merely a post-mortem scalpel cut made at the preceding autopsy?

Prior to analysis, administrative thought must focus on the forensic specialists likely to examine the evidence, their methodology and the ideal sequence of that work. Documentation of that progression, the methodology employed and sound record-keeping are needed to clarify issues that emerge. Such planning and thought begins with recovery at the scene because, at such an early stage in case development, the crime scene investigator is not likely to be aware of what issues will emerge as the case analysis progresses.

The maintenance of high-quality analysis and evidence protection involves appropriate and accurate technology, sound laboratory practices, employment of qualified and well-trained personnel and sound science. Many factors relate to such quality considerations, but paramount are laboratory accreditation and personnel certification.

Although laboratory accreditation does not ensure high quality, it certainly helps. The processes involved in accreditation relate to safety issues, the establishment and documentation of laboratory procedures and the steps to ensure the protection of evidence. Professional certification offers evidence that personnel involved in this work have appropriate training and experience, as judged by their peers and the forensic science community. Those who have participated in both accreditation and certification procedures mostly relate that they have been strengthened professionally by the experiences.

15.18 Technology

Nearly all of the chapters in this book relate the value of technology in the evolution of their discipline. The definition of modern forensic science clearly has a technological component. From DNA analysis to radiography within forensic pathology and remote sensing in anthropological recovery efforts, technological developments drive progress in the forensic fields. Although some of these developments have been engineered specifically for forensic applications, most have been adapted from other scientific areas and uses. For this reason (and others), it is important that practitioners and researchers in the forensic sciences maintain awareness of broader scientific developments and contact with the innovators. Clearly, technology will play a major role in our path forward.

While technology has become instrumental in conducting quality forensic science, it varies considerably in the international practice of forensic science. Modern technology is relatively expensive to acquire and maintain. Training of personnel is also costly, and it requires long-term commitments. Putting technological knowledge aside, the simple cost of many technological components to the modern laboratory leads to variability of practice, especially globally. While a new technique based on sophisticated technological analysis may clearly improve case analysis, it remains out of reach if the cost exceeds the budget of the laboratory. Many of us who have advocated new technological approaches to old problems usually find the first question to be asked is, 'What does it cost?' While national funding agencies and grant programs might consider funding 'big ticket' technological items for units that could otherwise not afford them, even the funding sources must examine priorities.

Changes in technology are paramount for sections and forensic activities in which technology is the object of case attention. Our colleagues working in digital and multimedia sciences, questioned documents, trace evidence analysis and the engineering sciences must strive to keep up with an ever-changing industry. As noted above, technological advances have led to the formation of the digital and multimedia section. Dramatic reduction in the use of the typewriter has affected the daily work routine of those working in questioned documents, but new technologies bring new challenges.

15.19 Scientific working groups

Beginning in the early 1990s, United States federal agencies have supported the organization and activities of scientific working groups (SWGs). These groups are designed to address the needs of different scientific areas within the forensic sciences, and each is composed of colleagues representing diverse areas and organizations within the discipline. Guidelines are presented for membership, the roles and duties of officers, meeting schedules, definitions of subcommittee working groups, the process for approval of recommendations, and a process for amendments to the bylaws [2]. Each discipline organizes working groups and conducts related business to meet the needs of their scientific area. The SWGs define standards, guidelines and best practices and promote effective communication and discussion within their disciplines. As noted in many chapters of this volume, the SWGs have represented a key development in the advancement of forensic science in the past two decades.

The following is a list of the existing SWGs in 2011 [3]:

1. Scientific Working Group for Forensic Anthropology (SWGANTH)
2. Scientific Working Group on DNA Analysis Methods (SWGDAM)
3. Scientific Working Group for Forensic Document Examination (SWGDOC)
4. Scientific Working Group on Friction Ridge Analysis, Study and Technology (SWGFAST)

5. Scientific Working Group on Firearms and Toolmarks (SWGGUN)
6. Scientific Working Group on Materials Analysis (SWGMAT)
7. Scientific Working Group on Shoeprint and Tire Tread Evidence (SWGTREAD)
8. Scientific Working Group on Toxicology (SWGTOX)
9. Scientific Working Group on Gun Shot Residue (SWGGSR)
10. Scientific Working Group on Bloodstain Pattern Analysis (SWGSTAIN)
11. Scientific Working Group on Medicolegal Death Investigation (SWGMDI)
12. Scientific Working Group for Fire and Explosives Scenes (SWGFEX)
13. Scientific Working Group on Disaster Victim Identification (SWGDVI)
14. Scientific Working Group on Dogs and Orthogonal Detection Guidelines (SWGDOG)
15. Scientific Working Group for the Analysis of Seized Drugs (SWGDRUG)
16. Scientific Working Group on Digital Evidence (SWGDE)
17. Scientific Working Group on Imaging Technology (SWGIT)
18. Scientific Working Group for Wildlife Forensics (SWGWILD)
19. Facial Identification Scientific Working Group (FISWG)

Although the organization, mission and products of the different scientific working groups display some variation, they share the common accomplishment of promoting dialogue within the forensic sciences. Simply meeting regularly to discuss the issues and needs of each discipline represents a major contribution to the field. This activity supplements discussion and presentations at larger scientific meetings and provides a forum for analysis of fundamental issues within each discipline.

15.20 Research

As with all sciences, the forensic sciences represent a dynamic endeavor, constantly evolving with new technology, enhanced methodologies and other forms of innovation. While the procedures of the forensic sciences are time-tested, their general scientific foundation is not static. Improvement is constantly needed and is strived for by all involved.

Research represents the key ingredient in any path forward within the forensic sciences. Stimulated by the challenges of casework and the critical climate of legal applications, professionals concerned with progress recognize the need for thoughtful, targeted research. Although research interest is strong today, it is not new.

Throughout their history, the forensic sciences have depended on scientific research and the knowledge gained from carefully executed experimentation and problem-solving. Research is international in scope and is frequently accomplished by interdisciplinary global teams. Such teams are formed from federal agencies, universities, laboratories, student initiatives and a wide variety of other sources.

Within the United States, the National Institute of Justice (NIJ) represents the primary funding source for formal sponsored research in forensic science. Created four decades ago within the Department of Justice, NIJ seeks to advance scientific research and development within the forensic sciences, especially at the state and local levels. Its general research agenda is formulated from input derived from the criminal justice community and is aimed at identifying high-priority needs. Technology working groups (TWGs) are composed of community-based practitioners, who advise on the most important needs and discipline-specific priorities, providing the information needed for priority formulation.

As funding resources become available each year, NIJ incorporates the community-based advice received to define major funding goals. For example, in the fiscal year of 2011, two general research solicitations were articulated – applied research and development in forensic science for criminal justice purposes, and basic scientific research to support forensic science for criminal justice purposes.

The applied research and development solicitation was aimed at funding projects that would guide policy and practice and/or produce useful products to enhance forensic applications. This general goal was to stimulate applications of basic research toward new and more effective methodologies.

The 2011 basic scientific research initiative recognized the need for innovative, 'fundamental' research in the physical, life and cognitive sciences that would increase knowledge underlying the forensic sciences. This aspect of the program recognized the need for basic knowledge that can be drawn upon for forensic applications. Proposals were sought that articulated research designs aimed at new knowledge with very broad forensic science impact.

In the three-year period from 2009 through 2011, NIJ approved 143 research awards totaling 61.3 million dollars. The dollar amounts were 19.7 million in 2009, 26.7 million in 2010 and 14.9 million in 2011. Awards recognized funding language targeting DNA research and other forensic activities. Many funded proposals directly address the needs articulated in the NAS report (4).

15.21 Awards

15.21.1 2009 Awards

Research and development on impression evidence

- A Proposal to Develop a Computer Program to Improve the National Integrated Ballistics Information Network

- Consecutive and Random Manufactured Semi-Automatic Pistol Breech Face and Fired Cartridge Case Evaluations
- Improve the NIBIN system by Providing Examiners a Capability to Match Infrared Images of Firing Pin Impressions and Deformed Bullets as Well as Accurate Large Database Searches
- Quantitative Measures in Support of Latent Print Comparison
- Application of Machine Learning to Toolmarks: Statistically Based Methods for Impression Pattern Comparisons
- Use of Magneto-Rheological Fluids for Collecting, Preserving, and Analyzing Toolmark and Impression Evidence

Trace evidence research and development

- Rapid Screening and Confirmation of Organic GSR using Electrospray Mass Spectrometry
- Significance of Elemental Analysis from Trace Evidence
- Method Development and Validation of Comparative Finished Fiber Analysis Using Nano-Sampling Cryomicrotomy and Time-of-Flight Secondary Ion Mass Spectrometry
- Developing a high throughput protocol for using soil molecular biology as trace evidence
- Improvements to Laser Ablation-Inductively Coupled Plasma-Mass Spectrometry (LA-ICP-MS) for Quantitative Forensic Analysis using a Short Pulse [100 Femtosecond] Ultraviolet Laser

Fundamental forensic science research awards

- Repeatability and Uniqueness of Striations/Impressions in Fired Cartridge Casings Fired in 10 Consecutively Manufactured Slides
- Quantified Assessment of AFIS Contextual Information on Accuracy and Reliability of Subsequent Examiner Conclusions
- Statistical Evaluation of Torn Duct Tape
- Experimental Study of the Validity and Reliability of Digital Forensic Tools
- Microscopic Analysis of Sharp Force Trauma in Bone and Cartilage
- Face Annotation at the Macro-scale and the Micro-scale: Tools, Techniques and Applications in Forensic Identification

- Establishing the Quantitative Basis for Sufficiency: Thresholds and Metrics for Friction Ridge Pattern Detail Quality and the Foundation for a Standard
- Application of Spatial Statistics to Latent Print Identifications: Toward Improved Forensic Science Methodologies
- Development of Synthetically Generated LEA Signatures to Generalize Probability of False Positive Identification Estimates
- Forensic Analysis of Ignitable Liquid Fuel Fires in Buildings
- Error Rates for Latent Fingerprinting as a Function of Visual Complexity and Cognitive Difficulty
- The Information Content of Friction Ridge Impressions as Revealed by Human Experts
- Statistical Assessment of the Probability of Correct Identification of Ignitable Liquids in Fire Debris Analysis
- Fundamental Research to Improve the Understanding of the Accuracy, Reliability, and Validity of Using the ENF Criterion for the Forensic Authentication of Digital Recordings
- Quantifying the Effects of Database Size and Sample Quality on Measures of Individualization Validity and Accuracy in Forensics
- Manipulative Virtual Tools for Tool Mark Characterization

DNA research and development

- Automated Processing of Sexual Assault Cases Using Selective Degradation
- Sperm Capture Using Aptamer Based Technology
- Development of Linkage Phase Analysis Software for Resolving mtDNA Mixtures
- Validation of Highly-Specific Protein Markers for the Identification of Biological Stains
- Development of a SNP Assay Panel for Ancestral Origin Inference and Individuals Somatic Traits
- Predicting the Biological Age of a Bloodstain Donor
- Microchip Analyzer for Forensic Short Tandem Repeat Typing of Single Cells
- Next Generation Sequencing-based STR Mixture Deconvolution and STR Profiling of Degraded Samples
- Application of Raman Spectroscopy for an Easy-to-Use, on-Field, Rapid, Nondestructive, Confirmatory Identification of Body Fluids

- Automated Processing of FTA Samples
- Improved Tools and Interpretation Guidelines for Examining Limited Low Copy Number DNA Obtained from Degraded Single Source Samples: Bones, Teeth, and Hairs
- A Low Cost Microfluidic Instrument for Typing SNPs
- Application of Proteinases for DNA Isolation of Challenged Bone Specimens Traits
- Identification and Separation of Evidence Mixtures Using SNP-Based FISH Techniques and Laser Microdissection
- Identification of Forensically Relevant Fluids and Tissues by Small RNA Profiling
- DNA Analysis of LCN Samples: Towards Fully Integrated STR Profiling
- Development of an Automated Holographic Optical Trapping Method for Rape Kit Analysis

15.21.2 2010 Awards

Crime scene and medicolegal death investigation research and development awards

- Novel Computer-Assisted Identification Method Using Radiograph Comparison
- Day and Night Real Time Signature Enhanced Crime Scene Survey Camera
- Utility of Post-mortem X-ray Computed Tomography (CT) in Supplanting or Supplementing Medicolegal Autopsies
- Estimating Post-mortem Interval: a Molecular Approach
- Establishing Blow Fly Development and Sampling Procedures to Estimate Post-mortem Intervals
- Development of a Sampling System to Stabilize Ignitable Liquid Residues in Fire Debris
- Estimation of Age at Death using Cortical Bone Histomorphometry

Research and development on pattern and impression evidence

- Acquisition of Fingerprint Topology using Columnar Thin Films

- Research and Development on Pattern and Impression Evidence- Evaluating high dynamic range (HDR) processing with regards to the presence of individualizing characteristics in shoeprint/tireprint impressions
- Digitizing Device to Capture Track Impressions
- Statistical Examination of Handwriting Characteristics Using Automated Tools
- Development of a Science Base and Open Source Software for Bloodstain Pattern Analysis
- Developing Methods to Improve the Quality and Efficiency of Latent Fingermark Development by Superglue Fuming
- Quantitative Analysis of High Velocity Bloodstain Patterns: A Double Blind Investigation of Impact Velocity Assessment
- Significance of Association in Tool Mark Characterization

Research and development on instrumental analysis for forensic science applications

- Raman Spectroscopy with Multi-component Searching for Complex Clandestine Laboratory Sample Analysis
- Expansion of a Cheminformatic Database of Spectral Data for Forensic Chemists and Toxicologists
- Use of Scanning Electron Microscopy/Energy Dispersive Spectroscopy (SEM/EDS) Methods for the Analysis of Small Particles Adhering to Carpet Fiber Surfaces as a Means to Test Associations of Trace Evidence in a Way that is Independent of Manufactured Characteristics
- Forensic Investigation Techniques for Inspecting Electrical Conductors Involved in Fire
- LA-ICP-MS and LIBS analysis of paper, inks, soils, cotton and glass
- Research and Development on Instrumental Analysis for Forensic Science Applications
- Replication of Known Dental Characteristics in Porcine Skin: Emerging Technologies for the Imaging Specialist
- Fundamentals of Forensic Pigment Identification by Raman micro-spectroscopy: A practical identification guide and spectral library for forensic science laboratories
- Development and Validation of Standard Operating Procedures for Measuring Microbial Populations for Estimating a Post-mortem Interval

- Validation of Forensic Characterization and Chemical Identification of Dyes Extracted from Millimeter-length Fibers
- On-Site Confirmatory Test for Tissue Type and Specimen Age
- Universal IR Fluorescent Latent-Print Detection Method

Fundamental forensic science research awards

- Quantified Assessment of Contextual Information in Latent Friction Ridge Impression Analysis Related to Accuracy and Reliability of Subsequent Examiner Suitability Determinations
- Independent Validation Test of Microscopic Analysis of Saw Marks in Bone
- Gunshot Residue in a Non-Firearm-Related Detainee Population
- The effects of acquisition of post-mortem blood specimens on drug levels and the effects of transport conditions on degradation of drugs
- Validity, Reliability, Accuracy, and Bias in Forensic Signature Identification
- Determination of Unique Fracture Patterns in Glass and Glassy Polymers
- Improving the Understanding and the Reliability of the Concept of 'Sufficiency' in Friction Ridge Examination
- Reducing Uncertainty of Quantifying the Burning Rate of Upholstered Furniture
- Improving Investigative Lead Information and Evidential Significance Assessment for Automotive Paint and the PDQ Database
- Frequency Occurrence of Handwriting and Hand Printing Characteristics
- Reliability of Forensic Data from Networked Process Control System
- Evaluation of Statistical Measures for Fiber Comparisons: Interlaboratory Studies and Forensic Databases
- Development and Quantitative Evaluation of Steganalysis and Digital Forgery Detection System
- Testing the Validity of Radiographic Comparisons in Positive Identifications
- Reliability Measures for Current Methods in Bloodstain Pattern Analysis
- Development of Individual Handwriting Characteristics in $\sim$1800 Students: Statistical Analysis and Likelihood Ratios that Emerge Over an Extended Period of Time
- Miami-Dade Repeatability and Uniqueness of Striations/Impressions in Bullets Fired in 10 Consecutively Manufactured Glock EBIS Barrels

- Miami-Dade Research Study for the Reliability of the ACE-V Process: Accuracy, Precision, Reproducibility, and Repeatability in Latent Fingerprint Examination
- Interpretation of Ignitable Liquid Residues in Fire Debris Analysis: Effect of Competitive Adsorption, Development of an Expert System and Assessment of the False Positive/Incorrect Assignment Rate
- Electrical Fires' Forensic Signatures of Fire Cause Events
- Investigation of a Novel Approach to Forensic Analysis Using Neutron Imaging Techniques

Forensic DNA research and development

- Tools for improving the quality of aged, degraded, damaged, or otherwise compromised DNA evidence
- Effective Long-Term Preservation of Biological Evidence
- Detection and remediation of PCR inhibition using real time PCR melt curves as a diagnostic tool
- Trace DNA from Fingernails: Increasing the Success Rate of Widely Collected Forensic Evidence
- Assessing Deep Sequencing Technology for Human Forensic mitochondrial DNA
- A Microfluidic Microarray Instrument to Type SNPs for Physical Appearance
- Targeted Non-Destructive Evidence Detection and Collection
- Resolution of DNA Mixtures and Analysis of Degraded DNA Using the 454 DNA Sequencing Technology
- Genetic Markers Associated with Sudden Unexplained Death or Sudden Infant Death
- Development of a Proteomic Assay for Menstrual Blood, Vaginal Fluid and Species Identification
- Evaluating the Use of DNA and RNA Degradation for Estimating the Post-Mortem Interval
- Addressing Quality and Quantity; the Role of DNA Repair and Whole Genome Amplification in Forensically Relevant Samples
- Molecular Characterization of Trace Biological Evidence for the Optimized Recovery and Analysis of 'Touch DNA'

- Developing an Empirically-Based Ranking Order for Bone Sampling: Examining the Differential DNA Yield Rates between Human Skeletal Elements over Increasing Post Mortem Intervals
- Developing a Forensic Resource/Reference On Genetics knowledge base
- Further Development of SNP Panels for Forensics
- Characterization of X Chromosomal Short Tandem Repeat Markers for Forensic Use
- Maximizing mtDNA testing Potential with the Generation of High Quality of mtGenome Reference Data

15.21.3 2011 Awards

Applied research and development in forensic science for criminal justice purposes

- Development of a Modern Compendium of Microcrystal Tests for Illicit Drugs and Diverted Pharmaceuticals
- Raman Spectroscopy for Analyzing Body Fluid Traces, Stain Aging, Differentiation between Races, Genders and Species
- The Behavior of Contamination and Methods for Extraction and Capture of Low Copy Number and Degraded DNA
- Low-template DNA Mixture Interpretation: Determining the Number of Contributors
- Development of an Immuno-Magnetic Procedure for the Separation of Spermatozoa from Vaginal Epithelial Cells
- Forensic Identification of an Individual in Complex Mixtures Utilizing SNP Technology
- Automated Multianalyte Screening Tool for Classification of Forensic Samples
- Automated Sperm Detection in Forensic DNA Analysis: Implications for Rape Kit Analysis
- Rapid and Selective Extraction of Male DNA from Rape Kits and Other Forensic Evidence using Pressure Cycling
- Use of Pressure Cycling Technology to Enhance DNA Yield and Profile Success in Touch Samples
- Comprehensive Forensic Toxicological Analysis of Designer Drugs

- Smartphone Technology for Capturing Untreated Latent Fingerprints
- Computerized Reconstruction of Fragmentary Skeletal Remains for Purposes of Extracting Osteometric Measurements and Estimating MNI
- Pediatric Fracture Printing: Creating a Science of Statistical Fracture Signature Analysis
- Graphical User Interface for a Multi-Factorial Age-at-Death Estimation Method Using Fuzzy Integrals
- Degraded Ignitable Liquids Database: An Applied Study of Weathering and Bacterial Degradation on the Chromatographic Patterns of ASTM E 1618 Ignitable Liquid Classes
- Comparison of Microspectrophotometry and Room-Temperature Fluorescence Excitation-Emission Matrix Spectroscopy for Non-Destructive Forensic Fiber Examination
- Microfluidic System for Automated Dye Molecule Extraction and Detection for Forensic Fiber Identification
- Raman Spectroscopy of Automotive and Architectural Pigments: in situ Identification and Evidentiary Significance
- Accessing the Probative Value of Physical Evidence at Crime Scenes with Ambient Mass Spectrometry and Portable Instrumentation
- Examining the Effects of Environmental Degradation on the Optical Properties of Manufactured Fibers of Natural Origin
- Developing Guidelines for the Application of Multivariate Statistical Analysis to Forensic Evidence

Basic scientific research to support forensic science for criminal justice purposes

- Extending the Microbial Forensic Toolkit Through Whole Genome Sequencing and Statistical Phylogenomics
- Investigating Unexplained Deaths for Molecular Autopsies
- Characterization of Bacterial and Fungal Communities Associated with Corpse Decomposition Using Next Generation Sequencing
- Human Hair Proteomics-Improved Evidence Discrimination
- DNA Forensics using Single Molecule Technology: From DNA Recovery and Extraction to Genotyping Degraded and Trace Evidence without PCR
- Population Genetic Issues for Forensic DNA Profiles

- Analytical and Synthetic Studies on Designer Drugs of the Piperazine Class
- Separation of Identification of Drug Abuse Using ESI-IMS-MS
- Prediction of Drug Interactions with Methadone, Buprenorphine and Oxycodone from in vitro Inhibition of Metabolism
- 3-D Characterization and Comparison of Fracture Surfaces
- Isotope Analyses of Hair as a Trace Evidence Tool to Reconstruct Human Movements: Combining Strontium Isotope with Hydrogen/Oxygen Isotope Data

The above list represents the response of the forensic community to the NIJ announcements of general target research initiatives. Since these projects were funded, their inclusion here indicates that the proposals were reviewed by their peers and judged to be of high priority and worthy of support. Collectively, they make a statement regarding the commitment of NIJ to meet the needs of advancing forensic science, as well as to promote ideas that are on the cutting edge of research as forensic science moves forward in the United States.

The 143 research awards listed above for the years 2009, 2010 and 2011 reveal patterns in both funding and new initiatives. In regard to subject matter, most grants were awarded in support of DNA topics (32%). The DNA grants reflect the importance of molecular advances to the field of forensic science, as well as the growing number of laboratory-based scientists in this area. Approximately 10% of awards each were devoted to topics on trace evidence (15 awards), latent print analysis (14 awards) and general forensic topics (14 awards). Other topics included skeletal analysis (6%), fire debris (6%), chemistry and toxicology (5%), ballistics, tool marks, document examination, forensic pathology and bloodstain patterns (3% each), digital evidence (2%), gunshot residue (1%), facial imagery (1%), entomology (1%), track impressions (1%) and odontology (1%).

The problems addressed in the 143 funded proposals can be classified into five general categories. Most awarded grants (67%) targeted methodological advancement in various ways. Approximately 14% focused on new statistical approaches to the assessment of forensic information. Methodology evaluation comprised 10% of awards, followed by new computer programs and approaches (4%) and quantification factors (4%).

The award patterns described above suggest that, while the path forward in the forensic sciences has been debated by legislators and others in Washington, NIJ has been steadily offering direct grant support to those demonstrating research leadership. The funding priorities directly address the major issues articulated by many, and they lead the way toward significant advancement.

15.22 Global forensic science

The diverse and dynamic research projects outlined above that are currently pushing forensic science forward in the United States are complemented by many similar initiatives throughout the world. Forensic science really is globally transcending national borders. We are reminded constantly of the global nature of forensic science when we gather at international meetings, work together with colleagues from different countries on research and cases, and read our journals.

This volume summarizes much of the direction and activity of the forensic sciences. It also documents the dynamic, international nature of forensic applications. Through research, casework experience, new technology and innovation, forensic science advances on its path forward. It is an exciting time to be a forensic scientist. All those involved in writing this book are proud to be part of the field. I hope all who read the volume sense that excitement and their commitment to excellence.

References

1. Field KS. *History of the American Academy of Forensic Sciences, 1948–1998*. West Conshohocken, PA: ASTM; 1998.
2. Adams DE, Lothridge KL. *Scientific Working Groups. Forensic Science Communications* 2000 Jul; 2 (3). http://www.fbi.gov/about-us/lab/forensic-science-communications/fsc/july2000/swgroups.htm/ (accessed January 12, 2012).
3. National Institute of Justice. http://nij.gov/ (accessed January 12, 2012).
4. National Research Council. Strengthening Forensic Science in the United States: A Path Forward. Washington, DC: The National Academies Press, 2009.

Index

Forensic Science: Current Issues, Future Directions, First Edition. Edited by Douglas H. Ubelaker.